How to Achieve Sustainable Development

Aimed at practitioners and students in environmental studies, development, economics and geography, this book systematically investigates how sustainable development can be achieved at the local and global level. Instead of problems, it focuses on practical solutions for the reader to take for themselves, in their community, and in their career.

Based on the latest scientific research, case studies, and voices from the Global North and the Global South, *How to Achieve Sustainable Development* is divided into five key Parts. It reviews why business-as-usual is no longer an option; how to engage with governance, finance, and collective action; where transformation hotspots for achieving sustainable development exist; how local change happens in sustainable development; and finally, what happens next. Chapters take a deep dive into improved human capabilities, just economies, sustainable food and energy production, gender inequality and resilience. It moves beyond merely recognizing the challenges faced in achieving the UN Sustainable Development Goals (SDGs), instead focusing on real-world pathways to address these, beyond the 2030 timeline. Each solution is framed within this context and promotes taking complexity seriously, as opposed to ignoring it.

Alongside the book, appendix material provides guidance for personal action, and digital resources offer up-to-date opportunities for further engagement.

Samuel T. Ledermann is an Associate Professor of Practice in the International Development Studies program at the Elliott School of International Affairs at George Washington University, USA. He is an economic geographer with an extensive background in sustainable agricultural development in Africa.

Routledge Studies in Sustainable Development

This series uniquely brings together original and cutting-edge research on sustainable development. The books in this series tackle difficult and important issues in sustainable development including: values and ethics; sustainability in higher education; climate compatible development; resilience; capitalism and de-growth; sustainable urban development; gender and participation; and well-being.

Drawing on a wide range of disciplines, the series promotes interdisciplinary research for an international readership. The series was recommended in the *Guardian*'s suggested reads on development and the environment.

Greening Higher Education in Europe
Institutional Transitions to Sustainable Development
Magdalena Popowska

Implementing Sustainable Cities
Edited by Sylvie Albert, Jeremy Millard, and Manish Pandey

Sustainable Urban Development in the European Arctic
Dorothea Wehrmann, Michał Łuszczuk, Katarzyna Radzik-Maruszak, Jacqueline Götze, and Arne Riedel

A Nordic Smart Sustainable City
Lessons from Theory and Practice
Barbara Maria Sageidet, Daniela Müller-Eie, and Kristiane M.F. Lindland

Inside the Sustainable Development Goals
From Fighting Poverty to Saving the World
Adele Orosz

Circular Economy Solutions for Sustainable Development
Katarzyna Nowicka

How to Achieve Sustainable Development
Local Actions for Global Change
Samuel T. Ledermann

Innovation Strategies and the Sustainable Development Goals
A Blueprint for Sustainable Transformation
Amit Singh Khokhar, Punita Duhan and Anshu Lochab

For more information about this series, please visit: www.routledge.com/Routledge-Studies-in-Sustainable-Development/book-series/RSSD

"Ledermann's *How to Achieve Sustainable Development* is a manual for change designed for students intent on thoughtfully fostering a better world. This inspiring text not only helps one comprehend our daunting global challenges, but offers well-reasoned strategies for pursuing various forms of sustainability without sinking into jingoistic activism, token behavior or charity. With a rich array of illustrative case studies, from sustainable local food systems to just economies, this book is a very welcome addition to interdisciplinary sustainability studies."

William G. Moseley, *DeWitt Wallace Professor of Geography, Macalester College, President of Association of American Geographers (2026), USA*

"Prof. Ledermann's *How to Achieve Sustainable Development* offers a clear, cohesive narrative on sustainability's societal dimensions and the key changemakers. This essential textbook provides valuable insights and practical real-life experiences, benefiting students and concerned individuals alike."

Juha I. Uitto, *Environmental Law Institute, and Past-Director, Independent Evaluation Office, Global Environment Facility*

"I wholeheartedly endorse this groundbreaking textbook, which serves as an essential resource for anyone committed to fostering sustainable development in their community and across the globe. Let it be your companion in the quest for a viable future and in achieving UN Sustainable Development Goals (SDGs) by 2030 and beyond for transformative change."

Ambassador Liberata Mulamula, *Member of Parliament, Tanzania, Former Minister of Foreign Affairs, Tanzania and Ambassador to the UN and the USA, Tanzania*

"The comprehensive overview of sustainable development laid out in this impressive book is informed by a wealth of research results, from the natural and social sciences, as well as derived from development practice experience. It clearly shows that better governance and policy-making as well as our individual and collective actions are indispensable to keep the development process inside planetary boundaries."

Andreas Sicks, *CEO Liechtenstein Development Services, Liechtenstein*

How to Achieve Sustainable Development

Local Actions for Global Change

Samuel T. Ledermann

Routledge
Taylor & Francis Group
LONDON AND NEW YORK

Designed cover image: ©Peter Lüthi/Biovision Foundation

First published 2026
by Routledge
4 Park Square, Milton Park, Abingdon, Oxon OX14 4RN

and by Routledge
605 Third Avenue, New York, NY 10158

Routledge is an imprint of the Taylor & Francis Group, an informa business

British Library Cataloguing-in-Publication Data
A catalogue record for this book is available from the British Library

ISBN: 978-1-032-50867-2 (hbk)
ISBN: 978-1-032-50866-5 (pbk)
ISBN: 978-1-003-40002-8 (ebk)

DOI: 10.4324/9781003400028

Access the Support Material: www.routledge.com/9781032508665
Access the author's website: www.sustainabledevelopment.me

Typeset in Univers
by SPi Technologies India Pvt Ltd (Straive)

MIX
Paper | Supporting responsible forestry
FSC™ C013985
FSC
www.fsc.org

Printed in the United Kingdom
by Henry Ling Limited

Contents

Figures

Tables

Acknowledgments

For my 6-year old daughter, Séraphine, who taught me about upcycling. And my wife, Leena, who transforms our small world for the better every day.

At the heart of this book's origin is the UN's *Global Sustainable Development Report 2019: The Future is Now* – co-chaired by Peter Messerli and Endah Murniningtyas. Upon getting my hands on a copy, it struck me that their exceptional report offered a re-framing of sustainable development pedagogy with their conceptualization of levers of change and transformation hotspots. I was concerned that the authors' insights would be relegated to the internet archives where my students would never encounter them. I hence revised my previous graduate seminar, "Agriculture and Sustainable Development," that focused on how food is at the heart of the SDGs. Anchored around their report, I piloted my new graduate seminar – "Achieving Sustainable Development" – in 2022.

Based on the encouraging feedback received, this book (and our companion website, sustainabledevelompent.me) are the culmination of my subsequent teaching and development practice over the years. I specifically thank my graduate students in my "Achieving Sustainable Development" graduate course in 2023 and 2024, who have contributed to our book through class discussions and their inspiring analyses of sustainable development: in alphabetical order, Shreya Agarwal, Eleonara Arca, Carsyn Baxter, Nabihah Begum, Ben Bresnee, Leonard Cisneros, Emily Costello, Simran Dali, Joshua Denicoff, Imesha Ediriweera Patabendige, Chiara Evelti, Mariana Felix Angioletti, Rawand Hidmi, Lauriane Immongault, Renuka Iyer, Zoe Jafflin, Hajrah Jalil, Ryker Jensen, Lydia Kiros, Robert Lalle, Rubayet Khundoker, Zirong Li, Fatima Lunsford, Lois Maison, Olivia Morris, Mia Nielsen, Annie Olise-Aikins, Jaya Parashar, Yehwan Park, Michael Purcell, Amica Rapids, Kanika Sahai, Swad Sathe, Vinithra Sudhakar, Angela Tarana, Evergrace George Tibaijuka, Olyvia Tremper, Duncan Wegner, Mackenzie Wenig, Christine Wenzel, and John Wilson. Morgan Urlaub also dropped her crime novel to proofread Part I while on our study abroad course in Lilongwe, Malawi.

A special thank you to two George Washington University undergraduate Honors students: Emma Taylor, who helped with my citations, and Pritam Garcha, who supported our sustainabledevelopment.me web design.

The journey to get us here took longer than expected. I want to thank my editor, Grace Harrison, her editorial assistants, Rohita Divyanshu and Sarah Lewis, production editors Isobel Watson and Jennifer Hicks, and copy editor Susan Dunsmore for their support throughout.

I have benefitted from the insights, mentoring, and encouragement from my academic colleagues, including Nicholas Anderson, Christina Fink, Moses Kansanga, Robin Leichenko, William Moseley, Bob Orttung, Carl Pray, and Sean Roberts. I was privileged to have had a chance to learn from my extended family comprised of world-class development researchers and practitioners, including John Choptiany, Sunday Ekesi, Menale Kassie, Fatiya Khamis, Julius Manda, Samira Mohamed, Annah Njui, Jimmy Pittchar, and Andreas Sicks.

A warm thank you to all my former colleagues at Biovision Foundation who offered me an opportunity to make our world better, *naturally*.

I conclude with gratitude to all the farmers I've met and worked with across the globe – your resilience and ability to manage complexity inspire me today and for our tomorrows.

Acronyms and Abbreviations

AI	artificial intelligence
CGIAR	Consultative Group on International Agricultural Research
COP	Conference of the Parties of the UNFCCC
EPA	U.S. Environmental Protection Agency
ESB	earth-system boundaries
ESG	Environmental, social, and governance
FAO	UN Food and Agriculture Organization
FASA	Financing for Agricultural Small and Medium Enterprises in Africa
GDP	Gross Domestic Product
ha	hectare
HCD	human-centered design
icipe	International Centre for Insect Physiology and Ecology
ICT	Information and Communications Technology
JETP	Just Energy Transition Partnership
KAP	knowledge, attitude, and practice
MDGs	Millennium Development Goals
NDCs	Nationally Determined Contributions
NGO	non-governmental organization
ODA	official development assistance
OECD DAC	Organisation for Economic Co-operation and Development's Development Assistance Committee
ppp	purchasing power parity
R&D	research and development
SDGs	Sustainable Development Goals
SDSN	Sustainable Development Solutions Network
SHARP	Self-evaluation and Holistic Assessment of climate Resilience of farmers and Pastoralists
SLM	Sustainable Land Management
SME	small and medium-sized enterprises
UN	United Nations
UNCTAD	United Nations Trade and Development
UNDP	United Nations Development Programme
UNESCO	United Nations Educational, Scientific and Cultural Organization

UNFCCC	United Nations Framework Convention on Climate Change
USA	United States of America
USAID	United States Agency for International Development
WASH	water, sanitation and hygiene
WOCAT	World Overview of Conservation Approaches and Technologies

Part I
Unsustainable pasts

1 Introduction

Silo breakers

It is Monday, September 28, 2015. You open your email inbox and find a message from the United Nations with the subject line "Unanimously adopted: 2030 Agenda for Sustainable Development". Curious, you open the mail to find aspirational language around new global efforts to end poverty and hunger, combat inequalities, build peaceful and just societies, and protect the planet and its resources. Dubbed the Sustainable Development Goals (SDGs), the material includes a color wheel representing the 17 goals that would five years later become one of the most desirable fashion pins to wear (see Figure 1.1) (Dooley and Ueno 2022).

Figure 1.1
The SDG pin

DOI:10.4324/9781003400028-2

Adopted by the United Nations General Assembly three days earlier, the resolution titled "Transforming Our World" (United Nations 2015) captures goals such as "Ensuring healthy lives and promote well-being for all at all ages" under SDG 3: Good Health and Well-Being, complemented by specific targets.[1] As you scroll to the bottom of the email, you wonder, how do we achieve these lofty goals? Neither the resolution nor the 17 goals and 169 targets offer an answer. The many unread messages in your inbox and news notifications of an emission scandal that have trickled in on your phone over the last 20 minutes are slowly regaining your attention. Sinking back in your chair, you decide to tackle the to-dos of yesterday and today as 2030 slowly slips below your radar.

Fast-forward to the present: What has changed since 2015? After a global pandemic, climate disasters, food crises, and war, are we not heading in the wrong direction entirely? With challenges omnipresent, this book takes you on a journey to address how we can harness the momentum of the UN's 2030 Agenda for Sustainable Development and turn it into action to create positive change across the globe by 2030 and beyond. Now, how do we get there?

WELCOME SILO FIGHTERS

Our journey starts with silos: eight silos to be exact. These silos represent the eight Millennium Development Goals (MDGs) that preceded the SDGs. The MDGs came to represent the epitome of silo thinking in development circles. At an institutional level, silo thinking translates into a lack of policy integration and coherence. At a personal level, it translates into sticking to your own box and ignoring synergies. And for development practitioners, silo thinking means you are ignoring holistic solutions.

The *2030* Agenda for Sustainable Development is calling on all of us to break down these silos:

> The challenges and commitments identified at these major conferences and summits are *interrelated and call for integrated solutions*. To address them effectively, *a new approach is needed*. Sustainable development recognizes that *eradicating poverty in all its forms and dimensions, combating inequality within and among countries, preserving the planet, creating sustained, inclusive economic growth and fostering social inclusion are linked to each other and are interdependent*.
>
> (United Nations 2015: Paragraph 13 of A/RES/70/1, emphasis added)

As noted by observers of the negotiations, breaking down silos has become a mantra in recognition of the 2030 Agenda's "comprehensive, holistic and systematic approach" (Niestroy and Meuleman 2016). The UN Development Group (now the Sustainable Development Group) – a consortium of 36 UN funds – even started a blog exchanging tales of their efforts as silo fighters (see Figure 1.2). Over the course of this book, you will learn how to join other silo fighters in delivering sustainable development for your generation and those to come.

Figure 1.2 The Silo Fighters blog banner, UNDG 2016, used under CC BY 3.0 IGO (https://creativecommons.org/licenses/by/3.0/igo/), https://web.archive.org/web/20181027225913/https:/undg.org/about-silo-fighters/

EMBRACE COMPLEXITY

To start us off, we should take a step back and review what has changed from the MDGs to the SDGs. The eight MDGs – which shaped global development for 15 years as we entered the new millennium – captured efforts in the Global South to address critical challenges, such as eradicating hunger and improving maternal health (see Figure 1.3). To give you a sense of their breadth and depth, the eight goals were accompanied by 18 targets and 60 indicators. Three targets were achieved by 2015: cutting extreme poverty by half (under Goal 1), closing gender gaps in education (under Goal 3), and reducing rates of infections in malaria and tuberculosis (under Goal 6).

Analyzed from the three dimensions of sustainability – the economic, environmental, and social dimensions – a clear pattern emerges (see Figure 1.4). Except for strengthening global partnerships for development, the MDGs are all in the social dimension. Only Goal 1 (focusing on poverty and hunger) and Goal 7 (focusing on environmental sustainability) overlap with the economic and environmental dimensions, respectively. No single goal is at the intersection where the economic, environmental, and social dimensions meet.

Shifting to the SDGs, the 17 goals were accompanied by nearly ten times more targets (169) and around four times more indicators (230+). Using the same three dimensions, we find most goals (12 out of 17) present at the dimensions' intersections (see Figure 1.5). Five goals are at the heart of sustainable

Figure 1.3 The eight Millennium Development Goals (2000–2015), https://www.un.org/millenniumgoals/

How the Millennium Development Goals were biased towards the social dimension

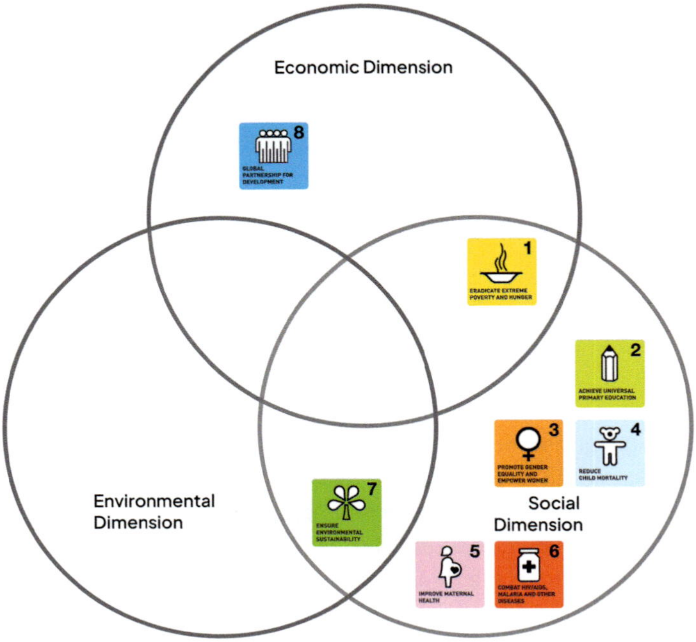

Figure 1.4
How the Millennium Development Goals were biased towards the social dimension, adapted from United Nations Economic Commission for Africa (2016)

How holistic development is at the center of the Sustainable Development Goals

Figure 1.5
How holistic development is at the center of the Sustainable Development Goals, adapted from United Nations Economic Commission for Africa (2016)

development: no poverty (SDG 1), zero hunger (SDG 2), decent work and economic growth (SDG 8), sustainable cities and communities (SDG 11), and life on land (SDG 15).

Reducing inequality (SDG 10), an essential new goal missing from the global agenda, has links to 12 other goals (see Figure 1.6). For example, unpaid care and domestic work are causing gender inequalities (SDG 5) to increase. Or uneven energy access (SDG 7) and a lack of universal healthcare (SDG 3) continue to stifle efforts at creating equal opportunities for all. The importance of inequality has further grown since the global pandemic emerged in 2020, dubbed by the United Nations the "inequality virus" (UN News 2021).

Irrespective of their importance, mapping these linkages might urge you to run away from them. They are difficult to digest. Our goal, however, is that we embrace complexity. As a silo fighter, you need to understand how the topic or cause you are passionate about relates to others so we can coordinate, collaborate and co-learn for a greater impact.

WHY THE SUSTAINABLE DEVELOPMENT GOALS MATTER

Beyond the North-South divide

The MDGs applied only to countries in the Global South. Figure 1.7 depicts the Brandt line delineating the Global North and the Global South adapted from the cover of the 1980 North-South report by the Independent Commission on International Development Issues (1980). Development was deemed mainly a problem to be solved in poorer economies, the Global South. Reflecting on your walk to class today or your last trip to the grocery store, you must have noted that issues like inequality, for example, are all too present in our neighborhoods, no matter how rich or poor. The SDGs recognize that to achieve sustainable development we need to move beyond a North-South divide.

This makes the SDGs genuinely *our* development goals, as both wealthy and poorer nations are asked to "conduct regular and inclusive reviews of progress at the national and sub-national levels, which are country-led and country-driven" (Paragraph 79, United Nations 2015). Look at the United Nations repository of the Voluntary National Reviews (United Nations 2025): I am confident your country of interest is among the more than 185 that reported progress made.[2] And beyond the country level, your own city or local municipality might be active in achieving the SDGs. In 2018, New York City was the first major city to report on its achievements in reaching the SDGs. Cities across the globe – from Accra in Ghana to Zürich in Switzerland – have since incorporated the SDGs into their urban development strategies.

Change in funding

Annual funding gaps between US$3 to US$5 trillion must be closed to achieve sustainable development. With funding for development efforts declining after the global COVID-19 pandemic, the total gap is expected to have increased well beyond US$100 trillion (Edgecliffe-Johnson and Neville 2022). Besides more

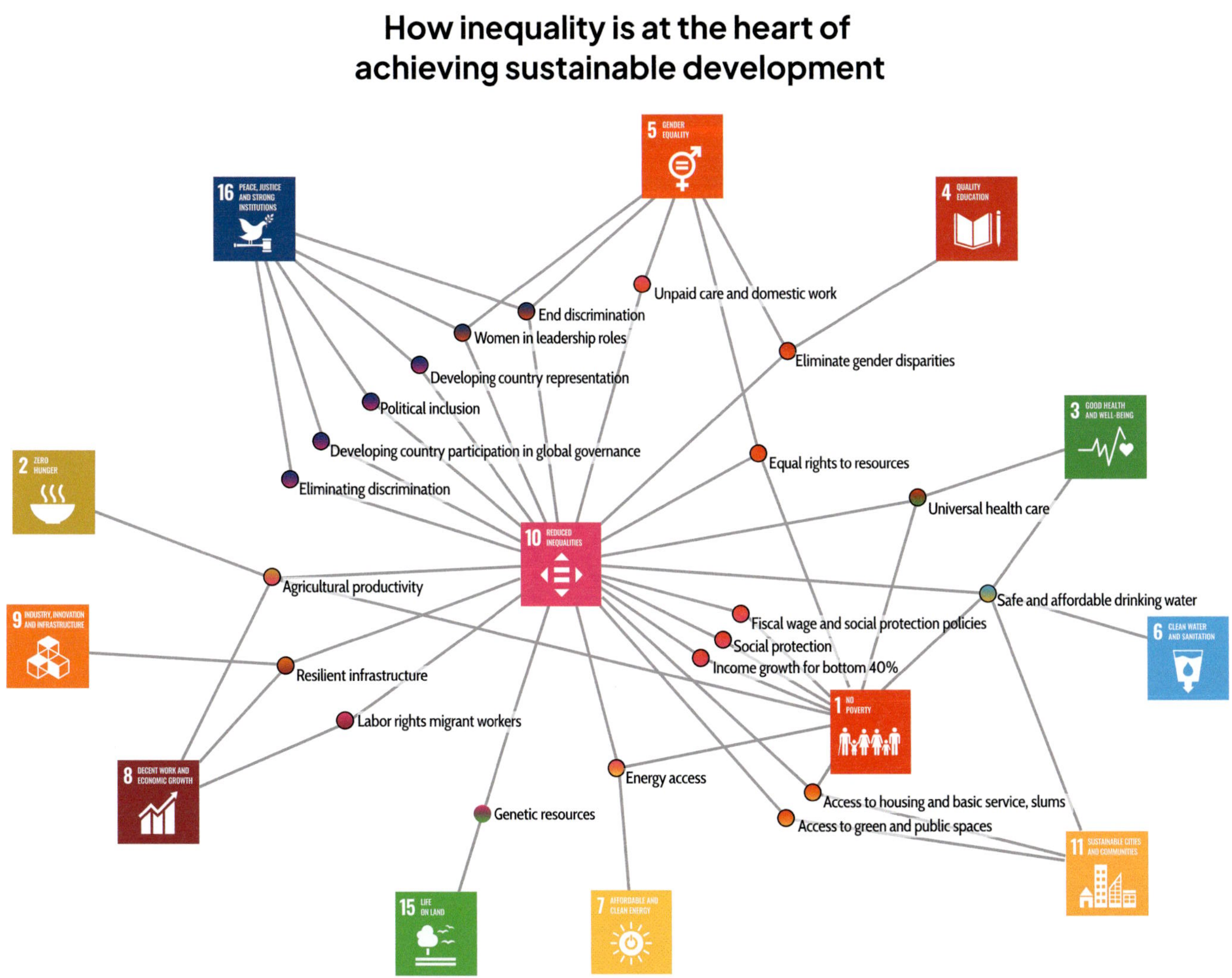

Figure 1.6
How inequality is at the heart of achieving sustainable development, adapted from Le Blanc (2015)

The Brandt Line: Global North and South

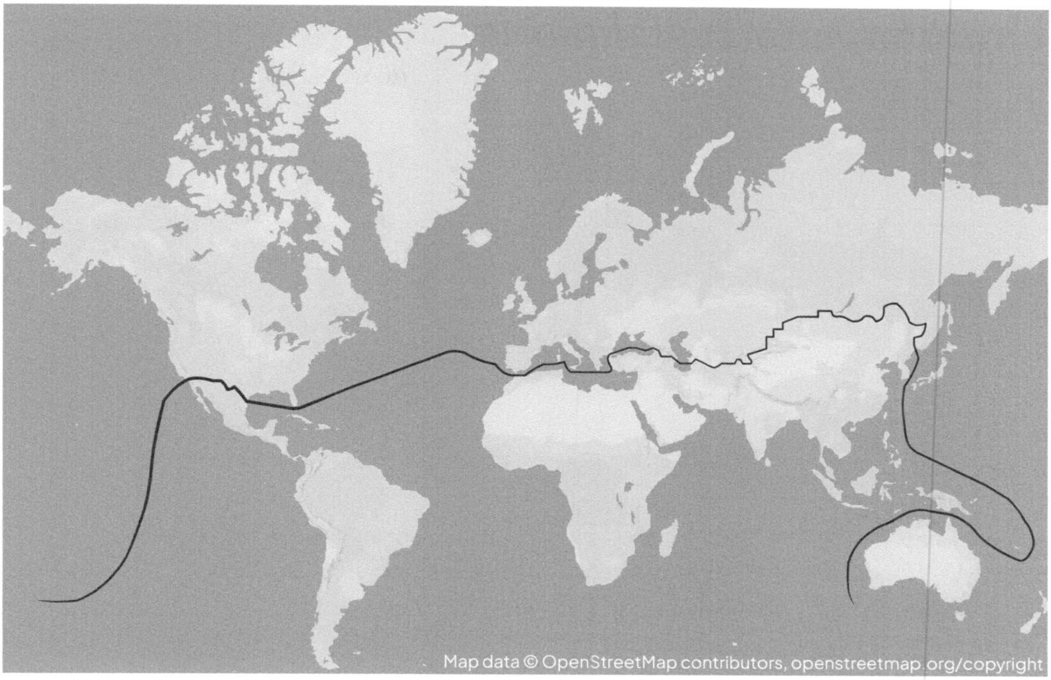

Figure 1.7 The Brandt Line depicting the Global North and the Global South, adapted from Independent Commission on International Development Issues (1980)

funding flows, increased accountability is needed to determine where existing and additional funds are heading.

The lesser-known SDGs – means of implementation, marked by a letter as opposed to a numeral – are usually hidden among the ambitious goals and targets.

For SDG 15 – Life on Land – they are crucial to increase funding flows toward sustainable development. Target 15.a, for example, states that we must "mobilize and significantly increase financial resources from all sources to conserve and sustainably use biodiversity and ecosystems." To review whether we achieve the target, countries will report their "[o]fficial development assistance and public expenditure on conservation and sustainable use of biodiversity and ecosystems" (Indicator 15.a.1). Countries use these means of implementation to align their development financing. Even non-state actors, such as foundations and private investors, orient their funding and investments to the SDGs.

Tipped toward environmental sustainability

Socio-economic development dominated the MDGs. The environment was relegated to a catch-all Goal 7 to "ensure environmental sustainability." This limited attention contrasts starkly to the 11 SDGs addressing environmental sustainability under the SDGs (Figure 1.8). Besides SDG 13 – Climate Action – focusing on building upon the Paris Agreement to address climate change, targets within the other ten SDGs specifically aim to improve environmental sustainability.

Figure 1.8 How the SDGs are tipped towards environmental sustainability

For example, SDG 2 – Zero Hunger – aims to "[e]nd hunger, achieve food security and improved nutrition and promote sustainable agriculture." Target 2.4 specifically captures that by 2030, we need to do the following:

> ensure *sustainable food production systems* and implement *resilient* agricultural practices that increase productivity and production, that help *maintain ecosystems*, that strengthen *capacity for adaptation to climate change, extreme weather, drought, flooding and other disasters* and that progressively *improve land and soil quality*.
>
> (emphasis added)

This shift toward environmental sustainability reflects the local and global challenges experienced by the climate crisis. The negative impacts from global warming on species loss, sea-level rise, and crop yields continue to rack up. The SDGs reflect the urgency communicated in widely read studies (including by policymakers) on how we have breached our planetary boundaries and are moving beyond the safe space for our planet (Rockström et al. 2009; Steffen et al. 2015).

MOVING BEYOND GREENER SILOS

At this stage, you might wonder what stops us from not just making our silos greener (Figure 1.9)? How can we effectively achieve sustainable development by taking complexity seriously and working towards more systematic change? What can be my role in improving sustainability and empowering others to bring about change?

Figure 1.9
The need to move beyond greening silos, 'A tree growing out of a silo' by Aaron Hall/ Flickr licensed under CC BY-SA 2.0

These are our guiding questions for the rest of this book. Yours is the generation that will bring novel approaches to the table that have been missing in the past. To set you up for success, there are five parts to our journey.

In Part I "Unsustainable pasts," you will discover why the tenet of "business-as-usual is no longer an option" has increasingly filled the halls of the UN and corporate boardrooms alike. Building on our "greener silo," we will use the case of agriculture to show how the environmental costs of productivity maximization are increasingly becoming apparent. We will also illuminate how the global pandemic that started in early 2020 has laid bare the vulnerabilities of millions and learn how it changed their livelihoods for the worse. The pandemic has also opened up opportunities for "building back better," bringing resilience and inequality to the forefront of policymakers' agendas, and spurring local action.

In Part II "Changemakers," we draw upon a framework developed by the Independent Group of Scientists appointed by the UN Secretary-General (2019) to introduce four changemakers: (1) governance and policymaking; (2) economy and finance; (3) individual and collective action; and (4) science and technology. In their combinations, they open up possibilities for transformations to achieve sustainable development, which are explored in Part III.

In Part III "Transformation hotspots," we investigate six domains adapted from the Independent Group of Scientists Appointed by the Secretary General (2019) where achieving sustainable development is warranted and possible: (1) sustainable food systems; (2) just economies; (3) human capabilities; (4) inclusive urban-rural development; (5) green energy access; and (6) regeneration of nature. Each offers significant opportunities for transforming the local and global. Case studies will illustrate their potential and challenges, and show how the changemakers enable them.

In Part IV "Change is local," we look more in-depth at how change happens. Challenging the common misconception that you and I can impose change top-down, we focus on more bottom-up efforts that empower. We take on a gendered lens to review how ignoring gender inequalities is detrimental to achieving sustainable development. We also focus on resilience's central role in ensuring the sustainability of any efforts made.

In Part V "Outlook beyond 2030," we take stock of the skill set you have gained and review where we go from here. As you progress in your careers, we reflect on how you can sharpen your role in achieving sustainable development. To guide you and your peers in this process, the appendices at the end of this book and our up-to-date companion website (sustainabledevelopment.me) offer guides and tools for further action.

YOUR TURN

Identify an SDG that you feel passionate about. Familiarize yourself with its targets and indicators by reviewing the UN Resolution "Transforming our world: the 2030 Agenda for Sustainable Development" (United Nations 2015) and the SDG indicators list (https://unstats.un.org/sdgs/indicators/indicators-list/).

Challenges to achieving sustainable development are all around us. Use this SDG as an entry point to explore your own neighborhood. Rejecting the notion that sustainable development is limited to the Global South, I encourage you to record one observation of a challenge. Identify how others in your community are taking local actions to address the challenge. You might be surprised to find how active your neighbors and community are in tackling sustainable development.

NOTES

1 The third target under SDG 3 – Target 3.3 – aims to achieve: "By 2030, end the epidemics of AIDS, tuberculosis, malaria and neglected tropical diseases and combat hepatitis, water-borne diseases and other communicable diseases" (United Nations 2015).

2 Or were you looking for the USA and had trouble finding anything before 2023? The US government was instrumental in negotiating the SDGs but prior administrations viewed the SDGs as non-binding guiding principles.

REFERENCES

Dooley, B., and H. Ueno. 2022. "Why Is This Colorful Little Wheel Suddenly Everywhere in Japan?" *New York Times*, August 25.

Edgecliffe-Johnson, A., and S. Neville. 2022. "Bill Gates Warns Ukraine War Is Sapping Europe's Foreign Aid Budgets." *Financial Times*, September 13.

Independent Commission on International Development Issues. 1980. *North-South: A Programme for Survival*. Cambridge, MA: MIT Press.

Independent Group of Scientists Appointed by the Secretary-General. 2019. *Global Sustainable Development Report 2019: The Future Is Now – Science for Achieving Sustainable Development*. New York: United Nations. Available at: https://sdgs.un.org/gsdr/gsdr2019.

Le Blanc, D. 2015. *Towards Integration at Last? The Sustainable Development Goals as a Network of Targets*. UN DESA. New York: United Nations. Available at: https://www.un.org/esa/desa/papers/2015/wp141_2015.pdf.

Niestroy, I., and L. Meuleman. 2016. "Teaching Silos to Dance: A Condition to Implement the SDGs." *SDG Knowledge Hub* (blog). Available at: http://sdg.iisd.org/commentary/guest-articles/teaching-silos-to-dance-a-condition-to-implement-the-sdgs/.

Rockström, J., W. Steffen, K. Noone, Å. Persson, F. S. Chapin III, E. Lambin, T. M. Lenton, M. Scheffer, C. Folke, H. J. Schellnhuber, B. Nykvist, C. A. De Wit, T. Hughes, S. R. Van der Leeuw, H. Rodhe, S. Sörlin, P. K. Snyder, R. Costanza, U. Svedin, M. Falkenmark, … J. Foley. 2009. "Planetary Boundaries: Exploring the Safe Operating Space for Humanity." *Ecology and Society* 14 (2).

Steffen, W., K. Richardson, J. Rockstrom, S. E. Cornell, I. Fetzer, E. M. Bennett, R. Biggs, S. R. Carpenter, W. de Vries, C. A. de Wit, C. Folke, D. Gerten, J. Heinke, G. M. Mace, L. M. Persson, V. Ramanathan, B. Reyers, and S. Sorlin. 2015. "Planetary Boundaries: Guiding Human Development on a Changing Planet." *Science* 347 (6223): 1259855. https://doi.org/10.1126/science.1259855.

UN News. 2021. "'Inequality Virus' Threatens 'Catastrophe for All' Unless Crisis Can Be Overcome Together." Available at: https://news.un.org/en/story/2021/04/1089872.

United Nations. 2015. "Transforming Our World: The 2030 Agenda for Sustainable Development." *General Assembly 70 Session*. New York: United Nations.

United Nations. 2025. "Countries Who Have Presented Their Voluntary National Reviews." *High-Level Political Forum on Sustainable Development*. Available at: https://hlpf.un.org/countries (accessed February 14, 2025).

United Nations Economic Commission for Africa. 2016. *From the Millennium Development Goals to Sustainable Development Goals: Path and Additional Efforts of West African Countries Towards Achieving the Objectives Set for 2030*. Available at: https://repository.uneca.org/handle/10855/23494.

2 "Business-as-usual is no longer an option"

TIPPING POINTS BEGGING FOR RADICAL CHANGE

We are reaching tipping points (see Figure 2.1; Zhong and Rojanasakul 2024 offer an animated version) – "a critical point in an Earth system component, around which small perturbations can trigger an irreversible transition from one stable state to another" (Independent Group of Scientists Appointed by the Secretary-General 2019, 13). UN Secretary-General António Guterres speaking on how we are on track to hit 2.8 degree Celsius global warming, stated it bluntly: "Humanity has opened the gates to hell" (Williams, 2023b, 6226). One of the most visible climate tipping points we are close to reaching are the melting of Arctic summer sea ice: a darker ocean area absorbs more sunlight, increasing warming and speeding up melting (Independent Group of Scientists Appointed by the Secretary-General 2019, 13). We could reach the first ice-free Arctic Sea by 2030 (Tauschinski 2024).

In Central Asia, glaciers are reaching their peak flow. After their tipping point, flow will reduce causing significant water shortages by 2050 (Fountain and Solomon 2019). Glacial retreat in Switzerland is already forcing adaptation to the centuries-old practice of transhumance – herding livestock in the Alps during summer (Porter and Steinmetz 2024). Other imminent tipping points include the abrupt thawing of the boreal permafrost (with resultant carbon release) or the die-off of coral reefs (Armstrong McKay et al. 2022).

What the climate tipping points have in common is that incremental change from global warming is having exponential effects. Their effects are not felt in isolation but expected to increase the likelihood of tipping others beyond the point of return (Armstrong McKay et al. 2022). As captured in Table 2.1 (Kaplinsky 2022), a 0.5 degree increase in global temperatures nearly triples the share of the population exposed to extreme heat every five years. Plant species losses are projected to double from 8 percent to 16 percent. Most dramatically, coral reefs are expected to decline by 99 percent (nearly extinguished) with 2 degrees of warming. Sea-level rise does not differentiate between the Global North and the Global South, affecting more than half the population in the Marshall Islands and more than a quarter of the population in the Netherlands by 2100 (Strauss et al. 2021).

DOI:10.4324/9781003400028-3

Our Global Tipping Points Mapped

Circles show projected tipping points that occur even if we stay below 2 degree Celsius threshold.

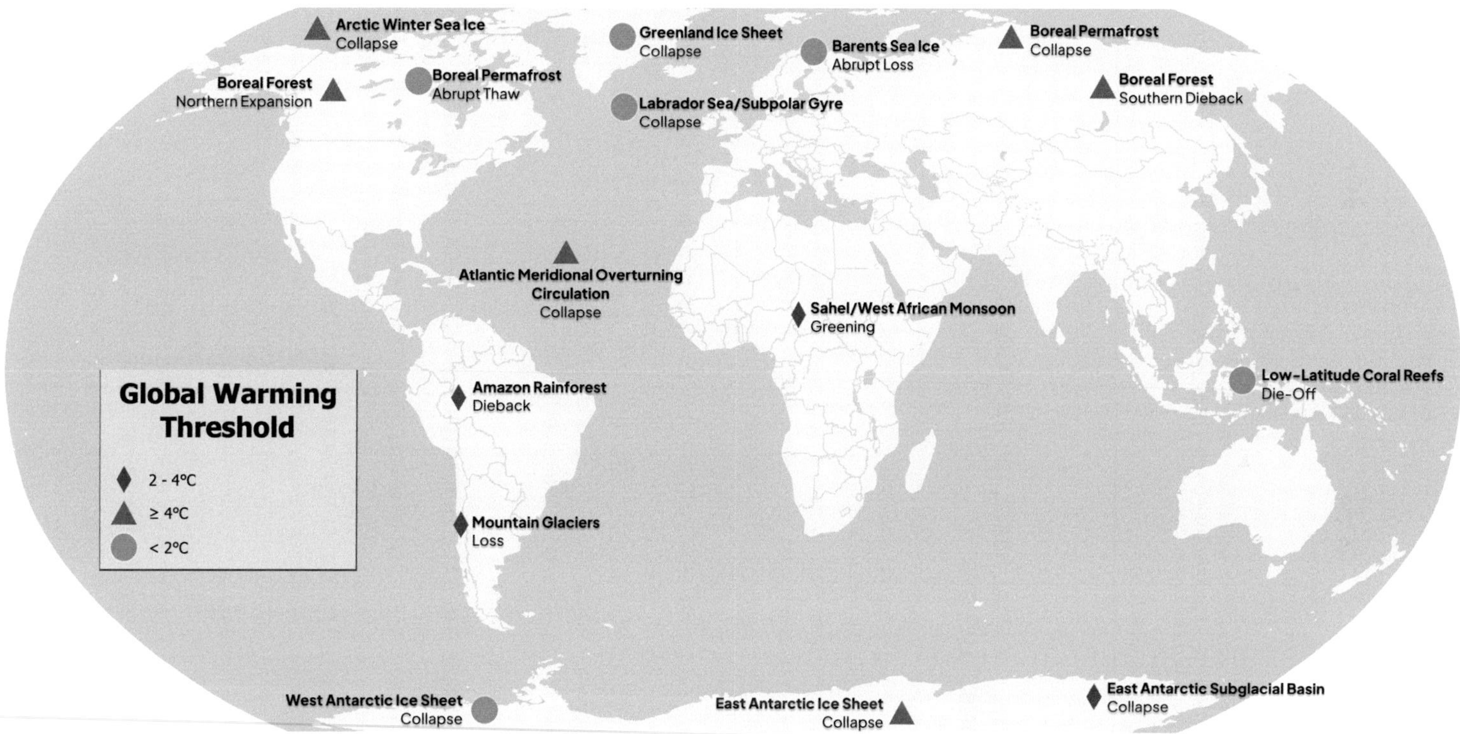

Figure 2.1
Our global tipping points mapped, adapted from Armstrong McKay et al. (2022)

Table 2.1 Projected impacts of 1.5 and 2.0 degree C of global warming, adapted from Levin (2018)

Impact	1.5° C	2.0° C
Extreme heat [global population exposed at least once every 5 years] (%)	14	37
Arctic ice [number of ice-free years]	1 every 100	1 every 10
Sea-level rise by 2100 (meters)	0.4	0.46
Vertebrate species loss (%)	4	8
Plant species loss (%)	8	16
Area of global land with shift in local ecosystems (%)	7	13
Amount of Arctic permafrost thaw (million sq kms)	4.8	6.6
Reduction in crop yields in tropics (%)	3	7
Decline in coral reefs (%)	70–90	99
Decline in marine fisheries (million metric tons)	1.5	3

VIGNETTE 1 CORAL REEFS, TIPPING POINTS AND CONSEQUENCES

Scientists project that tropical and subtropical coral reefs will cease to exist unless warming is halted well below 2° Celsius. At the heart of their tipping is an increase in water temperatures that results in a demise of the symbiotic algae that provides nutrients to the invertebrate coral and prevents them from bleaching. An increase of just 1° Celsius can cause mass bleaching events, which have increased in occurrence and reduce the timeframe for corals to recover (Great Barrier Reef Foundation 2023). Additional human-induced pressures include overfishing, direct damage (e.g., from recreation or coral harvesting for aquarium trade), and ocean acidification (Armstrong McKay et al. 2022; EPA 2023). The resultant loss in biodiversity will negatively impact the marine food web and millions of households depending on the oceans for their livelihoods.

LIVING OUTSIDE THE SAFE OPERATING SPACE OF OUR PLANET

We are entering the Anthropocene epoch, where human activities affect the Earth's systems and functioning. Scientists are increasingly worried that we are pushing the state of our planet too far to be still "conducive to the human welfare and societal development experienced during the [preceding] Holocene" (Richardson et al. 2023, 1). Human-induced climate change is attributed to increasing the likelihood of natural disasters, such as the flooding events in Western Europe during July 2021 by 1.2 to nine-fold (Tradowsky et al. 2023). Using an integrated approach analogous to the SDGs, the nine *planetary boundaries* introduced earlier (Rockström et al. 2009; Steffen et al. 2015) assess spaces "critically affected by anthropogenic [human] activities and relevant to Earth's overall state" (Richardson et al. 2023, 2). Six out of the nine planetary boundaries have already exceeded their safe operating space set at pre-industrial levels.[1]

Figure 2.2
We crossed three additional boundaries from 2009 to 2023, adapted from https://www.stockholmresilience.org/research/planetary-boundaries.html

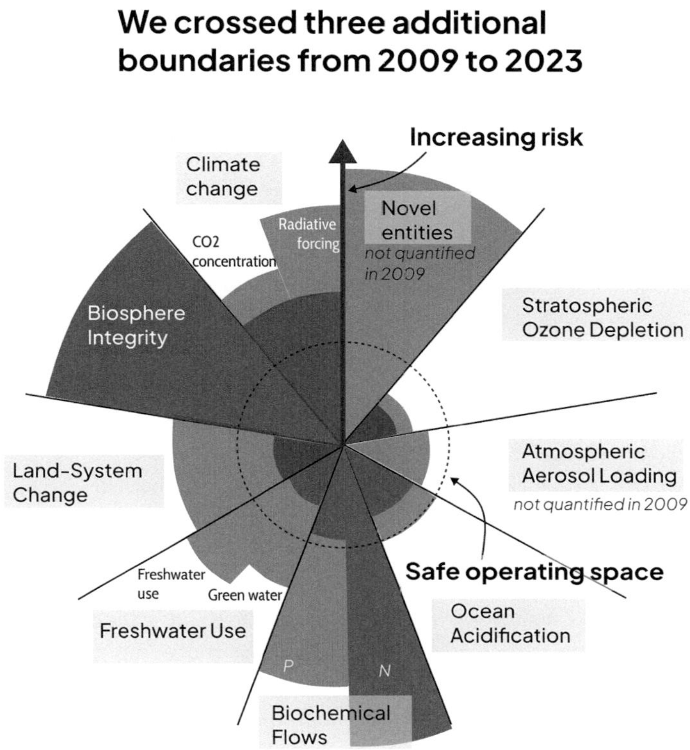

Out of those, four are approaching the highest risk zone (Richardson et al. 2023):

1. *Biogeochemical flow*, whereby nitrogen and phosphorous run-offs from fertilizer use lead to eutrophication, causing algal blooms, and ultimately ocean anoxia (depletion of oxygen) (Watson, Lenton, and Mills 2017).
2. *Biosphere integrity*, whereby a loss of genetic diversity through extinction of plants and animals limits the adaptive capacity of the biosphere.
3. *Climate change*, whereby the CO_2 concentration in the atmosphere accelerates global warming and the melting of the ice cap.
4. *Novel entities*, whereby synthetic chemicals,[2] microplastics[3] or organic pollutants that have not undergone an adequate safety assessment carry new risks (Richardson et al. 2023).

Since its original inception in 2009, three additional boundaries have been crossed thanks to improved analytics and continued worsening of our global state (see Figure 2.2).

WE LIVE BEYOND OUR MEANS SOONER EACH YEAR

On July 24, 2025 we reached the point where we are taking more from Earth than is available. This is 14 days earlier than in 2015, 55 days earlier than in 2000, and nearly four months earlier than in 1980. *Earth Overshoot Day* is calculated as the amount of resources the Earth can generate annually divided by humanity's demand that year.[4] The annual date vividly illustrates that even with all the progress made at the level of technological innovation and efficiency we are still not reducing our footprint enough (see Figure 2.3). Our current lifestyles still demand 1.8 Earths (Earth Overshoot Day 2025). You can calculate your own overshoot day at www.footprintcalculator.org; thanks to my air travel, mine occurs already on April each year.

How we are taxing our Earth led Indigenous activist and scholar Ailton Krenak to declare that sustainability is a myth:

> invented by corporations to justify their theft of our idea of nature ... But the fact is, it's dishonest to use a term like *sustainability* when we're on the verge of being expelled from Gaia ... No community that is in debt to the land can call itself sustainable, because we take out more than we can put back in.
>
> (Krenak and Doyle 2020, 21)

We will encounter more of his voice throughout our chapters.[5]

As argued by Raphie Kaplinsky (2022, 66ff.) and depicted in Figure 2.4, we are living in the era of the *Great Acceleration*: we make more (e.g., GDP per capita), we take more (e.g., forest area loss), we use more (e.g., electric power consumption), and we risk more (e.g., climate-related-disaster frequency).

Our material use has *only* increased from 7 to 10 metric tons per capita between 1970 to 2010. Taking into consideration our increase in global population we are now using over three times the materials for sustaining our livelihoods

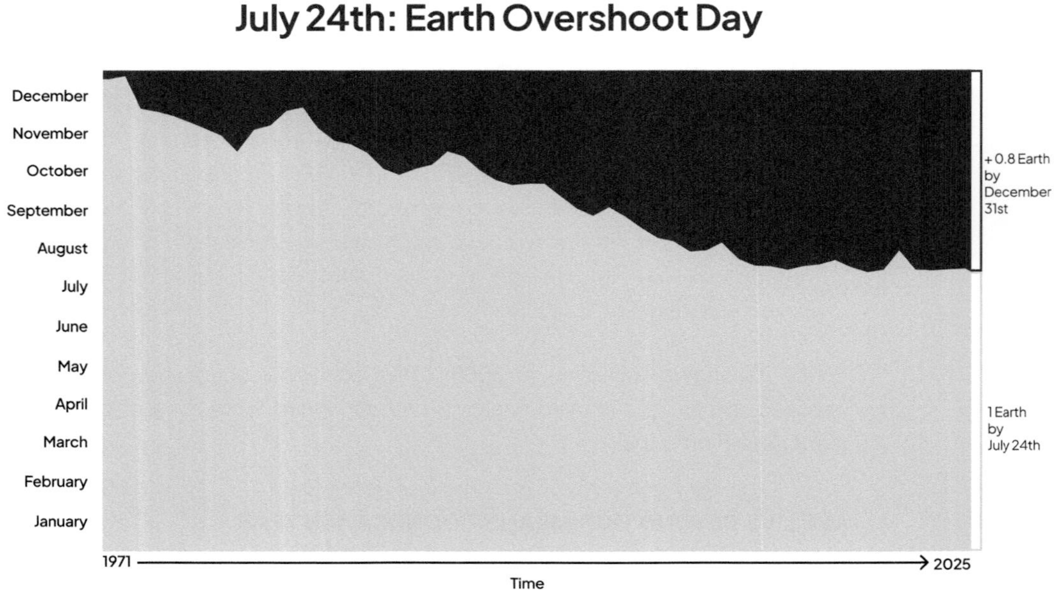

Figure 2.3 Earth Overshoot Day in 2025, adapted from Earth Overshoot Day (2025)

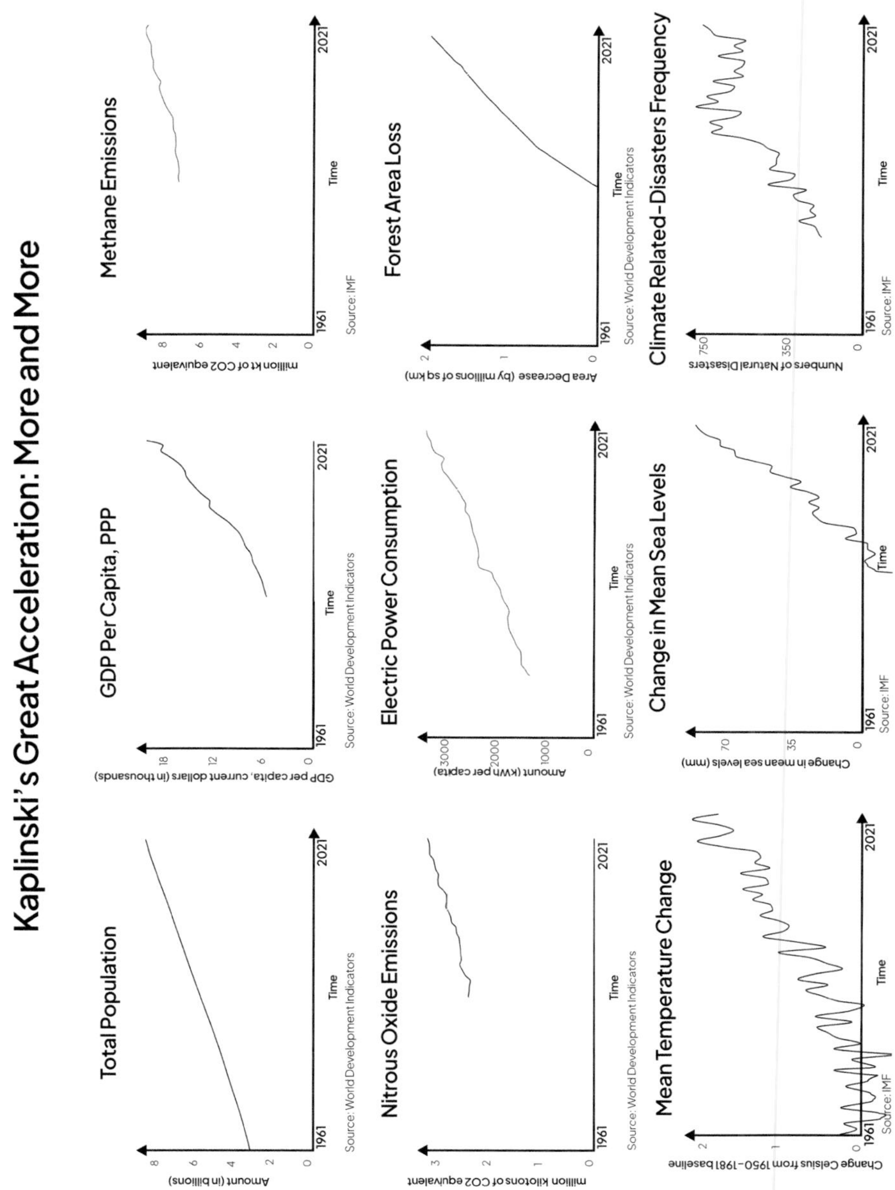

Figure 2.4 Kaplinsky's Great Acceleration: more and more, adapted from Kaplinsky (2022, 70)

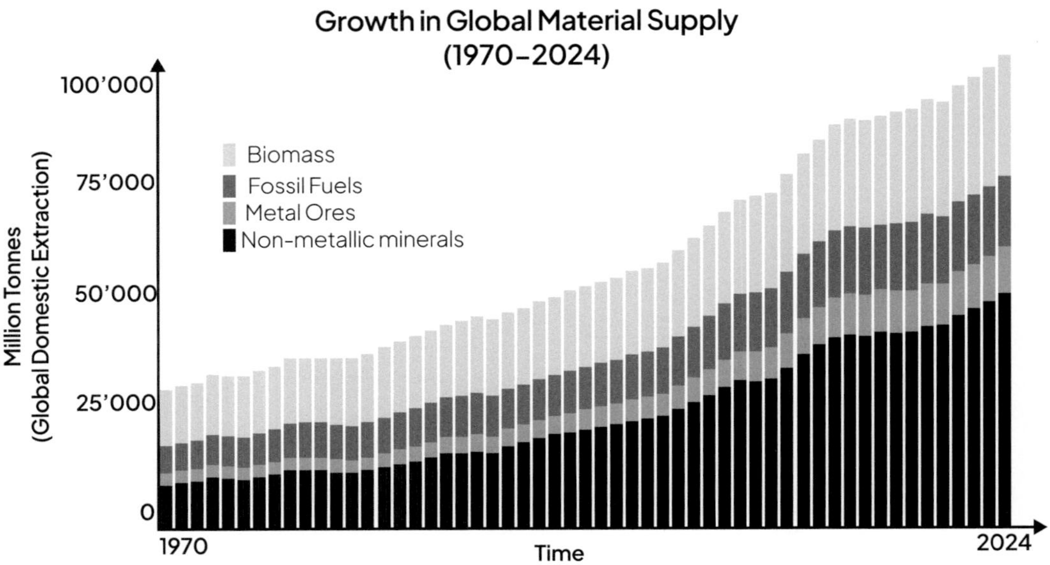

Source: UNEP (https://www.resourcepanel.org/global-material-flows-database)

Figure 2.5
The growth in global material supply, 1970–2024, adapted from Kaplinsky 2022, 77

(see Figure 2.5) (Kaplinsky 2022, 77). As of 2024, we are consuming twice the biomass since the 1970s, over 2.5 times more fossil fuels, four times more metal ores, and 5.5 times more non-metallic minerals. Technomass – structures built by humans, such as concrete, asphalt, or bricks — have surpassed in weight the combined biomass of Earth (Ménard and Shtarkman 2024).

But do not get hung up on the population increase: the high-income countries (representing 7 percent of global populations) account for nearly 50 percent of global consumption (Kaplinsky 2022, 81). The top 10 percent of the world's polluters account for nearly half of all greenhouse gas emissions, illustrating the inequalities (Hodgson 2023).

Sustaining that trend is conspicuous consumption and planned or perceived obsolescences. The Great Acceleration coincided with massive spending on advertisements to U.S. consumers (moving from around US$10 billion in 1960 to over US$250 billion by 2000) (Kaplinsky 2022, 86). Digital advertising revenue skyrocketed from 2000 onwards, moving from US$8 billion to US$225 billion in 2023 (Statistica Research 2024).

WORSENING TRENDS HIGHLIGHTING NEED FOR TRANSFORMATION

Inequality, climate change, biodiversity loss, and our material footprint are all trending in the wrong direction (Independent Group of Scientists Appointed by the Secretary-General 2019), undermining our ability to achieve sustainable development that supports safe and just living. Since the COVID-19 pandemic, further challenges in poverty, food security, and insecurity have trended negative (Independent Group of Scientists Appointed by the Secretary-General 2023). These challenges are not just limited to a single world region. European countries struggle to reduce their carbon emissions and inequalities. In South Asia, progress on gender inequalities is slowing. Conflicts in Gaza, Lebanon, Syria, and Yemen are increasing poverty and food insecurity across the Arab region. Poverty reduction is stagnating or decreasing in Latin America.[6] In sum, the potential of transformations is not limited to a single country or region, but omnipresent.

Below, we offer a deeper dive into youth unemployment, the effects of COVID-19 on inequality, and marginalization and backsliding.

Youth unemployment

In Africa, youth unemployment is rising (Independent Group of Scientists Appointed by the Secretary-General 2019, 15). In South Africa, unemployment among young individuals, for example, hovers around 50 percent. While educational quality was improved, a sizable rural-urban gap persists. Many highly qualified university graduates are struggling to find jobs. In East African countries, I would meet parents whose Ph.D. and M.A. graduates are struggling to find a job as the demand remains low for their skills. Even with higher skills, limits to social mobility persist. Youth unemployment in the Arab world has reached the highest level globally. I advised former students who worked with the UN Economic and Social Commission of West Asia (ESCWA) to analyze the effects of education as the great leveler across the Arab world. They found that "if not carefully managed, education can reinforce inequalities, as the wealthy have access to higher quality education systems than those living in poverty, which is reflected in the job opportunities available to them upon graduation" (UN ESCWA 2024). Especially in middle-income countries, there is increased pessimism, with nearly three out of four dissatisfied with their employment opportunities (UN ESCWA 2024). In lower-income countries, job opportunities are worsening, and satisfaction with the education system is declining (UN ESCWA 2024). And across the globe, the majority of parents think their children will be worse off (Doroszczyk 2025).

Inequality and COVID-19

Research by Perry et al. (2021) highlighted how shocks, such as the global pandemic, exacerbate pre-existing gender, race, age, and educational inequalities in the United States. In rural India, Singh (2020) reported on an increase in suicides as farmers' struggles deepened under the COVID-19 lockdown. Dubbed the *inequality virus* by the UN (UN News 2021), you will have experienced the challenges that came with shifting to remote education or work during COVID-19. These challenges were omnipresent across the globe, as students experienced increased levels of stress and anxiety (UNESCO 2022). Hundreds of millions of students could not access digital content, and innovations deployed – such as new digital learning platforms – are already falling apart post-COVID-19 (UNICEF 2022). If education was under-funded pre-COVID, it likely emerged with even less funding post-COVID. These challenges are not just limited to public institutions: even at my private university, we have yet to recover to pre-pandemic budgets. The results are devastating. As relayed by one of my graduate students who works with eighth graders at a local high school in Washington, DC, she can see the pandemic's effects as students fail to pass third-grade reading skills assessments. Across the United States, a new historic low was reached with 33 percent of eight graders scoring below basic reading skills (Goldstein 2025). While economies have recovered in the short run, we expect that the long-run effects of uneven access to opportunities exacerbated by shocks will be felt for a generation. As captured by the United Nations Educational, Scientific and Cultural Organization (UNESCO):

> Although connected technology supported the continuation of education for many learners, many more were left behind. Exclusion soared and inequities widened. Achievement levels fell, even for those with access to distance learning. Educational experiences narrowed. Physical and mental health declined. Privatization accelerated, threatening education's unique standing as a public good and human right. Invasive surveillance endangered the free and open exchange of ideas and undermined trust. Automation replaced human interactions with machine-mediated experiences. And technology production and disposal placed new strains on the environment.
>
> (West 2023, 18)

Marginalization and backsliding

In the words of Perry et al. (2021, 1), "crises lay bare the social fault lines of society." COVID-19 hit the most vulnerable hardest. About 15 percent of the world's population experience disabilities (Independent Group of Scientists Appointed by the Secretary-General 2019, 16). They were further isolated during the pandemic as inclusive access to healthcare was no longer a priority, remote education was inaccessible, and food was difficult to obtain (Kubenz and Kiwan 2022). Discrimination for over 370 million Indigenous peoples across 90 countries (Independent Group of Scientists Appointed by the Secretary-General 2019, 16) continued into the pandemic. State actors also increased repressions and human rights violations during the pandemic. Democratic backsliding came in the form of government-led disinformation campaigns (e.g., in Brazil[7] or Tanzania) and increased restrictions on media freedom (e.g., in Poland or India) (Gabbero et al. 2022; V-Dem 2022a). Select countries used the state of emergency to put discriminatory measures in place (V-Dem 2022a). At the beginning of the pandemic, Saudi Arabia, for example, deported thousands of Ethiopian migrant workers under the argument that they were "vulnerable to the coronavirus" (V-Dem 2022b). Human rights violations increased, with COVID-19 policies – such as limits on peaceful assembly – used to justify arrests of civil rights leaders and protestors. For example, in Niger, prominent members of the NGO Turning the Page Niger, which uncovered a misuse of public funds, were arrested (Gabbero et al. 2022). During the pandemic, attacks on defenders of Indigenous rights frequently increased, with COVID-19 restrictions limiting their ability to file police reports or access the judicial system (Gabbero et al. 2022).

VIGNETTE 2 HOW SUBSIDIES FUEL BUSINESS-AS-USUAL AND TOXIFY DEVELOPMENT

Subsidies historically represent the fuel to the *business-as-usual* engine.[8] In *Focal Point: SDG 2*, I further highlight how subsidies perpetuate business-as-usual in agriculture. In Chapter 8, I will dive deeper into my experience and the role of subsidies in agriculture. This vignette summarizes their detrimental effects across development, drawing on the World Bank's flagship report: *Detox Development: Repurposing Environmentally Harmful Subsidies* (Damania et al. 2023). The authors argue that "in an era of fiscal constraints and degrading natural capital, reform and repurposing of perverse and harmful subsidies offer an opportunity to promote greater sustainability, inclusion and shared prosperity" (Damania et al. 2023, xxi). In 2020, global debt reached the highest levels since a half-century ago at 263 percent of GDP (Damania et al. 2023). By 2022, subsidies for coal, oil and natural gas amounted to 7.1 percent of global GDP, a larger share than what governments around the world spend on education (Williams 2023a).

Direct fossil fuel subsidies amounted to US$577 billion in 2021, or three-quarters of all energy subsidies. They increase the consumption of fossil fuels and reduce incentives for transitioning to cleaner energy. The commitments under the Paris Agreement on Climate Change of financing US$100 billion are comparatively small (Damania et al. 2023). Even the COP29 decision raised the annual commitment to only US$300 billion by 2035 (UNFCCC 2024).

In agriculture, subsidies are expected to exceed US$1 trillion, most of which are coupled to production (e.g., input or crop-specific subsidies). Such subsidies "[distort] farmers' decisions, often reducing productivity and causing harmful environmental spillovers that encourage deforestation, pollute waterways, and deplete water supplies – often beyond national borders" (Damania et al. 2023, xxii). Subsidies in fishing directly threaten our nutrition security. They increase the fishing capacity on top of the already 34 percent of fisheries being overfished, putting at risk a critical nutritional source for over 3 billion people (Damania et al. 2023, xxi). Subsidized Chinese fishing fleets, for example, have spread as far as West Africa, aided by over US$20 billion in subsidies between 2011 and 2015 alone (Jacobs 2017). Estimates are that their ability to out-compete local boats cost West African nations annual fishing revenues as high as US$2 billion (Doumbouya et al. 2017). A new World Trade Organization (WTO) treaty aims to curb fisheries subsidies (WTO 2022), but subsidizing fleets in foreign waters are exempt from it (Jungwiwattanaporn 2024).

On top of explicit subsidies, implicit subsidies capture "the price difference between the 'undistorted' (socially optimal) price and the actual price that emerges after the subsidy is paid" (Damania et al. 2023, xxiii). They amount to US$6 to 10.8 trillion (Damania et al. 2023). Implicit subsidies in fossil fuel are estimated at US$ 5.4 trillion, including the effects of burning fossil fuels, which "emits harmful chemicals into the air, including fine particulate matter ($PM_{2.5}$) and sulfur dioxide, which have enormous impacts on health, as well as greenhouse gases such as carbon dioxide, which contribute to global climate change" (Damania et al. 2023, xxiii). Of the US$45.4 trillion, 75 percent is attributed to externalities affecting local air .pollution and global climate change.

In agriculture, environmental externalities are estimated to be more than US$5 trillion per year, derived from increased greenhouse gas emissions, loss of natural capital, pollution, overuse of pesticides, and antimicrobial resistance (Damania et al. 2023).

And who benefits the most from these subsidies? The richest. As subsidies encourage people to build bigger houses or purchase more cars, the wealthiest income group would lose 13 times more than the poorest if they were removed (Damania et al. 2023, xxvi).

Damania et al. identify the benefits accrued from special interest groups as an obstacle to change subsidy policies. We get used to business-as-usual: "economies and people adjust to [the subsidies'] presence, which builds inertia against change due to behavioral biases that favor the status quo" Damania et al. 2023, 265). Most harm by subsidies is also not visible or only appears after a significant time lag, "making attribution of damage difficult and weakening public pressure for reform" Damania et al. 2023, xxx).

With externalities manifesting themselves across national boundaries and nations often leveraging subsidies for their domestic politics, creating meaningful transnational action remains challenging.

The current use of subsidies captured in Vignette 2 runs counter to achieving sustainable development. While representing a persistent problem, their reform also offers a key avenue to achieving sustainable development. We will discuss potential reforms and the importance of engaging our changemakers throughout the book. Next, let us take a closer look at agriculture, a major sector fueled by subsidies where business-as-usual is no longer an option.

FOCAL POINT: SDG 2

Agricultural research and development have translated successively into higher yields and productivity globally. As global giants such as Brazil, China, and India increase their income per capita, so have their cereal yields. Starting from around 1,000 kilogram (kg) per hectare (ha) in the 1960s, they have reached between

3,000 to nearly 6,000 kg per ha by the 2000s. This period of growth is characterized as the Green Revolution, which marked the use of improved (i.e., hybrid) seed varieties combined with irrigation, synthetic fertilizers, and pesticides. As a result, the impressive productivity achievements followed the same pattern as the Great Acceleration described earlier.

Figure 2.6 captures the trends from 1961 to 2021. Meat production per capita increased nearly two-fold (from 23 kg per capita in 1961 to 43 kg per capita in 2021) and cereal production increased by around 25 percent by the 1980s (from 286 kg per capita in 1961 to 360 kg per capita in 1981). To support this growth, nitrogen use in agriculture quadrupled (from 45 million metric tons in 1961 to 184 million metric tons in 2021), with phosphorus use tripling (from 9 million metric tons in 1961 to 27 million metric tons in 2021). The area of land under irrigation more than doubled, reaching 353 million ha in 2021. Inflation-adjusted imports of agricultural pesticides have risen over 15-fold since 1961 (from US$893 million in 1961 to nearly US$17 billion in 2021). In summary, growth in agricultural production closely tracked the rise of capital-intensive inputs.

The Green Revolution, however, failed to reach large parts of sub-Saharan Africa (see Figure 2.7). In Asia, representing intensification, yields have more than doubled without increasing the area under cultivation. In sub-Saharan Africa, representing extensification, yield gains were achieved through an 80 percent increase in land under cultivation. African farmers largely lacked access to the four critical ingredients of hybrid seeds, irrigation, fertilizers, and pesticides.

I outline five reasons why these divergent agricultural development paths form the basis for calls that "business-as-usual is not an option" anymore.

First, the intensification model of agricultural development is reaching its limits (Figure 2.8). The substantial annual growth rates achieved at the beginning of the Green Revolution (as high as 6 percent for wheat and around 3 percent for maize and rice) are constantly declining. Using compound growth rates, the percent increase in yield gains is getting smaller over the years. By 2022, they are logging at 2.02 percent for wheat, 1.79 percent for maize (corn), and 1.52 percent for rice. Yield gaps – the spread between the optimal and attained yield – are also declining (e.g., rice; Yuan et al. 2022).

These declining yield increases showcase the limits to research and development (R&D) investments in improved breeding and hint at a reduced response to chemical fertilizer and pesticide use of crops grown. With chemical pesticides banned, facing lawsuits due to adverse health effects, or encountering lower efficacy due to pesticide resistance, biological pest control methods are on the rise (Cookson 2019).

Across Asia, rice farmers feel the brunt of climate change and challenges in water access. Adaptation is the name of the game: from shifting planting calendars, to identifying new rice varieties, to changes in irrigation practices (Sengupta et al. 2023). Given the heat-sensitive nature of rice, the seeds are highly susceptible to temperature increases, fueling fears of export bans and increased food inflation (Savage et al. 2023). Across Europe, wheat yields primarily stagnated in

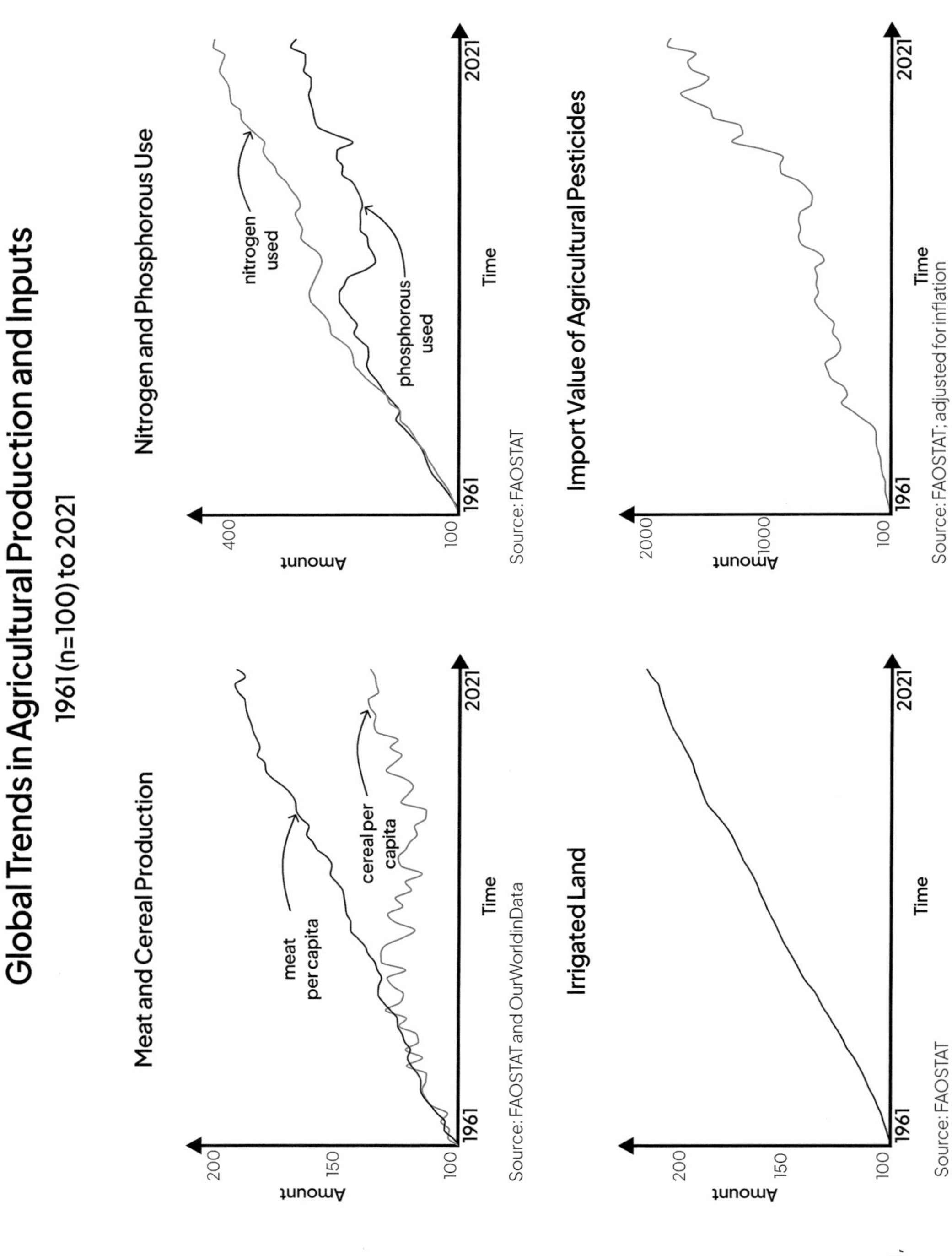

Figure 2.6
Global trends in agricultural production and inputs, 1961–2021, FAOSTAT (n.d.)
n = 100

Figure 2.7
Four decades of the Green Revolution: intensification in Asia and extensification in Africa from 1961 to 2004, adapted from World Bank (2007)

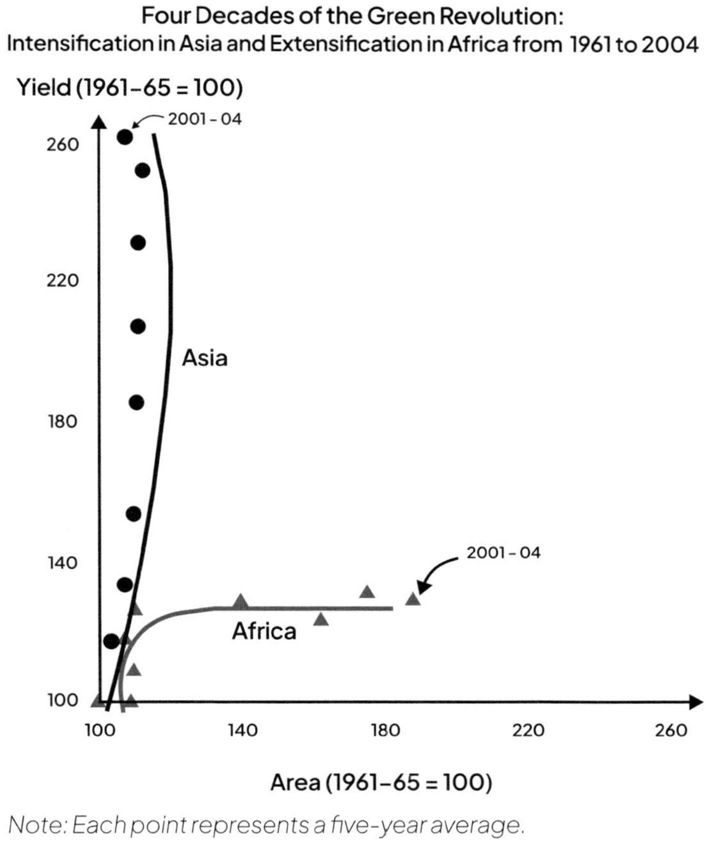

Figure 2.8
Declining growth rate of cereal yields, 1962–2022, FAOSTAT (n.d.)

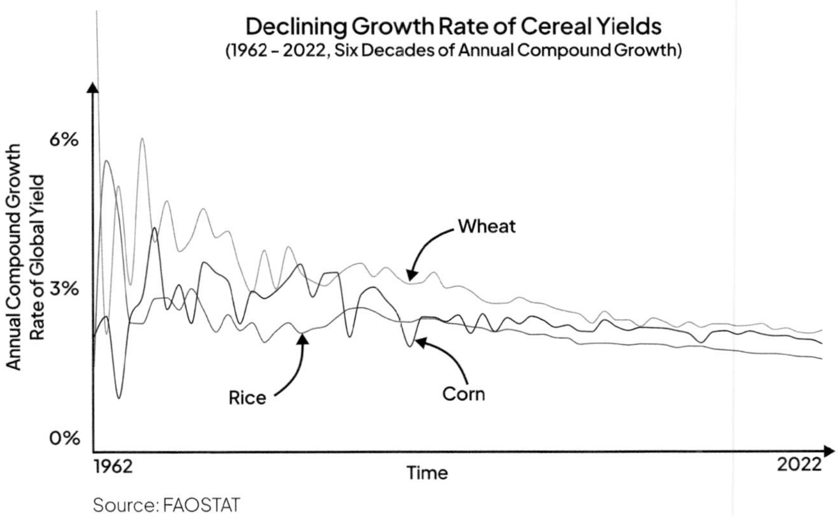

the 1990s, including Switzerland (1990), the United Kingdom (1996), and Germany (1999), resulting in diminishing returns as costs for inputs increased (Brisson et al. 2010). In South and East Asia, subsidized fertilizer stimulates excessive overuse detrimental to yields (Damania et al. 2023, xxvii).

Subsidies' record of improving productivity is mixed, with farm efficiency declining even with increased outputs (Damania et al. 2023, xxvi). At the low end, most fertilizer use in sub-Saharan Africa is between 3 to 6 kg per ha. In East Asia and the Pacific, the majority used over 154 kg per ha, around four times the global average (Damania et al. 2023, xxvii). Global fertilizer use could be reduced without a negative effect on yields outputs, as the crops can make use of less than half of all fertilizers applied. Every second calorie is produced in areas with excessive use of fertilizer (Damania et al. 2023, xxviii).. Excessive fertilizer use also creates runoffs into waterways, with input subsidies responsible for 17 percent of all nitrogen pollution (Damania et al. 2023, xxviii).

Second, conventional farming maximized yield gains through improved agricultural practices, including monocultures, at the cost of continued declines in soil fertility (Figure 2.9). Illustrating the power (and limitations) of reporting against the SDGs, SDG Goal 15 on *Life on Land*, Target 15.3 states that "[b]y 2030, combat desertification, restore degraded land and soil, including land affected by desertification, drought and floods, and strive to achieve a land degradation-neutral world." It is measured using a single indicator (15.3.1), which is the "annual change in degraded or desertified arable land (% or ha)." Based on the voluntary reporting done from 117 countries in 2015 and 2019, we have a picture of the degraded land share. Four regional hotspots emerge: (1) Central America (with Mexico having 64 percent of all land degraded); (2) the Sahel region in sub-Saharan Africa (with Benin reporting 38 percent of all land degraded);

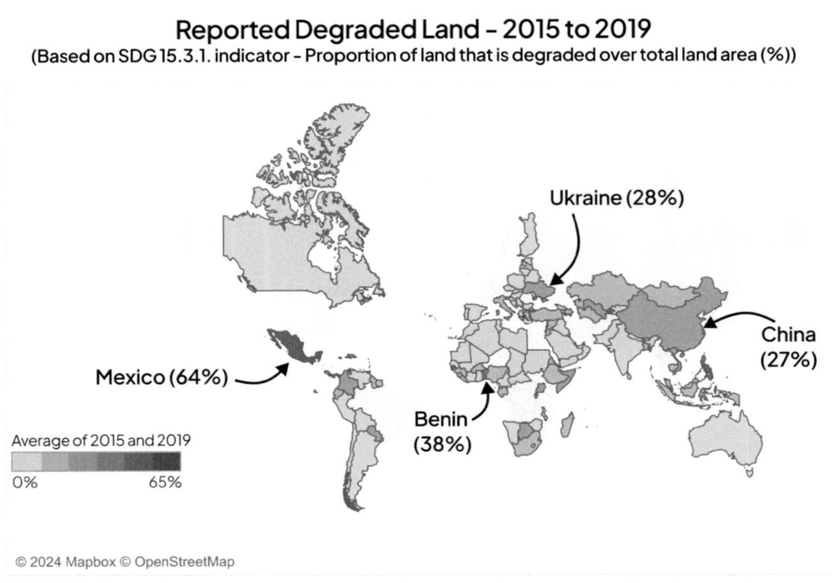

Figure 2.9 Reported degraded land, 2015–2019, https://unstats.un.org.sdgs/dataportal/database

(3) Eastern Europe (with Ukraine reporting 28 percent of all land degraded); and (4) East Asia (with China reporting 27 percent of all land degraded). The incomplete global picture, however, masks broader regional variations. Funding for soil research is notoriously limited. It was not until 1991 that the first global assessment on land degradation was completed (ISRIC n.d.). Our map in Figure 2.9 shows high or very high land degradation across the Sahel, Southern Africa, East Africa, and the Horn of Africa, extending into South Asia and East Asia. These regions overlap with severe food and nutrition insecurities.

Countries with very high land degradation include Brazil and the United States. Voluntary reporting also shows its limitations. For example, India states that degradation levels are at less than 7 percent, whereas the ISRIC assessment found large stretches of the country with high to very high land degradation. These land degradation challenges translate into declining soil fertility. We take capital out of our soil piggy bank without replenishing them with nurturing long-term investments or paying for the environmental costs at the check-out aisles (see True-Cost Accounting vignette).

VIGNETTE 3 TRUE-COST ACCOUNTING

The concept of True-Cost Accounting prices externalities (such as environmental costs of taxing the soil or pollution) into products. This method can yield vastly different prices for agricultural products (DePillis, Andreoni, and Einhorn 2024). At the low end, tofu, with a retail price of US$2.42 per pound, would cost less than 10 percent more (+US$0.21) if the estimated environmental costs were included. Water use is minimal (+US$0.01), with adverse ecosystem effects through the expansion of soybean cultivation accounting for nine extra cents and emissions from harvest to consumption for 11 extra cents. Beef – priced at US$5.34 per pound at a major US retailer – has an additional environmental cost estimated at US$22.02 (or over four times the current cost). Its lowest environmental cost is water use (+US$0.23), but emissions from manure and cow belching (from fermentation) contribute to significant methane emissions (+US$3.00). Its most considerable environmental cost relates to the inefficiency of converting agricultural production into animal feed into cow meat. The negative environmental impact of using agricultural land to grow cattle feed accounts for an additional US$18.79. Translating these costs into the grocery store risks further increasing uneven access to food. Better data on true costs, however, offer the potential to change consumers' and policymakers' decisions as they decide where to invest their finite funds to ensure we stay within our planetary boundaries (DePillis, Andreoni, and Einhorn 2024).

These challenges of degradation[9] and declining soil fertility are accelerated by climate change. Agriculture is now widely recognized as a crucial contributor to global warming (IPCC 2019). In addition to losing fertile topsoil due to wind and water erosion, plowing the soil releases carbon back into the atmosphere. The impacts are especially large when we repurpose nature for agricultural use with long-lasting effects. According to Tim Searchinger from the World Resources Institute, drained peatlands continue to emit greenhouse gases, accounting for one out of five gigatons released each year from land use changes (Flavelle 2019). Beyond draining peatland, deforestation is the largest driver of greenhouse gas emissions. Agricultural subsidies across the globe are fueling deforestation far away. For example, livestock subsidies in the USA or Switzerland increase demand for soybean as feed, driving further deforestation in Brazil (Damania et al. 2023, xxxviii).

Third, public funding for agriculture continues to decline, reaching a new normal in the last two decades (see Figure 2.10). As development priorities shift from agriculture to manufacturing and services industries, development assistance has dropped from nearly 19 percent in the early 1980s to less than 5 percent by the mid-2000s. Although funding levels have increased again following the global food crisis in 2007–2008, we do not expect a return of higher funding levels as long as it is primarily allocated based on expected growth. As discussed in Chapter 4, the war in Ukraine has shifted aid away from Africa. The dismantling of the U.S. Agency for International Development (USAID) in the first weeks of the second Trump administration also showed a new low point in support for aid among the dominant political class.

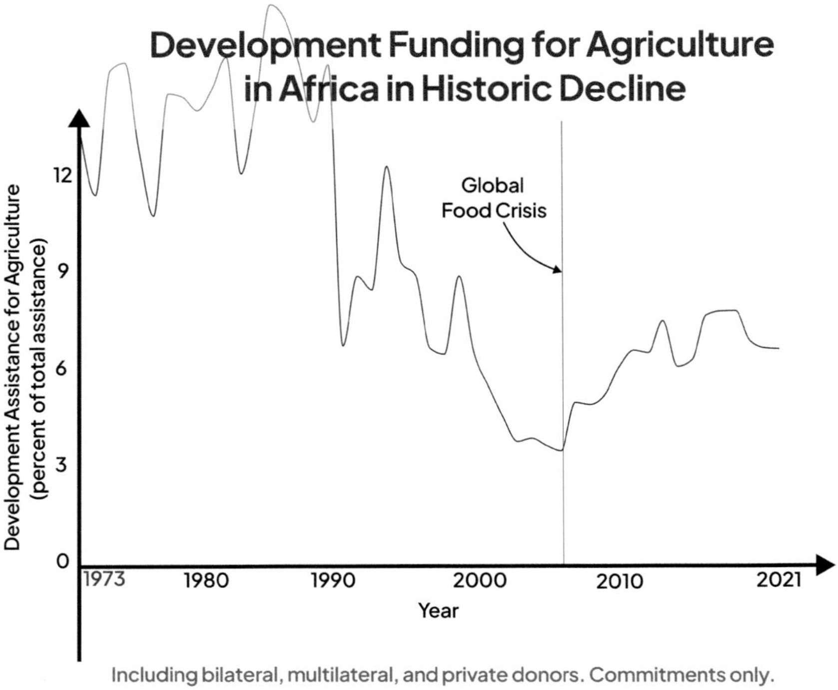

Figure 2.10 The new normal of development funding for agriculture in Africa, FAOSTAT (n.d.)

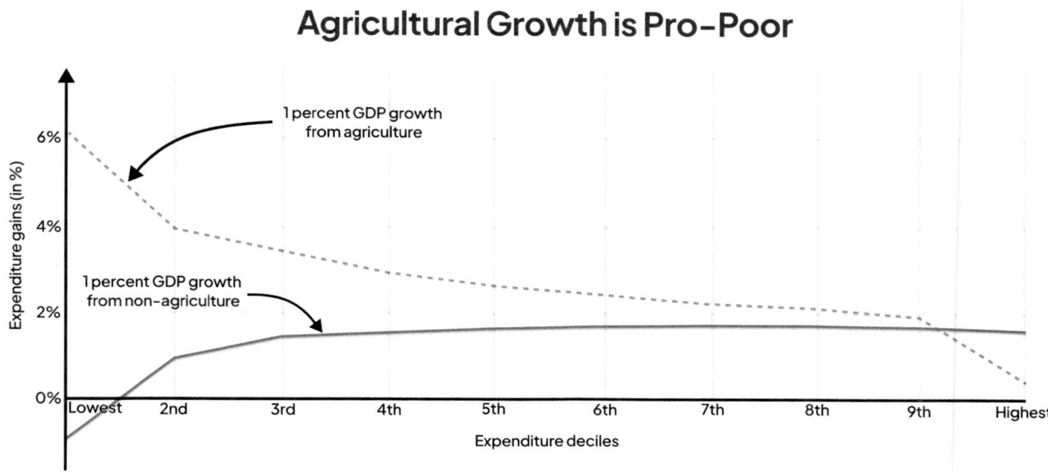

Figure 2.11 Agricultural growth is pro-poor, adapted from World Bank (2007) (see also Ligon and Sadoulet 2008)

Fourth, rural-to-urban inequalities have increased. Post-COVID-19 pandemic, there is an urgent need to reinvest in strategies to reduce these gaps by focusing on smallholder farmers and rural populations. Agriculture inherently is pro-poor (see Figure 2.11; Ligon and Sadoulet (2008)): a 1 percent growth in GDP in the agricultural sector raises expenditure levels the most for the poorest of the poor (by 6 percent), whereas a 1 percent growth in GDP in the non-agricultural sector results in around 1 percent lower expenditures by the poorest of the poor. In both scenarios, the wealthiest households increase their expenditures, with the growth from the non-agricultural sector disproportionally benefitting wealthier population segments.

While increasing income, agricultural subsidies might benefit the wealthiest farmers most.[10] The importance of focusing on smallholder or family farms is warranted given that 98 percent of all farms globally are family farms, cultivating over 50 percent of all farmland (Graeub et al. 2016).

Fifth, even with growth in agricultural productivity, vulnerabilities and fragilities have increased, especially in sub-Saharan Africa and Latin America. Undernourishment continues to loom large across sub-Saharan Africa and South Asia. Progress on achieving food security has also slowed, and we are still far from reaching the target by 2030 (Independent Group of Scientists Appointed by the Secretary-General 2023, xix). New vulnerabilities are emerging. The Intergovernmental Science-Policy Platform Biodiversity and Ecosystem Services (IPBES) estimated that the economic costs of invasive species amounted to US$423 billion a year, quadrupling every decade (Cookson 2023). Deforestation in the tropics is also furthering zoonotic and vector-borne diseases. Climate change is further pushing the range of insects, for example, expanding mosquito frontiers well north in both the United States and China (Peel 2024).

Agricultural subsidies are also linked to between 1.3 to 3.8 million additional malaria cases, creating annual economic costs up to US$19 billion (Damania et al. 2023, xxviii). Deforestation continues to affect marginalized and Indigenous groups most severely, especially women and rural youth. Beyond losing their livelihoods, forests served as their sources of food, medicine, and cultural meaning (Damania et al. 2023, xxviii).

These trends and developments highlight the need to transform agricultural production, as business-as-usual is no longer an option.

TRANSFORMATIONS AS AN EXIT RAMP

Transformations offer an off-ramp from the business-as-usual highway. Instead of a spur-of-the-moment change in lanes, they represent a conscious decision – based on societal and scientific support – to achieve substantial change (Independent Group of Scientists Appointed by the Secretary-General 2019).

Taking the exit lane to transformation will bring renewed hope and benefits. It will also come with costs and trade-offs. To better map out our journey ahead, Part II introduces an assessment of co-benefits and trade-offs before highlighting our changemakers.

YOUR TURN

Review the latest trend of your SDG using the updated SDG Dashboard (https://dashboards.sdgindex.org/map/goals/SDG1/trends). Capture why business-as-usual is no longer an option.

NOTES

1 For use of freshwater, for example, excess drawdown of groundwater for irrigation depletes water aquifers, "leading to reductions in surface-water flows and an increased risk of land subsidence [, the gradual settling or sudden sinking of land]" (Gupta et al. 2024, 24). Regarding air pollution, "[p]otentially controllable anthropogenic emissions contribute to a reduction of around 1.7 years of global average life expectancy, 1.1 years of which can be attributed to fossil fuel use" (Gupta et al. 2024, 29).

2 Novel research highlights the potentially devastating effects of forever chemicals — synthetical chemicals classified as per- and polyfluoroalkyl substances (PFAS). Contained in fertilizer made from city sewage, PFAS are contaminating US farm land and are suspected of causing deaths of livestock (Tabuchi 2024).

3 Ahuja (2024) points to a generational increase among young people of gastrointestinal cancers. In New Zealand, this increase mirrors the surge in microplastic pollution observed.

4 Large global variations exist. The United States has a bio capacity deficit of 3.7 ha per person, Switzerland has one of 2.6 ha per person. Canada, given its lower population density, has a bio capacity surplus of 6.6 ha per person (Global Footprint Network 2025).

5 Thanks to my former graduate student, Mariana Felix Angioletti, who introduced me to his work.

6 Sustainable Development Solutions Network (2024) allows you to explore the latest trends.

7 Then-president Bolsonaro minimized the consequences of COVID-19 and reduced budgets. With fatal consequences: remote Indigenous populations in the Amazon failed to have access to medical services, with 50,000 Indigenous people infected, and over 1,000 deaths recorded by January 2021 (Gabbero et al. 2022).
8 Subsidies can also fuel a transformation. In 2023, the United States passed the Inflation Reduction Act and the Chips and Science Act. Europe and Asia were racing to not get left behind in advances on semiconductors, hydrogen, or batteries (Chazan, Fleming, and Inagaki 2023).
9 Beyond soil degradation, the Millennium Ecosystem Assessment (2005) identified 15 out of 24 ecosystem services that "are being degraded or used unsustainably."
10 The biggest spenders on subsidies are in the Global North. The distorting effects of subsidies are also present in the Global South. In Malawi and Tanzania, for example, input subsidy programs are meant to improve access for poorer farmers to fertilizers, seeds or pesticides. They have a significant effect on their income, but also increase inequalities: for every single US dollar paid to the bottom quintile, 5 US dollars are paid to the top income quintile (Damania et al. 2023, xxvi).

REFERENCES

Ahuja, A. 2024. "Rising Rates of Cancer in Young People Prompt Hunt for Environmental Culprit." *Financial Times*, August 6. https://www.ft.com/content/491d7760-c329-4f57-9509-0da36bc9e7de.

Armstrong McKay, D. I., A. Staal, J. F. Abrams, R. Winkelmann, B. Sakschewski, S. Loriani, I. Fetzer, S. E. Cornell, J. Rockström, and T. M. Lenton. 2022. "Exceeding 1.5 Degrees C Global Warming Could Trigger Multiple Climate Tipping Points." *Science* 377 (6611): eabn7950. https://doi.org/10.1126/science.abn7950.

Brisson, N., P. Gate, D. Gouache, G. Charmet, F.-X. Oury, and F. Huard. 2010. "Why Are Wheat Yields Stagnating in Europe? A Comprehensive Data Analysis for France." *Field Crops Research* 119 (1): 201–212. https://doi.org/10.1016/j.fcr.2010.07.012.

Chazan, G., S. Fleming, and K. Inagaki. 2023. "A Global Subsidy War? Keeping Up with the Americans." *Financial Times*, July 12. https://www.ft.com/content/4bc03d4b-6984-4b24-935d-6181253ee1e0.

Cookson, C. 2019. "Farmers Reap Rewards of Biological Pest Control." *Financial Times*, September 25. https://www.ft.com/content/fd98fd9c-be7f-11e9-9381-78bab8a70848.

Cookson, C. 2023. "Biodiversity Body Warns of $423bn Annual Hit from 'Invasive Alien Species'." *Financial Times*, September 4. https://www.ft.com/content/e574795f-77a6-4407-a7af-dbafe01f9494.

Damania, R., E. Balseca, C. de Fontaubert, J. Gill, K. Kim, J. Rentschler, J. Russ, and E. Zaveri. 2023. *Detox Development: Repurposing Environmentally Harmful Subsidies*. Washington, DC: World Bank. Available at: https://www.worldbank.org/en/topic/climatechange/publication/detox-development.

DePillis, L., M. Andreoni, and C. Einhorn. 2024. "The Hidden Environmental Costs of Food." *New York Times*, September 19. https://www.nytimes.com/interactive/2024/09/19/climate/food-costs-protein-environment.html.

Doroszczyk, M. 2025. "Generational Concerns." *Finance and Development Magazine.*, IMF. March.

Doumbouya, A., O. T. Camara, J. Mamie, J. F. Intchama, A. Jarra, S. Ceesay, A. Guèye, D. Ndiaye, E. Beibou, A. Padilla, and D. Belhabib. 2017. "Assessing the Effectiveness of Monitoring Control and Surveillance of Illegal Fishing: The Case of West Africa." *Frontiers in Marine Science* 4. https://doi.org/10.3389/fmars.2017.00050.

Earth Overshoot Day. 2025. "Earth Overshoot Day 2025 Falls on July 24th." Available at: http://overshoot.footprintnetwork.org (accessed September 14 2025).

EPA. 2023. "Threats to Coral Reefs." Available at: https://www.epa.gov/coral-reefs/threats-coral-reefs.

FAOSTAT. n.d. Available at: www.fao.org/faostat

Flavelle, C. 2019. "Climate Change Threatens the World's Food Supply, United Nations Warns." *The New York Times*, August 8. https://www.nytimes.com/2019/08/08/climate/climate-change-food-supply.html.

Fountain, H., and B. C. Solomon. 2019. "Glaciers Are Retreating. Millions Rely on Their Water." *The New York Times*, January 16. https://www.nytimes.com/interactive/2019/04/17/climate/melting-glaciers-globally.html.

Gabbero, H., Y. Louanchi, D. Reculeau, and C. Ferrerons Galeano. 2022. *Human Rights Defenders and Covid-19: The Impact of the Pandemic on Human Rights Defenders and Their Work*. The Observatory for the Protection of Human Rights Defenders. Available at: observatoryfordefenders.org.

Global Footprint Network. 2025. "Calculate Your Footprint.". https://data.footprintnetwork.org/#/ (accessed February 14, 2025).

Goldstein, D. 2025. "American Children's Reading Skills Reach New Lows." *The New York Times*, January 29.

Graeub, B. E., M. J. Chappell, H. Wittman, S. Ledermann, R. B. Kerr, and B. Gemmill-Herren. 2016. "The State of Family Farms in the World." *World Development* 87. https://doi.org/10.1016/j.worlddev.2015.05.012.

Great Barrier Reef Foundation. 2023. "Coral Bleaching: What It Is, How It Happens and What We're Doing to Help." Available at: https://www.barrierreef.org/the-reef/threats/coral-bleaching.

Gupta, J., X. Bai, D. M. Liverman, J. Rockström, D. Qin, B. S. Koster, J. C. Rocha, L. Jacobson, J. F. Abrams, L. S. Andersen, D. I. Armstrong McKay, G. Bala, S. E. Bunn, D. Ciobanu, F. DeClerck, K. L. Ebi, L. Gifford, C. Gordon, S. Hasan, N. Kanie, ... G. Gentile. 2024. "A Just World on a Safe Planet: A Lancet Planetary Health–Earth Commission Report on Earth-System Boundaries, Translations, and Transformations." *The Lancet Planetary Health*. Available at: https://doi.org/10.1016/s2542-5196(24)00042-1.

HLPE. 2020. *Food Security and Nutrition: Building a Global Narrative Towards 2030*. Rome: High Level Panel of Experts on Food Security and Nutrition of the Committee on World Food Security.

Hodgson, C. 2023. "Global Elite Produce Almost Half Greenhouse Emissions, UN Says." *Financial Times*, January 30. https://www.ft.com/content/47db95b4-ce9f-42c7-8b37-86483c85267a.

IAASTD. 2009. *Agriculture at a Crossroads: Global Report*. Washington, DC: Island Press. https://wedocs.unep.org/20.500.11822/9569.

Independent Group of Scientists Appointed by the Secretary-General. 2019. *Global Sustainable Development Report 2019: The Future is Now – Science for Achieving Sustainable Development*. New York: United Nations. Available at: https://sdgs.un.org/gsdr/gsdr2019.

Independent Group of Scientists Appointed by the Secretary-General. 2023. *Global Sustainable Development Report 2023: Times of Crisis, Times of Change: Science for Accelerating Transformations to Sustainable Development*. New York: United Nations.

IPCC. 2019. "Land Is a Critical Resource, IPCC Report Says." Available at: https://www.ipcc.ch/2019/08/08/land-is-a-critical-resource_srccl/.

ISRIC. n.d. "Global Assessment of Human-Induced Soil Degradation (GLASOD)." Available at: https://www.isric.org/projects/global-assessment-human-induced-soil-degradation-glasod.

Jacobs, A. 2017. "China's Appetite Pushes Fisheries to the Brink." *The New York Times*, April 30. https://www.nytimes.com/2017/04/30/world/asia/chinas-appetite-pushes-fisheries-to-the-brink.html.

Jungwiwattanaporn, M. 2024. *These 5 Rules Would Strengthen WTO Deal to End Harmful Fisheries Subsidies*. Pew Research Center. Available at: https://www.pewtrusts.org/en/research-and-analysis/articles/2024/01/17/these-5-rules-would-strengthen-wto-deal-to-end-harmful-fisheries-subsides.

Kaplinsky, R. 2022. *Sustainable Futures: An Agenda for Action*. Cambridge: Polity Press.

Krenak, A., and A. Doyle. 2020. *Ideas to Postpone the End of the World*. Toronto: Anansi International.

Kubenz, V., and D. Kiwan. 2022. "'Vulnerable' or Systematically Excluded? The Impact of Covid-19 on Disabled People in Low- and Middle-Income Countries." *Social Inclusion* 11 (1). https://doi.org/10.17645/si.v11i1.5671.

Levin, K. 2018. "8 Things You Need to Know about the IPCC 1.5 Degree Report." World Resources Institute. Available at: www.wri.org.

Ligon, E., and E. Sadoulet. 2008. *Estimating the Effects of Aggregate Agricultural Growth on the Distribution of Expenditures*. Washington, DC: World Bank. Available at: https://openknowledge.worldbank.org/bitstream/handle/10986/9096/WDR2008_0021.pdf.

Ménard, B., and N. Shtarkman. 2024. "Biocubes: An Inventory of Biomass and Technomass." Available at: https://biocubes.net (accessed March 23, 2025).

Millennium Ecosystem Assessment. 2005. "Current State & Trends Assessment." Available at: https://www.millenniumassessment.org/en/Condition.html.

Peel, M. 2024. "The Race Against Time to Defeat Mosquito-Borne Diseases." *Financial Times*, June 17. https://www.ft.com/content/816c4a17-b938-49c2-b2c0-28b1497084de.

Perry, B. L., B. Aronson, and B. A. Pescosolido. 2021. "Pandemic precarity: COVID-19 Is Exposing and Exacerbating Inequalities in the American Heartland." *Proceedings of the National Academy of Sciences* 118. https://doi.org/10.1073/pnas.2020685118.

Porter, C., and G. Steinmetz. 2024. "As Switzerland's Glaciers Shrink, a Way of Life May Melt Away." *The New York Times*, January 21. https://www.nytimes.com/2024/01/21/world/europe/switzerland-glaciers.html.

Richardson, K., W. Steffen, W. Lucht, J. Bendtsen, S. E. Cornell, J. F. Donges, M. Druke, I. Fetzer, G. Bala, W. von Bloh, G. Feulner, S. Fiedler, D. Gerten, T. Gleeson, M. Hofmann, W. Huiskamp, M. Kummu, C. Mohan, D. Nogues-Bravo, S. Petri, … J. Rockstrom. 2023. "Earth Beyond Six of Nine Planetary Boundaries." *Science Advances* 9 (37). https://doi.org/10.1126/sciadv.adh2458.

Rockström, J., W. Steffen, K. Noone, Å. Persson, F. S. Chapin, E. Lambin, T. M. Lenton, M. Scheffer, C. Folke, H. J. Schellnhuber, B. Nykvist, C. A. De Wit, T. Hughes, S. Van der Leeuw, H. Rodhe, S. Sörlin, P. K. Snyder, R. Costanza, U. Svedin, … J. Foley. 2009. "Planetary Boundaries: Exploring the Safe Operating Space for Humanity." *Ecology and Society* 14 (2).

Savage, S., J. Singh, B. Parkin, and A. Adeoye. 2023. "The Return of the Rice Crisis." *The Financial Times*, October 23. https://www.ft.com/content/416736ec-7960-496d-b6c8-fd7a2fd99668.

Sengupta, S., T. Le Thuy, T. Nguyen, and R. Doyle. 2023. "Rice Gets Reimagined, from the Mississipi to the Mekong." *The New York Times*, May 20. https://www.nytimes.com/interactive/2023/05/20/climate/rice-farming-climate-change.html.

Singh, K. D. 2020. "'The Lockdown Killed My Father': Farmer Suicides Add to India's Virus Misery." *The New York Times*, November 11. https://www.nytimes.com/2020/09/08/world/asia/india-coronavirus-farmer-suicides-lockdown.html.

Statistica Research. 2024. "Online Advertising Revenue in the United States from 2000 to 2023." Available at: https://www.statista.com/statistics/183816/us-online-advertising-revenue-since-2000/.

Steffen, W., K. Richardson, J. Rockstrom, S. E. Cornell, I. Fetzer, E. M. Bennett, R. Biggs, S. R. Carpenter, W. de Vries, C. A. de Wit, C. Folke, D. Gerten, J. Heinke, G. M. Mace, L. M. Persson, V. Ramanathan, B. Reyers, and S. Sorlin. 2015. "Planetary Boundaries:

Guiding Human Development On A Changing Planet." *Science* 347 (6223). https://doi.org/10.1126/science.1259855.

Strauss, B.H., S. A. Kulp, D. J. Rasmussen, and A. Levermann. 2021. "Unprecedented Threats to Cities from Multi-Century Sea Level Rise." *Environmental Research Letters* 16 (11). https://doi.org/10.1088/1748-9326/ac2e6b.

Sustainable Development Solutions Network. 2024. "Sustainable Development Report: Overall Score." Available at: https://dashboards.sdgindex.org/map (accessed February 14, 2025).

Tabuchi, H. 2024. "Something's Poisoning America's Land: Farmers Fear 'Forever' Chemicals." *The New York Times*, August 31. https://www.nytimes.com/2024/08/31/climate/pfas-fertilizer-sludge-farm.html.

Tauschinski, L. 2024. "Climate Graphic of the Week: First Ice-Free Day in the Arctic Ocean May Come Before 2030, Study Shows." *Financial Times*, December 6. https://www.ft.com/content/63fbcf2b-9acc-4ccb-ac12-9f3b78e85b68.

Tradowsky, J. S., S. Y. Philip, F. Kreienkamp, S. F. Kew, P. Lorenz, J. Arrighi, T. Bettmann, S. Caluwaerts, S. C. Chan, L. de Cruz, H. de Vries, N. Demuth, A. Ferrone, E. M. Fischer, H. J. Fowler, K. Goergen, D. Heinrich, Y. Henrichs, F. Kaspar, G. Lenderink, ... N. Wanders. 2023. "Attribution of the Heavy Rainfall Events Leading to Severe Flooding in Western Europe during July 2021." *Climatic Change* 176 (7). https://doi.org/10.1007/s10584-023-03502-7.

UN ESCWA. 2024. "What Is Driving Inequality in the Arab Region?" Available at: http://www.unescwa.org/publications/what-driving-inequality-arab-region.

UN News. 2021. "'Inequality Virus' Threatens 'Catastrophe for All' Unless Crisis Can Be Overcome Together." Available at: https://news.un.org/en/story/2021/04/1089872.

UNESCO. 2022. *Resuming or Reforming?: Tracking the Global Impact of the COVID-19 Pandemic on Higher Education After Two Years of Disruption.* New York: Scientific and Cultural Organization United Nations Educational and UNESCO International Institute for Higher Education.

UNFCCC. 2024. "COP29 UN Climate Conference Agrees to Triple Finance to Developing Countries, Protecting Lives and Livelihoods." Available at: https://unfccc.int/news/cop29-un-climate-conference-agrees-to-triple-finance-to-developing-countries-protecting-lives-and.

UNICEF (United Nations Children's Fund). 2022. "Pulse Check on Digital Learning," Available at: https://www.unicef.org/reports/pulse-check.

V-Dem. 2022a. "Pandemic Backsliding: Democracy During COVID-19 (March 2020 to June 2021)." Available at: https://www.v-dem.net/pandem.html (accessed November 10 2024).

V-Dem. 2022b. "Saudi Arabia." Available at: https://github.com/vdeminstitute/pandem/blob/master/by_country/Saudi%20Arabia.md (accessed February 14, 2025).

Watson, A. J., T. M. Lenton, and B. J. W. Mills. 2017. "Ocean Deoxygenation, The Global Phosphorus Cycle and the Possibility of Human-Caused Large-Scale Ocean Anoxia." *Philosophical Transactions Series A, Mathemathics and Physical Sciences* 375 (2102). https://doi.org/10.1098/rsta.2016.0318.

West, M. 2023. *An Ed-Tech Tragedy?* Paris: UNESCO.

Williams, A. 2023a. "Global Fossil Fuel Subsidies and Costs Hit Record $7tn in 2022, IMF Reports." *Financial Times*, August 24. https://www.ft.com/content/fa607c72-7810-4ed5-98a3-d27d004c15bd.

Williams, A. 2023b. "UN Chief Denounces 'Foot-Dragging' and 'Naked Greed' for Lack of Climate Action." *Financial Times*, September 20. https://www.ft.com/content/6199e651-5559-4320-abd8-865892407809.

World Bank. 2007. *World Development Report 2008: Agriculture for Development*. Washington, DC: World Bank. Available at: https://openknowledge.worldbank.org/handle/10986/5990.

WTO (World Trade Organization). 2022. *Agreement on Fisheries Subsidies*. Available at: https://www.wto.org/english/tratop_e/rulesneg_e/fish_e/fish_e.htm.

Yuan, S., A. M. Stuart, A. G. Laborte, J. I. Rattalino Edreira, A. Dobermann, L. V. N. Kien, L. T. Thúy, K. Paothong, P. Traesang, K. M. Tint, S. S. San, M. Q. Villafuerte, E. D. Quicho, A. R.P. Pame, R. Then, R. J. Flor, N. Thon, F. Agus, N. Agustiani, N. Deng, … P. Grassini. 2022. "Southeast Asia Must Narrow Down the Yield Gap to Continue to Be a Major Rice Bowl." *Nature Food* 3 (3): 217–226. https://doi.org/10.1038/s43016-022-00477-z.

Zhong, R., and M. Rojanasakul. 2024. "How Close Are the Planet's Climate Tipping Points?" *The New York Times*, August 11. https://www.nytimes.com/interactive/2024/08/11/climate/earth-warming-climate-tipping-points.html.

Part II
Changemakers

Part II introduces our guiding framework for achieving sustainable development, adapted from Independent Group of Scientists Appointed by the Secretary-General (2019). It focuses on the four key changemakers that play a role across all transformations. Six examples across the SDGs are used to illustrate how technological fixes are not enough to transform development. We learn a method for assessing co-benefits and trade-offs across the SDGs, including my own research on holistic development in Ethiopia. Part II prepares you for reviewing the six transformation hotspots that are the focus of Part III.

REFERENCE

Independent Group of Scientists Appointed by the Secretary-General. 2019. *Global Sustainable Development Report 2019: The Future is Now – Science for Achieving Sustainable Development*. New York. United Nations. Available at: https://sdgs.un.org/gsdr/gsdr2019.

DOI:10.4324/9781003400028-4

Part II Overview

INTRODUCING OUR CHANGEMAKERS

To achieve a transformation in sustainable development, we need to activate four key changemakers.[1] As illustrated in *Figure II.1* with drawings from South African artist Marizanne Linde, the four changemakers are *governance and policymaking, economy and finance, individual and collective action* and *science and technology. Governance and policymaking* is represented by a town hall that needs all of us to finish its construction. *Economy and finance* captures how the infamous bull on Wall Street in New York City, symbolizing economic growth in the stock market, should not just be filled up like a piggybank. We need to mobilize the funds to support sustainable development. The three signs in *individual and collective action* proclaiming *Stand together for Change* represent how our individual engagement is needed to create a resounding tenet for a common cause. The gears and lightbulb in *science and technology* illustrate how innovation and scientific discovery can lead to change, including through problem-solving that relies on human ingenuity and our adaptive capacities.

To achieve a sustained change across one of our six transformation hotspots – *sustainable, local food systems, just economies, human capabilities, inclusive urban-rural development, green energy access*, and *regeneration of nature* – each changemaker needs to be engaged. Failing to leverage the changemakers will lead to inaction at worst, or at least undermine the sustainability and effectiveness of the transformation.

We can use the example of *science and technology* to illustrate this crucial point. The archives of development projects are crammed with "silver bullet" technologies that offer a cure-all – a so-called *panacea* – to the world's problems. I introduce a brief vignette of a silver bullet for each hotspot and why we need all four changemakers to work together to achieve a transformative change.

DOI:10.4324/9781003400028-5

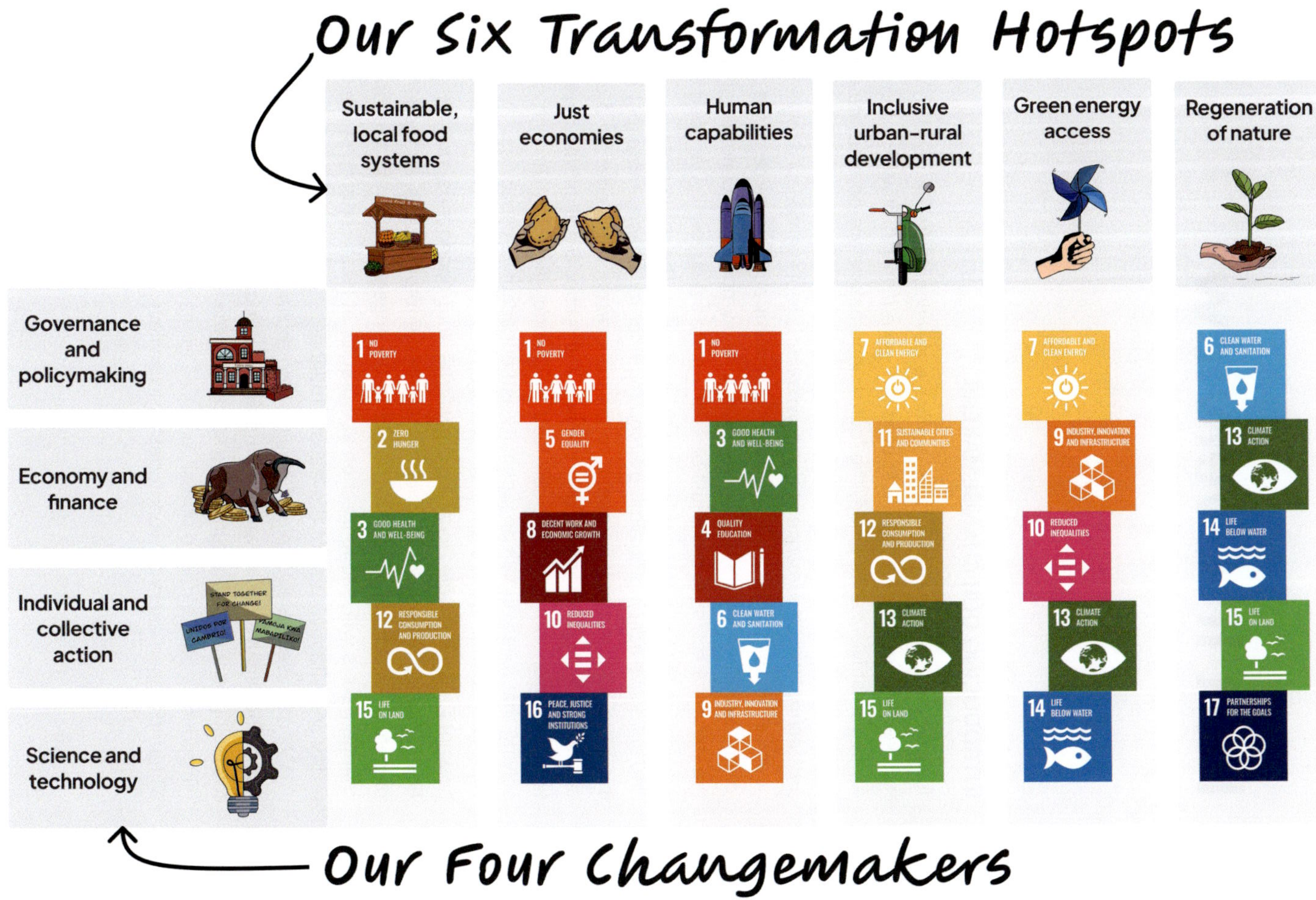

Figure II.1
Our changemakers and transformation hotspots, adapted from Independent Group of Scientists Appointed by the Secretary-General (2019)

VIGNETTE 1 SUSTAINABLE, LOCAL FOOD SYSTEMS AND VITAMIN-ENRICHED SEEDS

To improve nutrition (SDG 2 and SDG 3), scientists developed a vitamin-enriched rice variety (known as Golden Rice) piloted across South and Southeast Asia. I followed Golden Rice over the years since it emerged as a key technology targeting the Global South. For several decades, however, it failed to take off, as government policies refused to permit the genetically modified seeds for widespread consumption because of regulatory concerns, including fear of cross-contamination of the existing gene pool (SDG 15). According to Adrian Dubock, executive secretary of the Golden Rice Humanitarian Board, Golden Rice is more cost-effective than the other silver bullet alternatives of Vitamin-A food fortification or providing Vitamin A capsules (Dubock 2019). Farmers, however, lack access to locally adapted varieties (SDG 1). As the seed only addressed a specific segment of malnutrition, it struggles to result in larger improvements in nutrition. Programming efforts were limited in involving individuals or communities to change their nutrition habits toward healthier diets and lacked promotion of more local, complementary approaches, such as kitchen gardens (SDG 12).

VIGNETTE 2 JUST ECONOMIES AND VACCINE BREAKTHROUGH

With the arrival of COVID-19 in early 2020, a coordinated global effort was launched to identify viable vaccines. With several viable candidates mass-produced later in the year, access to such innovations was readily available in the Global North. Public health campaigns were launched to overcome people's skepticism and increase vaccination rates. Where funding for such campaigns was lacking, or governments were outright feeding vaccine skepticism (e.g., in Tanzania; Pilling and Schipani 2021), individual and collective actions to get vaccinated were limited. A lack of funding also meant that access to vaccines was much more limited in the Global South. These disparities were accentuated by rural-to-urban gaps.

VIGNETTE 3 HUMAN CAPABILITIES AND DIGITAL TEACHING PLATFORMS

In 2020, governments also raced to find digital solutions to overcome the challenges of social distancing in education (SDG 4). Existing information and communication technology (ICT) platforms, such as Zoom, were leveraged, and students were supplied with laptops and tablets to ensure continued educational progress. These technological innovations hit a major stumbling block: teachers lacked the expertise to teach in virtual environments, and students faced digital literacy gaps and a lack of access to essential services like electricity. Poor educational quality was the result, with digital divides across regions exacerbating uneven outcomes (UNESCO 2022). Because of a lack of funding, digital learning platforms have since stalled or become unusable (UNICEF 2022). A similar fate has hit the computer equipment handed out, creating environmental waste (Rockett Gutterman 2023). While attractive as a political quick fix, the global pandemic has laid bare the crisis in education as years of underfunding and low prioritization by policymakers had left the educational system vulnerable (West 2023). And our move to remote teaching came at the cost of ignoring lower-tech and more equitable options (West 2023).

VIGNETTE 4 INCLUSIVE URBAN-RURAL DEVELOPMENT AND LIGHT-RAILS

The Ethiopian capital city, Addis Ababa, opened its light rail to great fanfare in 2015, serving one of the fastest-growing cities in the world. I observed its construction when visiting colleagues at the nearby International Livestock Research Institute (ILRI) campus and followed its development with excitement. The first light rail in sub-Saharan Africa represented a significant milestone in sustainable urban and peri-urban development. The US$475 million infrastructure project was built at record speed by the China Railways Group Limited and funded through loans from the Export-Import Bank of China. As a prestige project, colleagues shared with me concerns about the quality of the materials used. The lack of involvement of Ethiopian workers in the construction also led to an initial mistrust by the local population of the safety of the train, as they had only seen Chinese workers using it. Fast forward a decade, the light rail is in high demand. Unfortunately, the lack of local capacity building for repairs and dedicated funds for their maintenance have resulted in only 8 out of 31 trains remaining functional. With limited spare parts and insufficient local capacity, the government of Ethiopia depends on its Chinese counterparts to keep the trains running (Tadesse 2023).

VIGNETTE 5 GREEN ENERGY ACCESS AND WIND TURBINES

Offshore wind turbines – invented decades ago – offer a proven avenue to increase access to green energy. Wind is a crucial energy source at night when solar cannot power the energy transformation. So why did the world's biggest offshore wind developer – Ørsted – cancel two projects off the coast of New Jersey in the USA in 2023 (DiGangi 2023)? Ørsted was one of the early pioneers of national energy companies going green, committing in 2009 to produce 85 percent of its energy from renewables by 2040 (Millard 2024). It turns out that Ørsted faced significant economic challenges, including supply chain shortages, rising interest rates, and lower-than-expected tax incentives from new government policies. The company also struggled to obtain construction permits, causing further delays. Despite a positive environmental impact assessment, collective opposition by residents and local leaders concerned about the negative impact on tourism and their property values further hastened the demise of these two energy transformation projects (Antonucci 2024). Meanwhile, policymakers were working to make it easier to build renewable energy projects by limiting local government's ability to restrict wind projects. With the change in administration in 2025 and resultant failure to mobilize all four changemakers, the USA continues to fall behind its transformation as devised under the Inflation Reduction Act (IRA), the largest climate funding in US history (nearly US$400 billion) (Plumer and Popovich 2024).

VIGNETTE 6 REGENERATION OF NATURE AND CONSERVATION

National park conservation in sub-Saharan Africa presents several instances where newly introduced conservation practices are ineffective in protecting wildlife without harming human livelihoods. Known as "fortress-style conservation," new parks were established across East and Southern Africa, and local populations were relocated outside park boundaries. Guillaume Blanc (2022, 126–127) captured the contradictions inherent in the promise of conservation: "give nature to the people; prevent the people from living in it."

While effective in improving wildlife habitat in Europe or the USA, the colonial and post-colonial transfer of this park management practice in the Global South led to less desirable outcomes (Adams and Hutton 2007; Di Marzo and Espinosa 2023). Changemakers of individual and collective action, the economy, and governance were ignored. Regarding the economy, as household members depended on access to the park area for their livelihoods (e.g., collecting herbs or fuelwood), their relocation (to less fertile grounds) worsened their livelihoods. In response, individuals and communities living near park boundaries ignored the new rules, encroaching on parkland to sustain at least part of their livelihoods. Local inhabitants of park areas were also frequently deemed invisible to policymakers. Their lack of visibility resulted in the creation of parks on falsely perceived *virgin land* (Blanc 2022). Poor governance further failed to account for losses incurred by wildlife incursions beyond the new park boundaries. The spectacular cases include elephants who raid farmers' plots or, worse, their housing. Monkeys who decided that garden vegetables and grain are tasty are less infamous but more frequent cases, as witnessed by my first fieldwork in Kibale National Park, Uganda in 2006 (Hartter 2007). Where they existed, economic compensation schemes often lacked transparency or were inadequately funded to compensate for losses. Over the past decade, more bottom-up approaches have been explored, with conservationists bringing social scientists and communities aboard to more holistically assess and plan conservation efforts (Brockington and Wilkie 2015). In this vignette in particular, space, power, and whose knowledge counts are key determinants of what sustainable development is supposed to look like.

ON CO-BENEFITS AND TRADE-OFFS

To achieve sustainable transformations, we want to harness synergies (co-benefits) between interventions and understand trade-offs that you can address. Given the extensive nature[2] of the SDGs, achieving each of the 17 goals brings co-benefits or trade-offs. Based on research by the Independent Group of Scientists Appointed by the Secretary-General (2019) and their review of 65 global assessments and 112 scientific articles, the three highest-scoring interactions for each SDG were selected. These trade-offs and co-benefits were scored from a -3 to +3 scale by Nilsson, Griggs, and Visback (2016) as outlined in Table II.1.

Table II.1 Scoring of co-benefits and trade-offs, adapted from Nilsson, Griggs, and Visback (2016)

Interaction	Name	Explanation	Example	Related chapter
+3	Indivisible	Inextricably linked to the achievement of another goal	Ending all forms of discrimination against women and girls is indivisible from ensuring women's full and effective participation and equal opportunities for leadership	14. Focus I: Gender Inequality
+2	Reinforcing	Aids the achievement of another goal	Strengthening the capacity to adapt to climate-related hazards reduces losses caused by disasters	15. Focus II: Resilience
+1	Enabling	Creates conditions that further another goal	Providing electricity access in rural homes enables education, because it makes it possible to do homework at night with electric lighting	11. Green Energy Access
0	Consistent	No significant positive or negative interactions	Ensuring education for all does not interact significantly with infrastructure development or conservation of ocean systems	-
-1	Constraining	Limits options on another goal	Improved water efficiency can constrain agricultural irrigation	7. Sustainable, local food systems
-2	Counteracting	Clashes with another goal	Boosting consumption for growth can counteract waste reduction and climate mitigation	8. Just economies
-3	Canceling	Makes it impossible to reach another goal	Full protection of natural reserves excludes public access for recreation	12. Regeneration of nature

Synergies and co-benefits are at the heart of the SDGs. Except for SDG 5 (Gender Equality), each goal has trade-offs. Among the 51 top three interactions captured in Figure II.2, co-benefits dominate (48 out of 51, 94 percent). Only three SDGs have trade-offs ranked in the top three. Achieving SDG 10 (Reduced Inequalities) positively influences the achievement of poverty reduction (SDG 1), hunger (SDG 2), and health and well-being (SDG 3). SDG 2 (Zero Hunger) is the most synergistic goal, influenced by nine other goals. Looking across all interactions, the SDG that influences the most goals is SDG 7 on affordable and clean energy, with the top three co-benefits being poverty (SDG 1), health and well-being (SDG 3), and clean water and sanitation (SDG 6). That same co-benefit of SDG 6 represents a trade-off negatively impacted by achieving SDG 1 and SDG 2. Together with SDG 14 (Life below Water), they are the only ones that rank in the top three interactions.

As motioned in Chapter 1, we need to break down silos and move outside our comfort zone. You can apply this scale in your own work and deeper dives, including going beyond the SDG goal and moving to the targets. See suggestion to move at the beginning of paragraph. Let us review the example of SDG 13 (Climate Action) to illustrate both the complexity that ensues, but also how you can use co-benefits and trade-offs to make sense of it.

As illustrated in Figure II.2, for SDG 13, co-benefits exist at the level of achieving zero hunger (SDG 2) and good health and well-being (SDG 3). Greater climate action will improve the viability of farming, as well as reduced stress on health. While significant co-benefits exist as well with Life below Water (SDG 14), Targets 13.1–13.3[3] create trade-offs with achieving Targets 14.2–14.6. As argued by Schmidt et al. (2017, 206):

> Policies and measures taken to adapt to climate change might counteract or even cancel SDG 14 targets aiming at the protection and conservation of coastal ecosystems, such as if technical protection measures fail to provide enough space for saltmarshes to keep up with sea-level rise (coastal squeeze).

These interactions are scored as counteracting (-2; making it more difficult to achieve progress) or even canceling (-3; automatically leading to a negative impact). To address these trade-offs, Schmidt et al. (2017, 206) recommend we should "ensure coherent and integrated coastal zone management and coastal protection management" as well as "develop nature-based solutions that promote both coastal and marine conservation and sustainable urban development in an integrated way." We will return to nature-based solutions in *Chapter* 12.

It would be false, however, to assume that the above trade-offs mean that achieving SDG 13 and 14 are in conflict with each other. As classified by Schmidt et al. (2017), there are both enabling (+1) and reinforcing (+2) interactions between the targets of SDG 13 and 14 that create co-benefits to achieve the SDGs. Table II.2 captures their analysis.

Having more closely reviewed a trade-off within SDG 13, let us review how achieving SDG 13 on Climate Action positively influences the achievement of SDG 2 on Zero Hunger, and any possible counteracting relationships (see Table II.3).

Top Three Interactions across each SDG Goal

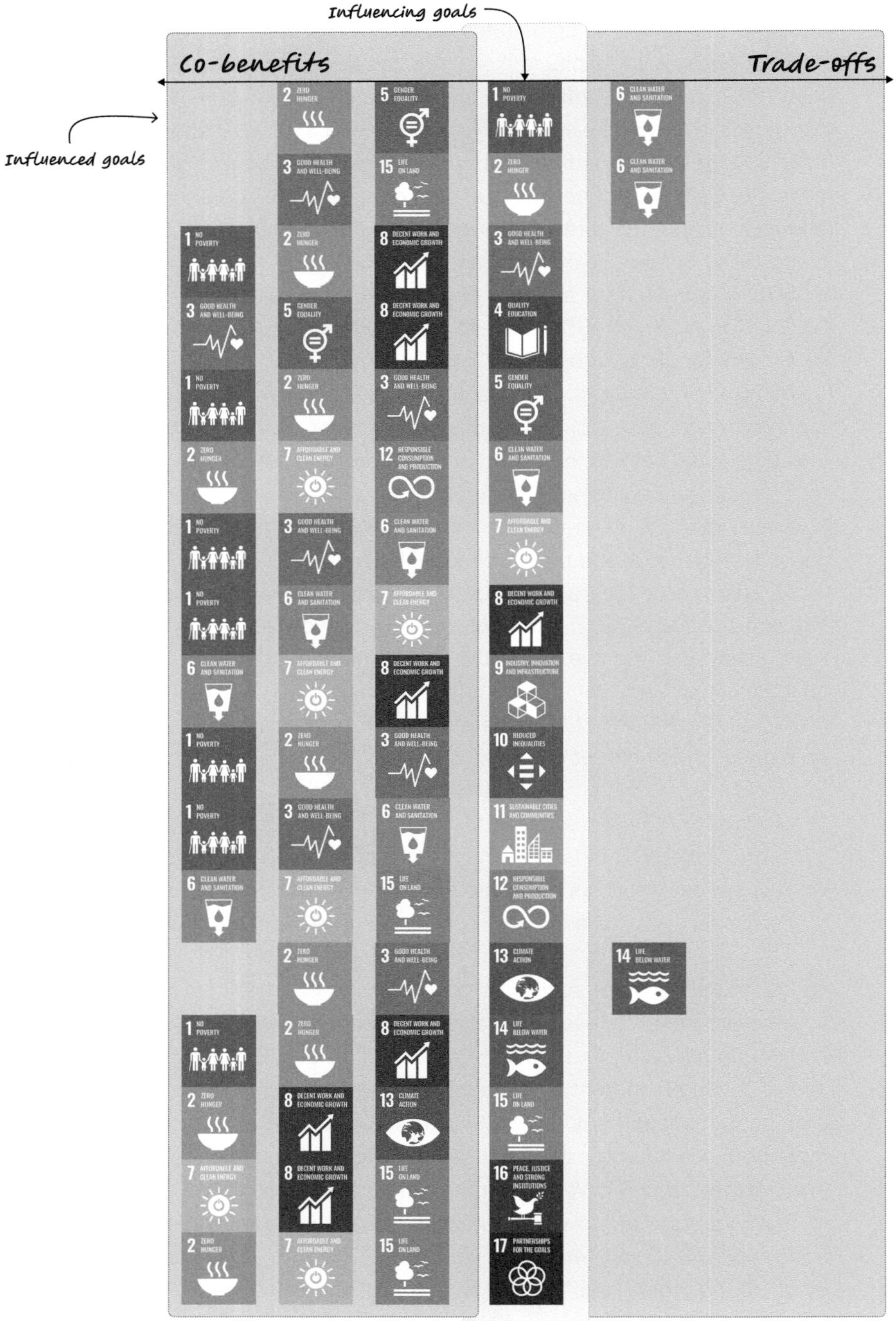

Figure II.2
Top three interactions across each SDG, adapted from Independent Group of Scientists Appointed by the Secretary-General (2019)

Table II.2 Reinforcing and enabling interactions between SDG 13 and SDG 14. Source: Schmidt et al. (2017)

SDG 13 targets	SDG 14 targets	Type of interaction	Description of interaction
13.1	14.1, 14.2, 14.3, 14.4, 14.5, 14.6	Reinforcing (+2)	Action taken to strengthen the health of coastal and marine ecosystems including fish stocks will reinforce the strengthening of environmental and societal resilience and adaptive capacities to climate change, and vice versa
13.2, 13.a, 13.b	14.2, 14.3, 14.5	Enabling (+1) and Reinforcing (+2)	Action taken for promoting healthy oceans and coastal systems will also enable or even reinforce the development and integration of climate exchange measures into policies, planning and management, such as by promoting coastal ecosystems servicing as blue carbon systems, and vice versa
13.3, 13.b	14.a	Enabling (+1) and Reinforcing (+2)	Increasing marine scientific knowledge, research capacities and technologies will enable or even reinforce awareness raising and capacity building for climate change mitigation measures, planning and management, and vice versa

Table II.3 Interactions between SDG 2 and SDG 13. Source: Mollier et al. (2017)

SDG 2 targets	SDG 13 targets	Type of interaction	Description of interaction
2.4, 2.5	13.1	Reinforcing (+2)	Resilient agricultural practices and maintaining and giving access to seeds/plant/animal genetic diversity should reinforce adaptation to climate change
2.a	13.2, 13.3, 13.b	Reinforcing (+2)	Enhancing international cooperation in agriculture research, science, and services should enable climate change measurements and raise awareness on climate challenges, and promote mechanisms to address them
2.3, 2.4, 2.5	13.b	Reinforcing (+2)	Positive feedback from raising awareness and capacity on climate change impacts (mitigation and adaptation) to setting up sustainable and productive agriculture practices, and maintaining biodiversity
2.3	13.1	Counteracting (-2)	Unsustainable agriculture focusing solely on productivity may counteract climate adaption by increasing climate instability and extreme events

As captured in *Table II.3*, the trade-offs are not a given but something to monitor and address through programming (Mollier et al. 2017, 63). If approached holistically, for example, the productivity goal in Target 2.3 is contextualized by Target 2.4 which emphasizes more sustainable agricultural practices that improve climate resilience.

Regarding gender, empowering women and achieving greater gender equality is a no-brainer as it yields only co-benefits. To use the above example of agriculture, three key interactions have been identified in Table II.4.

In summary, the motivation for undertaking an analysis of co-benefits and trade-offs is to harvest the synergies where they exist and address trade-offs before they can occur. This framework presents a key puzzle piece toward a greater understanding of how to achieve transformative local and global change. To illustrate how synergies pay off, I next apply the framework in a case of holistic development programming in rural Ethiopia.

Table II.4 Interactions between SDG 2 and SDG 5. Source: Mollier et al. (2017)

SDG 2 targets	SDG 5 targets	Type of interaction	Description of interaction
2.1, 2.2	Overall SDG 5	Reinforcing (+2)	Ensuring food and nutrition security reinforces women's empowerment. In turn, women's empowerment is enabling nutrition security due partly to their role in food production and preparation and their greater inclination to spend resources they control on family nutrition and health.
2.3	5.5, 5.a	Reinforcing (+2)	Promoting investment in rural infrastructure, securing equal access to productive resources (including land), and increasing income strengthens women's empowerment and gender equality.
2.a	5.b	Enabling (+1)	Access to technology is an important lever to enable women's empowerment in agriculture and overall – the two means of implementation mutually reinforce each other.

LEVERAGING CO-BENEFITS TO ACHIEVE HOLISTIC DEVELOPMENT

In my work at Biovision Foundation for Ecological Development, under the lead of my colleague, Dr. Menale Kassie from the *International Centre for Insect Physiology and Ecology* (icipe), we investigated if combining complementary development interventions that generate co-benefits will yield greater impact (measured at the level of livelihood improvements) than the sum of their individual parts (Kassie et al. 2020).

Smallholder farmers in rural Ethiopia face multiple challenges: malaria, animal diseases, pest infestation of crops, or climate change (just to name a few). Categorized by *icipe* as issues across animal, human, plant, and environmental health, their impacts on livelihoods are far-reaching. Their interconnected nature reduces our ability to achieve sustained benefits from one-dimensional development interventions. For example, an intervention focused on reducing the pest infestation of crops might fail if draft animals (i.e., oxen who can pull a plow) fall ill and cannot support land preparation. Farmers' abilities to substitute a lack of draft animals with their own labor is further limited by incidences of malaria. Families that have experienced malaria in the past will take fewer risks as they expect to lose labor capacity over the course of a growing season, further reducing the overall farm productivity (Diiro et al. 2022).

Starting in 2005 and continuing over the next decade, *icipe*, in collaboration with the Ethiopian government and donors, such as the foundation I worked for, piloted four interventions in Chora Boter district in Ethiopia. The first intervention was the deployment in 2005 of biconical (Ngu) traps to control tsetse and trypanosomiasis to improve animal health. The second intervention between 2009 and 2015 was to reduce malaria through an integrated vector management package (IVM) aimed at "nipping the mosquito in the bud." It improved human health by using biological control methods (*Bacillus thuringiensis*) to reduce the malaria larva population in breeding grounds, in combination with bed nets and an education campaign. The third intervention focused on reducing the pest pressure from the stem borer and improving resultant productivity for maize and sorghum, using an ecological intercropping strategy: Push-Pull (see also Chapter 7). It works by growing desmodium, a leguminous crop, between the maize or sorghum that creates plant volatiles that push the main pest (i.e., the stem borer) away, and a border crop, usually Napier grass, that attracts the stem borer. The fourth intervention involved the establishment of tree nurseries combined with improved beekeeping (using modern Langstroth hives) to improve environmental health and generate additional income.

Applying the Nilsson, Griggs, and Visback (2016) framework, we can identify the three positive expected synergies across the interventions (Table II.5).

Each intervention is assumed to contribute to economic livelihoods through increased productivity and yields (plant and animal health), increased income through the sale of honey (environmental health), and reduced medical expenditures (animal and human health).

Based on data collected from 544 randomly selected households across 19 villages in the district, the total discounted net income over eight years was calculated. The working hypothesis was that the sum of all interventions combined is greater than the stacked interventions individually, yielding a synergy effect. Put differently: a household that benefitted from all four interventions experienced a higher total benefit compared to four households that each benefited individually from a single intervention.

Figure II.3 illustrates the four individual interventions and their combined impact at the level of income gained. Starting from the left, if a rural household was not a beneficiary of any intervention, they would hold an average household

Table II.5 Interactions between development interventions in rural Ethiopia

First intervention(s)	Second intervention(s)	Type of interaction	Description of interaction
Trypanosmiasis intervention [animal health]	Push-Pull intervention [plant health]	Enabling (+1)	Improvements in the health of the oxen population increases the ability of farmers to prepare land for sustainable intensification practices. Push-Pull also provides nutritious fodder grass, which can support livestock health.
Push-Pull intervention [plant health] and malaria intervention [human health]	Beekeeping intervention [environmental health]	Reinforcing (+2)	The ability to control pests using ecological means reduces the use of chemical pesticides, which increases the bee populations. Increased bee populations also offer pollination services to other crops (excluding maize or sorghum which are primarily wind- or self-pollinated).
Malaria intervention [human health]	Push-Pull intervention [plant health]	Enabling (+1)	Establishing a Push-Pull plot in the first year is labor-intensive. Reduced instances of malaria increase available labor supply for households to take risks and establish a plot.

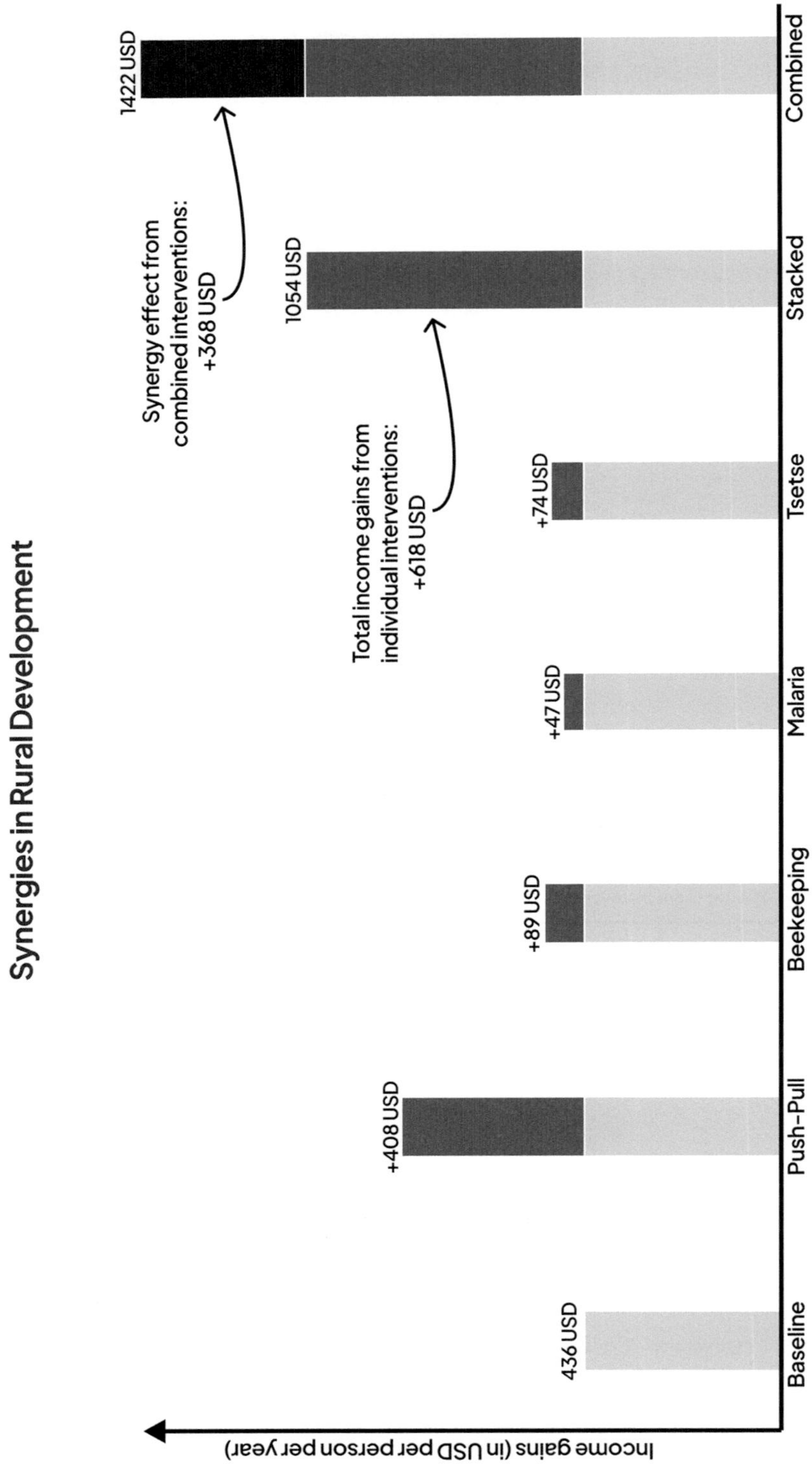

Figure II.3 Synergies in rural development, adapted from Kassie et al. (2020). Note: USD = US dollar

income per family member of US$436, or around US$1.20 per day. Implementing the Push-Pull intervention creates the largest income gains (an additional US$408 per person per year; or around US$1.10 per day), as it directly increases yields of staple crops. Additional income from beekeeping adds another US 37 cents (for an annual total of US$89 per person). Combating malaria-carrying mosquitoes and Tsetse flies adds US$47 and US$74, respectively. These are the benefits attained by a household member who has benefitted from a single intervention. If we stack all the total income gains from individual interventions on top of the baseline, we record additional income gains of US$618 (or an additional US$1.69) for a combined sum of US$1,054. Now, if these households benefitted from all four interventions, we record combined income gains of US$1,422. The difference between the combined and stacked totals is US$368 or one additional US dollar per person per day per household. This significant gain in household income represents the synergy effect that one can obtain by building on co-benefits and managing trade-offs. This gain demonstrates that holistic development pays off (Kassie et al. 2020).

Two important notes: First, it is false to conclude that individual development interventions do not *work*. Each sustainable intervention increases income between 11 percent and 96 percent. Second, these increases, sustained over eight years, occur even as some interventions are not focused explicitly on increasing incomes. For example, an intervention on malaria focuses on improving human health, and affects income only indirectly.

While our case illustrates the power of holistic interventions, it was limited by the choice of interventions in the first place. The interventions were drawn from a bucket of *icipe*'s expertise available. We will dive deeper into fostering local change in *Chapter* 13. Holistic effects from development also extend well beyond income. We will assess non-monetary effects by searching for the most significant changes in a case study in *Chapter* 15.

YOUR TURN

The Independent Group of Scientists Appointed by the Secretary-General (2019, 28) provides a helpful example of how addressing childhood obesity (SDG 2.2.) requires multiple changemakers to step up:

> For instance, changes in food habits towards more healthy diets may result from individual and collective action, which is informed by scientific knowledge that can directly influence choices made by families, while supporting governance initiatives such as mandatory food labelling and schools' limiting students' access to sugary drinks.

Identify an example in your SDG of how multiple changemakers might make a difference.

NOTES

1 Both the changemakers and transformation hotspots are adapted from (Independent Group of Scientists Appointed by the Secretary-General 2019).

2 There are criticisms that areas, such as migration, are not having their own SDG Goal. As a result, refugees are not visible enough in country-led efforts to monitor progress toward sustainable development (Grossman and Post 2019; Denaro and Giuffré 2022).

3 See United Nations (2015) or our companion website for the detailed wording of SDG targets and indicators.

REFERENCES

Adams, W. M., and J. Hutton. 2007. "People, Parks and Poverty: Political Ecology and Biodiversity Conservation." *Conservation and Society* 5 (2): 147–183.

Antonucci, P., 2024, "Congressman Van Drew Slams BOEM's Environmental Impact Statement of Offshore Wind," May 24. Available at: https://vandrew.house.gov/news/documentsingle.aspx?DocumentID=721#.

Blanc, G. 2022. *The Invention of Green Colonialism*. Cambridge: Polity Press.

Brockington, D., and D. Wilkie. 2015. "Protected Areas and Poverty." *Philosophical Transactions of the Royal Society of London B Biological Sciences* 370 (1681). https://doi.org/10.1098/rstb.2014.0271.

Denaro, C., and M. Giuffré. 2022. "UN Sustainable Development Goals and the 'Refugee Gap': Leaving Refugees Behind?" *Refugee Survey Quarterly* 41 (1): 79–107. https://doi.org/10.1093/rsq/hdab017.

Di Marzo, D., and C. Espinosa. 2023. "Conservation Conflict: A Political Ecology Meta-Synthesis of East Africa." *Society & Natural Resources* 1–19. https://doi.org/10.1080/08941920.2023.2253744

DiGangi, D. 2023. "Ørsted Cancels Two Offshore Wind Projects Along New Jersey Coast." *Utility Dive*. November 1. https://www.utilitydive.com/news/orsted-cancel-ocean-wind-offshore-project-development-new-jersey/698464/.

Diiro, G. M., M. Kassie, B. W. Muriithi, and C. M. Mutero. 2022. "Gender Heterogeneous Effects of Malaria Risk on Agriculture Productivity: Empirical Evidence from Rural Ethiopia." *The Journal of Developing Areas* 56 (1): 73–91.

Dubock, A. 2019. "Golden Rice: To Combat Vitamin A Deficiency for Public Health." In Z. L. Queiroz, R. V. de Vera, and J-L. Eduardo (Eds.), *Vitamin A*. Rijeka: IntechOpen.

Grossman, A., and L. Post. 2019. *Missing Persons: Refugees Left Out and Left Behind in the Sustainable Development Goals*. IRC. Available at: www.rescue.org.

Hartter, J. N. 2007. "Landscape Change around Kibale National Park, Uganda: Impacts on Land Cover, Land Use, and Livelihoods." PhD dissertation, Geography, University of Florida.

Independent Group of Scientists Appointed by the Secretary-General. 2019. *Global Sustainable Development Report 2019: The Future is Now – Science for Achieving Sustainable Development*. New York: United Nations. Available at: https://sdgs.un.org/gsdr/gsdr2019.

Kassie, M., Z. Abro, T. Wossen, S. T. Ledermann, G. Diiro, S. Ballo, and L. Belayhun. 2020. "Integrated Health Interventions for Improved Livelihoods: A Case Study in Ethiopia." *Sustainability* 12(6). https://doi.org/10.3390/su12062284.

Millard, R. 2024. "How the World's Biggest Offshore Wind Company Was Blown Off Course." *Financial Times*, December 4. https://www.ft.com/content/5130251d-06dd-40f5-a5d7-edf24ec1e571.

Mollier, L., F. Seyler, J.-L. Chotte, and C. Ringler. 2017. "SDG 2: End Hunger, Achieve Food Security and Improved Nutrition and Promote Sustainable Agriculture." In International

Council for Science (Ed.), *A Guide to SDG Interactions: From Science to Implementation*. Available at: https://council.science/wp-content/uploads/2017/05/SDGs-Guide-to-Interactions.pdf.

Nilsson, M., D. Griggs, and M. Visback. 2016. "Map the Interactions between Sustainable Development Goals." *Nature* 534: 320–322. https://doi.org/10.1038/534320a.

Pilling, D., and A. Schipani. 2021. "Tanzania Looks to Divine Providence as Government Shuns Covid Vaccines." *Financial Times*, February 18. https://www.ft.com/content/5dd89bb4-2efb-4195-a38a-ece11993f3b5.

Plumer, B., and N. Popovich. 2024. "As Solar Power Surges, U.S. Wind Is in Trouble." *The New York Times*, June 4.

Rockett Gutterman, L. 2023. "Chromebook Churn." U.S. PIRG Education Fund. Available at: https://publicinterestnetwork.org/wp-content/uploads/2023/05/PIRG-Chromebook-Churn-Full-Report-May-1.pdf.

Schmidt, S., B. Neumann, Y. Waweru, C. Durussel, S. Unger, and M. Visbeck. 2017. "SDG 14: Conserve and Sustainably Use the Oceans, Seas and Marine Resources for Sustainable Development." In International Council for Science (Ed.), *A Guide to SDG Interactions: From Science to Implementation*. Available at: https://council.science/wp-content/uploads/2017/05/SDGs-Guide-to-Interactions.pdf.

Tadesse, F. 2023. "A Crumbling Metro Reveals Failed Promise of China's Billions in Africa." *Bloomberg*, 2023.

UNESCO. 2022. *Resuming or Reforming?: Tracking the Global Impact of the COVID-19 Pandemic on Higher Education After Two Years of Disruption*. New York: Scientific and Cultural Organization United Nations Educational and UNESCO International Institute for Higher Education.

UNICEF. 2022. "Pulse Check on Digital Learning." Available at: https://www.unicef.org/reports/pulse-check.

United Nations. 2015. "Transforming Our World: The 2030 Agenda for Sustainable Development." *General Assembly 70 Session*. New York: United Nations.

West, M. 2023. *An Ed-Tech Tragedy?* Paris: UNESCO.

3 Governance and policymaking

THE CASE FOR GOVERNANCE AND POLICYMAKING TO ACHIEVE THE SDGs

The 2030 Agenda for Sustainable Development was adopted by consensus of all UN member states in 2015. Each country committed to contributing to achieving the SDGs. Using goal-setting as opposed to rules as a governance mechanism, the SDGs are non-legally binding. The 2030 Agenda depends on broad support from diverse stakeholders to generate momentum in achieving the targets (Kanie et al. 2019; Smith, Ehren, and Grek 2024). To be effective, the 17 SDGs need to be translated into action by countries (and their civil society, academia, and businesses). The 2030 Agenda increased demands placed on governance and policymaking. Investments in the capacity-building of stakeholders is a central (although from a donor's perspective, less attractive) part of achieving sustainable transformations (Independent Group of Scientists Appointed by the Secretary-General 2019).

As local, national, and global stakeholders collaborate to achieve the SDGs, there is a recognized need to create safe physical and virtual spaces for civil society to engage with governments and international organizations. Increasingly, virtual spaces are being silenced, as governments crack down on (opposition) movements by blocking internet access (Searcey and Essomba 2017). Effective policymaking requires efforts to fight corruption and improved transparency in budget planning. Inclusive governance relies on the robust participation of civil society and an engaged private sector (Independent Group of Scientists Appointed by the Secretary-General 2019). Both stakeholder groups are active at the local and global levels. We will revisit their roles at the end of this chapter and when we examine the other three changemakers.

Let us look first at four SDGs to identify the role of governance and policymaking in their achievement.

SDG 1: No Poverty

Reducing poverty (and, by extension, inequality) requires a profound structural transformation of social, political and economic relations. Policymakers need a better accounting of the local realities. Ensuring that the SDGs can be adapted to the local context, including target setting, was a less visible but necessary step in this direction. Target 1.2, for example, states "[b]y 2030, reduce at least by half the proportion of men, women and children of all ages living in poverty in

DOI:10.4324/9781003400028-6

all its dimensions *according to national definitions*" (emphasis added).[1] Around 40 countries have deployed a multidimensional poverty index (MPI) to assess the incidence and intensity of poverty according to their values, generating improved data to feed into policymaking initiatives (OPHI and University of Oxford 2024). These national definitions of poverty move well beyond income. They usually include dimensions of health, education, water and sanitation, housing, and employment. I even applied the MPI at the local level for my research in Tanzania (Ledermann 2012).

SDG 5: Gender Equality

To achieve gender equality, securing the political will to sustain any progress toward more equality is imperative. In a context where political will is present, making strides at the national level is possible. Rwanda pioneered in its 2003 Constitution a 30 percent quota for women in elected positions. Two decades later, Rwanda became the first country in the world where women hold the majority of positions in parliament, with 61 percent in the Chamber of Deputies and 37 percent in the Senate (Parliament of Rwanda 2024). Driving the change is civic education for all genders and the recognized leadership of women after the genocide in 1994 as Rwanda steered toward a path of reconciliation. Creating an environment where violence and threats against women are reduced is crucial. In Latin America, for example, female political candidates continue to experience abuse and threats, both online and in person, because of their gender, reducing the likelihood that other women will want to follow their path. To address challenges in gender equality, we also need greater visibility and data. Unless discrimination and violence against women are documented and their causes understood, it will be difficult for policymakers to develop effective policies to improve them. The Global Women's Health Index (Hologic 2025), developed in collaboration with Gallup, is one such attempt to shine more light on understudied, sensitive gender topics.

SDG 7: Affordable and Clean Energy

Governance and policymakers play a crucial role in navigating trade-offs that come with a sustainable energy transformation. As elaborated further in Case Study 11.1 in Chapter 11, on South Africa, moving away from coal toward renewable energy production demands reskilling and offering prospects to workers whose jobs will have become redundant. The same fear of reduced workloads in transitions from combustion to electric engines for cars in the U.S. demands effective policy solutions beyond the auspices of the private sector alone. Europe also faces a skills shortage to achieve its transition to net zero emissions by 2050 (Hancock 2023). As seen in the panacea of offshore windmills in Part II Overview, it is crucial to ensure that governance and policymaking are aligned with transformation initiatives, including finding the right balance between local concerns and national ambitions.

SDG 12: Responsible Consumption and Production

Accountability, stability, and continuity in local and national governments further responsible consumption and production. Case Study 3.2, on the Brazilian city of Belo Horizonte, speaks directly to the importance of continuity combined with popular support. A weak state that fails to hold actors to account, and lacks capacity for monitoring safety and working conditions, will struggle to improve responsible production. The case of the Rana Plaza disaster in 2013, where a collapsed garment factory in Dhaka, Bangladesh, killed over 1,000 people, illustrated the failures of a government overseeing compliance. As new policies, such as Bangladesh's National Occupational Safety and Health Policy, are established, trade unions and labor inspectors need to be empowered to monitor policy implementation (ILO 2023). Beyond increased policy commitments, building the capacity of these actors is equally vital.

ADDRESSING WICKED PROBLEMS THROUGH GOVERNANCE

For select transformations, we need to move beyond the boundaries of a single state: current wicked problems demand governing transboundary flows of goods, capital, information, and people (Independent Group of Scientists Appointed by the Secretary-General 2019).

One of the best-known examples of such wicked problems is CO_2 emissions. These emissions affect non-polluters in the form of climate change, as noted by Albert Abel Williams, Director of the Department of Environmental Protection and Conservation of the island nation of Vanuatu in the South Pacific:

> Climate change has already brought loss and damage to Vanuatu that could not be avoided through mitigation efforts, nor could be avoided through adaptation. While much of our adaptation has been successful, some of Vanuatu's losses are permanent, unavoidable, irreversible and unrecoverable and such that no adaptation measure can adequately address them.
>
> (Williams 2023, 2)

Emitters are asked to compensate for losses incurred via the financial mechanisms of the UN Framework Convention on Climate Change (UNFCCC) and the Green Climate Fund.

Small island nations have increasingly strengthened their positioning in global governance concerning sustainable development and climate change. The UN Conference on Small Island Development States (2014) led to the establishment of the SAMOA Pathway: the Small Island Nations Accelerated Modalities of Action, amplifying their collective calls for achieving sustainable development in the face of rising sea levels and climate change. The UN's mandate is to mobilize support and engage in advocacy to address global challenges in these most vulnerable local contexts. Wicked problems highlight the dependencies and co-dependencies within global governance, as no nation-state alone can tackle sustainable development challenges in isolation.

Climate change has also exacerbated the push factors of migration, a long-standing governance challenge. In Central America, the African Sahel, and Southeast Asia, smallholder farmers are pushed off their land in search of better futures that are more difficult to find (Semple 2019; Lustgarten 2020a). Displacement by climate change also affects populations in the Global North: wildfires in California, heatwaves in the Southwest and Midwest, crop productivity decline in Texas, sea-level rise in the Carolinas, and hurricanes in Florida (Lustgarten 2020b).

Given the shared challenges, the importance of governance is omnipresent. We next turn our focus on how sub-national governance can make a difference, especially as we are experiencing the whiplash from national policy shifts retreating from global cooperation efforts.

HOW LOCAL, NATIONAL, AND GLOBAL GOVERNANCE MAKE A DIFFERENCE

To provide you with a better sense of the role of governance and policymaking, I highlight the following three case studies from the Global South and the Global North, where governance and policymaking are making a difference. Our cases use food as an entry point and address SDG 2, SDG 3, SDG 10, SDG 16, and SDG 17. Case Study 3.1 provides a global overview of the role of governance and policymaking in causing food insecurities. Case Study 3.2 presents a local success story that has been scaled across Brazil. Case Study 3.3 highlights pre-war efforts in Ukraine to address inequalities in land ownership. It includes a review of how the war in Ukraine has caused global ripple effects, revealing a shake-up in global governance.

CASE STUDY 3.1 FAMINES ARE MAN-MADE

Amartya Sen stated famously that famines do not occur in democracies, as democratic systems are accountable to the public and policymakers fear being voted out of power if they fail to avert famines (Sen 1999). Food prices spiked during the global food crisis of 2007 to 2008. They were a driver behind the Arab Spring, mobilizing people to go onto the streets and forcing government changes (PBS Newshour 2011). Alex de Waal's research (de Waal 2018) finds that catastrophic famines are man-made and politics matter. Climate change or conflict exacerbate the effects. A *Financial Times* headline from 2019 captures as much: "*Drought and mismanagement push Zimbabwe to brink of famine"* (Cotterill 2019). In Gaza, in the immediate aftermath of the October 7, 2023 attack by Hamas, Israel deployed a siege on the region, blocking or severely limiting access to food, water, fuel, and electricity. Combined with the ongoing military actions, food insecurity has reached catastrophic levels, with doctors treating undernourished patients or finding starved children in a region that historically has been food secure (United Nations 2024). An official famine declaration by the United Nations was delayed until August 2025 (United Nations 2025), as they "convey significant political and emotional weight and the expectation of action from the international community" (Welsh, Kirschenbaum, and Swanson 2024). So even the definition and subsequent declaration of famine – akin to a declaration of genocide – matter. As captured by de Waal (2018, 194), "[w]hat politicians have created, politicians — under pressure from their publics — can remedy." This vignette on governance and SDG 2 demonstrates the importance of commitment by governments and their policymakers. Beyond the case of man-made famines, a household's hunger is affected by a breakdown of global governance. The war in Ukraine illustrated how changes in the global supply of both essential staples (e.g., cereals) and inputs (e.g., synthetic fertilizers) can have substantial ripple effects on food availability and production, to the detriment of local populations (see also Case Study 3.3). During the war, Turkey led diplomatic efforts with monitoring by the UN to support cereal shipments out of ports in Odessa, Ukraine (Bengali 2022). The Nobel Committee recognized the importance of global governance by awarding the UN's World Food Program the Nobel Peace Prize in 2020 for its work during the pandemic and in fragile places (Specia and Stevis-Gridneff 2020).

CASE STUDY 3.2 FOOD SECURITY IN THE CITY OF BELO HORIZONTE

In the sixth largest city of Brazil, Belo Horizonte, home to 2.8 million people in the mining state of Minas Gerais, hunger and malnutrition have been successfully addressed through effective governance and policymaking (Rocha and Lessa 2009). Around 12,000 locals can get subsidized meals from municipal workers in city-owned cafeterias. These nutritious meals – costing around 1 US dollar – are subsidized by taxpayers at a cost of 5 cents per week. According to Jose Enrique Serata, a cafeteria manager, "[t]hose of us who work here consider us defenders of the right to eat. Anyone, rich or poor, can eat here. For us, food is a basic right" (Mauro 2010). Their cafeterias offer affordable meals with dignity, with lawyers and laborers sharing tables. Launched in 1993 through a mayoral initiative that aimed to tackle urban hunger by improving food access, the currently named *Municipal secretariat for food and nutrition security* (SMASAN) serves around 800,000 citizens per year at a total cost of less than 2 percent of the city's budget (Chappell 2018, 2024).

Beyond providing access to meals (including around 200,000 school meals every day), Belo Horizonte's policies also improve urban food production and distribution. Neighborhood community gardens are maintained to provide fellow citizens with access to vegetables, fruits, and medicinal herbs. Local, privately run supermarkets receive subsidized access to land if they offer fixed-price produce. The municipal policies also support adult cooking, nutrition education programs and training in agroecology (Chappell 2018, 2024).

A testament to the program's success is its survival over three decades, outlasting seven different mayors. The secret to its success is that the intervention is too effective, cost-efficient, and popular to be canceled. The program has increased food availability and accessibility, ensured food acceptability and adequacy by engaging nutritionists and local cooks, strengthened food safety, and engrained the right to food in its policies. Each mayor engages with civil society and draws on participatory budgeting and reviews to build legitimacy and ensure the effectiveness of the food programming (Chappell 2018, 2024).

The model of Belo Horizonte has been scaled across Brazil and beyond. Its food policymaking and governance have contributed to a 70 percent drop in infant mortality, a 61 percent decrease in infant hospitalization for malnutrition, and a 33 percent drop in diabetes-related hospitalizations in the first two decades of the program (Chappell 2018). Beyond serving as a case study for scholarship, Belo Horizonte brought inspiration for change into the halls of think tanks (e.g., CSIS 2018) and the UN (e.g., HLPE 2020). The case serves as a powerful example of how local governance, when combined with the changemakers from civil society, academics, and the private sector, can make a lasting difference. Popular programs can better withstand national or global shifts, as sub-national and local actors find avenues to sustain them.[2]

CASE STUDY 3.3 LAND GOVERNANCE IN UKRAINE

The importance of agriculture in Ukraine's culture[3]

Agriculture serves as a place of pride in Ukraine and is ingrained in its culture. Ukraine has over a quarter of the world's most fertile soil: *chernozem* meaning "black earth" in Ukrainian and Russian. Rich in humus and covering more than half the country, not only is Ukraine's production highly sought after – with vital global exports being cereals and sunflowers – but so is this soil: estimates of the black market for illegally acquired *chernozem* in Ukraine reached US$900 million per year in the early 2010s (Kraznozhon 2011). Ukrainians' sensibility to their land, food production, and politics is furthermore grounded by their experience of the *Holodomor*, or the Ukrainian Famine, from 1932 to 1933. Recognized by some scholars as a genocide against the Ukrainian people conducted by the Soviet regime under Joseph Stalin, up to 10 million people died of starvation-related illness in one of the most agriculturally productive areas in the world. Fearing the power of the peasants and potential counterrevolutions, the aim of the Soviet regime was not only mass starvation but also to erase the Ukrainian culture (National Park Service 2023). As stated by Victor Karvchenko, a former Soviet official who defected:

> Here I saw people dying in solitude by slow degrees, dying hideously, without the excuse of sacrifice for a cause. They had been trapped and left to starve, each in his own home, by a political decision made in a far-off capital around conference and banquet tables. There was not even the consolation of inevitability to relieve the horror.
>
> (National Park Service 2023)

Challenges in equality and growth during the post-Soviet era

When looking at the distribution of resources[4] since the 1970s in Figure 3.1, we witness a marginally more equal distribution of resources across the globe. Eastern Europe and Central Asia were the exceptions, where after the fall of the Soviet Union, access to resources became less equal. Ukraine gained independence in 1991 and ratified its new constitution in 1995. The country witnessed a rapid decline, dropping by 35 percent from near-maximum equality (0.99 out of 1 in 1990) to reach close to the global average (0.64 in 2023 for Ukraine, versus 0.57 for the world) (Coppedge et al. 2024).

Ukraine has not experienced the same growth in GDP per capita compared to its neighbors, Poland and Russia. Ukraine's continued status as a non-EU member state has further stifled growth potential and a more equal distribution of resources. By 2023, Poland's equal distribution of resources was over 20 percent higher than Ukraine's, reflecting the gaps between the neighbor's development trajectories (Coppedge et al. 2024). Poland's membership in the EU, for example, has afforded it improved trade and development opportunities, whereas Ukraine's challenges with corruption and the rise of an oligarchic class have hindered its growth potential. Given the deficiencies of the Ukrainian state and economy, civil society has played a crucial role in the past decades in driving reforms toward improved governance (Solonenko 2015).

Outsized role of rural Ukraine in development

As a driver of rural development, agriculture plays a large role. In the Soviet Union, farming was organized in collective state-owned farms, so-called *kolhkohz*. After independence in 1991, workers on these state-owned farms received land titles as part of a post-independence privatization process. The workers received certificates of ownership, usually around 3 to 4 hectares, as shares of the former state-owned farms. Private companies later leased the land from these new landowners, frequently retirees hoping to sustain their income through lease payments. With increased calls for further privatization of the land markets, a 1999 policy permitted the sale of land. By 2001, however, a moratorium had to be put in place as wealthy Ukrainian oligarchs were buying up large swaths of land at undervalued prices (including the fraudulent sale of state-owned land). Contingent on IMF funding, the land market was reopened in 2021 with an initial prohibition of foreign ownership in place. Land was trading at a significant discount due to the war (Matuszak 2023). Expectations are that smaller farmers who cannot maintain their operations will be pushed out of agriculture and integrated into the larger landholdings (Amosov 2019; Reicher and Mousseau 2021).

To provide a sense of the land ownership inequality in Ukraine, the five agricultural conglomerates hold over two million hectares of agricultural land – close to four times the size of Switzerland's arable land. Kernel, the largest of the five, held over 500,000 hectares alone, with plans to increase to 700,000 hectares (Latifundist.com 2021). To put it in comparison with US agricultural landholdings, it is equivalent to approximately 4,000 farming households in the state of Iowa alone and around seven times more than the largest US farm. These landholdings put in question productivity and the potential to efficiently manage farms of this size, as the diversified production by family farmers is assumed to be producing more output per hectare (Graeub et al. 2016). Land plays a crucial role in fueling geopolitical strategies beyond Ukraine, with a US private equity company owning 430,000 hectares, a conglomerate owned by the sovereign wealth fund of Saudi Arabia owning 195,000 hectares, and a French agricultural company owning 120,000 hectares. These large farms can also wield considerable political influence in Ukraine, garnering reduced taxes and considerable support through external development finance. Kernel, for example, is reported to have received US$248 million in loans from the European Bank for Reconstruction and Development (EBRD) since 2018 (Reicher and Mousseau 2021).

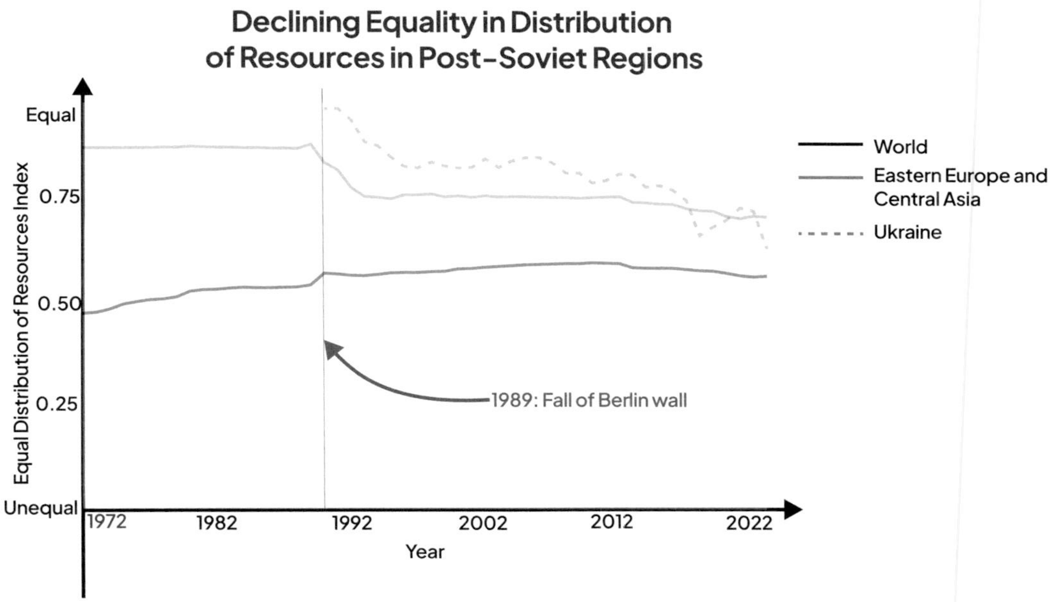

Figure 3.1 Declining equality in distribution of resources in post-Soviet regions, Equal Access Index, Coppedge et al. (2024)

LAND REFORM FOR SUSTAINABLE DEVELOPMENT

The case of Ukraine shows how rural development efforts are influenced by good governance and policymaking. It also showcases how agriculture extends beyond food, demonstrating its relevance to the political and social dimensions of society. SDG 16, focusing on good governance, aims at "promoting peaceful and inclusive societies for sustainable development, providing access to justice for all, and building effective, accountable and inclusive institutions at all levels." For land reform efforts to work, they need to build upon effective, accountable, and inclusive institutions and a justice system that provides access to all, aligned with the UN's *Voluntary Guidelines on the Responsible Governance of Tenure of Land, Fisheries and Forests* (Amosov 2019). The legacy structures left behind from the Soviet era and subsequent challenges in the transition to a market-oriented democracy have created a lack of accountability, with Ukrainian institutions perceived as corrupt, servicing the oligarchy. The land markets themselves lacked inclusivity, with 64 percent of people opposing the creation of a land market in 2021 before the onset of the war. Ukraine is also challenged by a corrupt judiciary, failing to provide justice for all (Reicher and Mousseau 2021).

Good governance matters in achieving the SDGs: uneven land ownership can negatively impact the achievement of agricultural production (SDG 2). Corruption can reduce economic growth (SDG 8). Lower or weakened accountability can provoke increased environmental externalities or pollution, affecting our ability to improve biodiversity and life on land (SDG 15) and overall leading to an increase in inequality (SDG 10). This case also highlights the importance of diplomacy and international relations, as the US government, for example, has been encouraging its European counterparts to strengthen Ukraine's case for accession into the EU. The lifting of the moratorium on land sales has also taken place in concert with IMF demands based on conditional loans. Good governance on land reform can make a difference in supporting the achievement of sustainable development if it yields a more equal distribution of land and affords greater societal benefits with reduced environmental externalities. The history of land governance in Ukraine demonstrates some of the challenges in achieving these goals.

THE BIGGER PICTURE AND ITS RIPPLE EFFECTS

The rural space in sustainable development is contested. Beyond our Ukrainian case, four questions emerge that can guide you in other contexts:

1. Who has access to land, capital, or technologies?
2. What is the role of government in facilitating more inclusive access to the above?
3. How do we manage inevitable trade-offs from improved technologies with lower employment?
4. What are the social and environmental dimensions of development that I am missing with my economic focus?

The initial Russian invasion of Ukraine caused not only an estimated loss of US$2 billion in lost harvest and widespread food insecurity in Ukraine but also incited global ripple effects, including increased fertilizer and grain prices pushing food inflation higher in vulnerable regions (Schipani, Terazono, and Saleh 2022; Vos 2024). The war in Ukraine also put into question the dominance of the Western sphere of influence on the global world order. With the BRICS countries – Brazil, India, China, and South Africa – failing to throw their support behind the West's alliance against their member, Russia, our picture of global governance has been shaken. South Africa accused Israel of apartheid and genocide at the international criminal court (AP 2024). West African states in the Sahel region rejected French and UN peacekeeping forces in favor of Russia (Chafer, Bertrand, and Stoddard 2024). As noted by Olander and Ramani (2022), leaders from the Global South articulate that "ideology and principle [are] just something we cannot afford."

STEPS TO IMPROVE THE EFFECTIVENESS OF GOVERNANCE AND POLICYMAKING

To address sustainable development challenges more effectively, it is increasingly clear that we need to improve how we build consensus and move past an impasse to reach the hard decisions while mobilizing the necessary resources

(Independent Group of Scientists Appointed by the Secretary-General 2019). On the global stage, it includes reforming the UN system to deal with the new geopolitical power dynamics (including in the Security Council) (Mahbubani 2024). At the UN *Summit of the Future* in 2024, UN Secretary-General Guterres "expressed concern that the UN's authority to tackle mounting global challenges is fast eroding, and without reform of its composition and working methods, 'it will lose all credibility'" (IISD 2024).

Even as the sustainability agenda has become more political, major global powers were not retreating from its relevance: the current and past chairs of the G20, South Africa, Brazil and India, have put the SDGs at the center (Hornidge and Berger 2024). After the Biden Administration released in September 2024 its *US Strategy on Global Development* drawing on the SDGs, the transition of power in the United States in 2025 re-emphasized the importance of local and sub-national governance. Necessary efforts include empowering grassroots organizations, including those human rights defenders referenced in Chapter 2, and strengthening an independent judicial system to counterbalance the democratic backsliding that gained traction during COVID-19 (Carothers 2023).

Development actors can contribute to these efforts by advancing their localization agenda and providing greater (purchasing) power to local partners who are more adaptive to change.[5] Calls for localization are not new – networks such as NEAR (the *Network for Empowered Aid Response*) with its aim to "reshape the top-down humanitarian and development aid system to one that is locally driven and owned" have just become more visible in the process (NEAR 2024). Momentum for change in the Global South has gathered in speed as it resonated with the Black Lives Matter movement in the Global North (Editorial Board 2021). It is being tested in the final years of the SDGs, as economic growth, inflation, and domestic security are trumpeting socio-environmental considerations. These trends reinforce a focus on sustainable development beyond 2030.

Beyond aid, greater recognition and support for non-traditional alliances – such as between academia, local citizens, and subnational actors (e.g., cities) – can power change through subnational governance (Independent Group of Scientists Appointed by the Secretary-General 2019). On January 1, 2025, the Dutch city The Hague became the first city in the world to use local laws to ban fossil-fuel advertising to counteract greenwashing (C40 Cities 2024). Besides drawing on local economies and their small and medium-sized enterprises, non-traditional alliances can also draw on the members of the *World Business Council for Sustainable Development* or the *We Mean Business Coalition* to influence a shift in policymaking (Nelson and Ingram 2023). We will explore further their role as changemakers in Chapter 4.

YOUR TURN

What is the role of governance and policymaking in achieving your SDG? What are the roles of local, national, or global stakeholders in shaping your SDG?

NOTES

1 Indicator 1.2.2. draws on that national definition by looking at the "[p]roportion of men, women and children of all ages living in poverty in *all its dimensions according to national definitions*" (emphasis added).
2 A case in point playing out in early 2025 was the attempt to dismantle the U.S. Inflation Reduction Act, which had large support in Republican-controlled states and among local constituencies.
3 The following case is based on a conversation with a former US foreign service officer who was stationed in Ukraine during the mid-2010s. They generously guest-lectured in my course over the years.
4 The index is composed of "measurements of power distribution by socioeconomic position, social group and gender; equality in health and education; and the breadth of welfare program coverage" (V-DEM 2016).
5 In the humanitarian sector, the "Grand Bargain" agreement was reached in 2016 to provide "more support and funding tools for local and national responders" (United Nations 2016, 5), with regular updates on progress made on its website (Inter-Agency Standing Committee 2025).

REFERENCES

Amosov, M. 2019. "The Land Question." TNI, *Commons: Journal of Social Criticism*. https://longreads.tni.org/the-land-question-ukraine.

AP. 2024. "South Africa Tells Top UN Court That It's Accusing Israel of Apartheid Against Palestinians." Available at: https://apnews.com/article/israel-palestinians-court-africa-apartheid-9eb5f75f05f636b41ba2873b0b296fce.

Bengali, S. 2022. "The U.N. and Turkey Work to Restore the Suspended Grain Deal." *New York Times*, October 30. https://www.nytimes.com/live/2022/10/30/world/russia-ukraine-war-news.

C40 Cities. 2024. "How Cities Can Restrict Carbon-Intensive Advertising." Available at: https://www.c40knowledgehub.org/s/article/How-cities-can-restrict-carbon-intensive-advertising?language=en_US (accessed February 14, 2025).;

Carothers, T. 2023. "Is the Global Tide Turning in Favor of Democracy?" Carnegie Endowment for International Peace. Available at: https://carnegieendowment.org/research/2023/05/is-the-global-tide-turning-in-favor-of-democracy?lang=en&center=middle-east.

Chafer, T., E. Bertrand, and E. Stoddard. 2024. "France's Strategic Failure in Mali: A Postcolonial Disutility of Force?" Available at: https://www.rusi.org/explore-our-research/publications/commentary/frances-strategic-failure-mali-postcolonial-disutility-force.

Chappell, M. J. 2018. *Beginning to End Hunger: Food and the Environment in Belo Horizonte, Brazil, and Beyond*. Berkeley, CA: University of California Press.

Chappell, M. J. 2024. "Beginning to End Hunger: Lessons from Belo Horizonte Brazil." Available at: https://www.youtube.com/watch?v=N_qUI8ovrtA.

Coppedge, M., J. Gerring, C. Henrik Knutsen, S. I. Lindberg, and J. Teorell. 2024. V-Dem [Country—Year/Country—Date] Dataset v14. edited by Varieties of Democracy (V-Dem) Project.

Cotterill, J. 2019. "Drought and Mismanagement Push Zimbabwe to Brink of Famine." *The Financial Times*, October 8. https://www.ft.com/content/011b4534-dacc-11e9-8f9b-77216ebe1f17.

CSIS. 2018. "When the City Does the Feeding: The Role of Local Governance in Urban Food Security." Available at: https://www.csis.org/events/when-city-does-feeding-role-local-governance-urban-food-security.

de Waal, A. 2018. "The End of Famine? Prospects for the Elimination of Mass Starvation by Political Action." *Political Geography* 62: 184–195. https://doi.org/10.1016/j.polgeo.2017.09.004.

Editorial Board. 2021. "Foreign Aid Is Having a Reckoning." *The New York Times*, February, 13. https://www.nytimes.com/2021/02/13/opinion/africa-foreign-aid-philanthropy.html.

Graeub, B. E., M. J. Chappell, H. Wittman, S. Ledermann, R. B. Kerr, and B. Gemmill-Herren. 2016. "The State of Family Farms in the World." *World Development* 87. https://doi.org/10.1016/j.worlddev.2015.05.012.

Hancock, A. 2023. "EU Workers Lack Skills to Green the Economy, EIB Poll Finds." *Financial Times*, February 27. https://www.ft.com/content/ec47ede3-773a-4bdd-a1e9-1a7f48037099.

HLPE. 2020. "Food Security and Nutrition: Building a Global Narrative Towards 2030." Rome: High Level Panel of Experts on Food Security and Nutrition of the Committee on World Food Security.

Hologic. 2025. "Global Women's Health Index." Hologic, Inc. Available at: https://hologic.womenshealthindex.com (accessed February 14, 2025).

Hornidge, A.-K.., and A. Berger. 2024. "After the Pact Is Before the Future: From the Summit of the Future to the Hamburg Sustainability Conference." German Institute of Development and Sustainability (IDOS) (The Current Column). 30 September. Available at: https://www.idos-research.de/fileadmin/user_upload/pdfs/publikationen/aktuelle_kolumne/2024/German_Institute_of_Development_and_Sustainability_EN_Hornidge-Berger_30.09.2024.pdf.

IISD. 2024. "Summit of the Future Highlights: Sunday, 22 September 2024." Available at: https://enb.iisd.org/summit-future-22sep2024.

ILO. 2023. "The Rana Plaza Disaster Ten Years On: What Has Changed?" Available at: https://webapps.ilo.org/infostories/en-GB/Stories/Country-Focus/rana-plaza#safety-challenges.

Independent Group of Scientists Appointed by the Secretary-General. 2019. *Global Sustainable Development Report 2019: The Future is Now – Science for Achieving Sustainable Development*. New York: United Nations. https://sdgs.un.org/gsdr/gsdr2019.

Inter-Agency Standing Committee. 2025. "The Grand Bargain (Official website)." Available at: https://interagencystandingcommittee.org/grand-bargain (accessed March 31, 2025).

Kanie, N., D. Griggs, O. Young, S. Waddell, P. Shrivastava, P. M. Haas, W. Broadgate, O. Gaffney, and C. Körösi. 2019. "Rules to Goals: Emergence of New Governance Strategies for Sustainable Development." *Sustainability Science* 14 (6): 1745–1749. https://doi.org/10.1007/s11625-019-00729-1.

Kraznozhon, L. 2011. "Black Market for Rich Black Earth." *KyivPost*, November 9.

Latifundist.com. 2021. "Kernel to Increase Its Leasehold Farmland Bank to 0.7 mln ha: Verevskiy." Available at: https://latifundist.com/en/novosti/56910-kernel-nameren-narastit-zemelnyj-bank-do-700-tys-ga--verevskij.

Ledermann, S. 2012. "Organic Revolution: Cotton and Its Impact on Poverty, Inequality and Sustainability." PhD dissertation, Geography Department, Rutgers University.

Lustgarten, A. 2020a. "The Great Climate Migration Has Begun." *New York Times Magazine*, July 23.

Lustgarten, A. 2020b. "How Climate Migration Will Reshape America." *New York Times Magazine*, September 15.

Mahbubani, K. 2024. "UN Credibility Depends on Adjusting Veto Rights to Match Shift in Global Power." *Financial Times*, August 11. https://www.ft.com/content/abf9a9bb-c88c-4faa-9a86-6ec500bcd5db.

Matuszak, S. 2023. "Ukraine: Two Years' Free Trading of Arable Land." OSW – Centre for Eastern Studies. Available at: https://www.osw.waw.pl/en/publikacje/analyses/2023-07-17/ukraine-two-years-free-trading-arable-land.

Mauro, C. 2010. "Fighting Hunger in Brazil." Al Jazeera. Available at: https://www.youtube.com/watch?v=6Lxg2gY4RiU.

National Park Service. 2023. "Holodomor Memorial." Available at: https://www.nps.gov/nama/planyourvisit/holodomor.htm.

NEAR. 2024. "Who We Are." Available at: https://www.near.ngo/who-we-are-copied (accessed March 23, 2025).

Nelson, J., and G. Ingram. 2023. "Scaling Private Sector Engagement in the SDGs." In Center for Sustainable Development at Brookings (Ed.), *The SDG Second Half: Ideas for Doing Things Differently*. Washington, DC: Center for Sustainable Development at Brookings.

Olander, E., and S. Ramani. 2022. "The Ukraine War and Great Power Politics in the Global South." *The China in Africa Podcast*. China Global South Project.

OPHI, and University of Oxford. 2024. "National MPI and Policy." Available at: https://ophi.org.uk/national-mpi-and-policy.

Parliament of Rwanda. 2024. "Women Representation." Government of the Republic of Rwanda. Available at: https://www.parliament.gov.rw/women-representation (accessed November 17, 2024).

PBS Newshour. 2011. "Did Food Prices Spur the Arab Spring?", September 7. (accessed June 6, 2024).

Reicher, B., and F. Mousseau. 2021. "Who Really Benefits from the Creation of a Land Market in Ukraine?" Oakland Institute, CA.

Rocha, C., and I. Lessa. 2009. "Urban Governance for Food Security: The Alternative Food System in Belo Horizonte, Brazil." *International Planning Studies* 14 (4): 389–400. https://doi.org/10.1080/13563471003642787.

Schipani, A., E. Terazono, and H. Saleh. 2022. "Soaring Fertiliser Prices Deepen Africa's Food Crisis." *Financial Times*, August 21. https://www.ft.com/content/4d746aa5-29e3-4796-b9e6-0b64c6865389.

Searcey, D., and F. Essomba. 2017. "African Nations Increasingly Silence Internet to Stem Protest." *The New York Times*, February 10.

Semple, K. 2019. "Central American Farmers Head to the U.S., Fleeing Climate Change." *The New York Times*, April 13. https://www.nytimes.com/2019/04/13/world/americas/coffee-climate-change-migration.html.

Sen, A. 1999. *Development as Freedom*. New York: Knopf.

Smith, W. C., M. C. M. Ehren, and S. Grek. 2024. "Global Governance of Education: The Sustainable Development Goals as a Product and Mechanism?" *International Review of Education* 70 (4): 531–545. https://doi.org/10.1007/s11159-024-10108-3.

Solonenko, I. 2015. "Ukrainian Civil Society from the Orange Revolution to Euromaidan: Striving for a New Social Contract." In *OSCE Yearbook 2014*. Baden-Baden: IFSH, pp. 219–235.

Specia, M., and M. Stevis-Gridneff. 2020. "World Food Program Awarded Nobel Peace Prize for Work During Pandemic." *The New York Times*, October 9. https://www.nytimes.com/2020/10/09/world/2020-nobel-peace-prize.html?login=email&auth=login-email.

UN Conference on Small Island Developing States. 2014. "Small Island Developing States Accelerated Modalities of Action. Sustainable Development Knowledge Platform." Available at: sustainabledevelopment.un.org/sids2014

United Nations. 2016. "The Grand Bargain – A Shared Commitment to Better Serve People in Need." Available at: https://interagencystandingcommittee.org/sites/default/files/migrated/2016-10/grand_bargain_final_22_may_final-2.pdf.

United Nations. 2024. "Over 1.8 Million in Gaza Face Extreme Hunger." Available at: https://news.un.org/en/story/2024/10/1155836.

United Nations. 2025. "Gaza: Famine "Irrefutably" Confirmed, UN Humanitarians Unite in Plea for Aid Access – OCHA, OHCHR, WFP and WHO." Available at: https://www.un.org/unispal/document/ocha-ohchr-wfp-who-press-briefing-22aug25/

V-DEM. 2016. "Equal Distribution of Resources in Colombia." Available at: https://v-dem.net/weekly_graph/equal-distribution-of-resources-in-colombia.

Vos, R. 2024. "The War in Ukraine Continues to Undermine the Food Security of Millions." IFPRI. Available at: https://www.ifpri.org/blog/war-ukraine-continues-undermine-food-security-millions.

Welsh, C., A. Kirschenbaum, and Z. Swanson. 2024. "Famine in Gaza." *CSIS*, 2024.

Williams, A. A. 2023. Speech by the Republic of Vanuatu at the High-Level Segment of the 19th Conference of the Parties to the United Nations Framework Convention on Climate Change Warsaw, Poland.

4 Economy and finance

THE CASE FOR ECONOMY AND FINANCE TO ACHIEVE THE SDGs

Economic policy – be it fiscal, monetary, or trade policy – matters. Given the transnational challenges in sustainable development, cross-border trade and global financial flows play a large role in either limiting or expanding the impacts.[1] Stable financial flows ensure that markets, economies, and people are more resilient to shocks, especially through a dependable social safety net from government programming (Independent Group of Scientists Appointed by the Secretary-General 2019). Increasingly, remittances play a large role in supporting development (Independent Group of Scientists Appointed by the Secretary-General 2019), reaching an estimated US$656 billion in 2023 – predominantly from the Global North to the Global South – and recovering from their COVID-19 lows (World Bank 2024b).

Growth has come at the cost of the environment. Incentives from the state or activist campaigns from civil society have pushed us toward "developing new ideas that led not just to more growth, but a different type of growth as well" (Susskind 2024, 208). How governments set incentives makes a difference: at an individual level, providing tax credits, for example, for individuals to adopt more environmentally friendly technologies, such as solar energy, can significantly spur demand (e.g., the Inflation Reduction Act). Creating more effective tax policies can generate additional revenue for increasing public expenditures without increasing inequality.[2] On the flip side, not putting any costs on the negative environmental externalities fails to create disincentives for companies to move toward more sustainable production. False incentives, such as subsidies discussed earlier, can further expand these harmful externalities.

Official development assistance (ODA) has ebbed and flowed over the past decades (see Figure 4.1). In sub-Saharan Africa, aid has increased seven-fold, from US$8.08 billion in 1972 to US$57.77 billion 50 years later in 2022. Periods of increased aid flow were the COVID-19 pandemic and the aftermath of the global food crisis. In Central America and the Caribbean, aid has increased only four-fold, from US$1.48 billion to US$5.65 billion over the same period. In 2022, with the invasion of Ukraine by Russia, ODA funds were redistributed from the Global South to address this crisis in the Global North, with aid to Ukraine increasing more than ten-fold in a single year (from US$2.12 billion in 2021 to US$28.70 billion in 2022).

To put these numbers into perspective, if you invested US$8.08 billion in 1972 into the stock market, you would have around a 40-fold increase (adjusted

DOI:10.4324/9781003400028-7

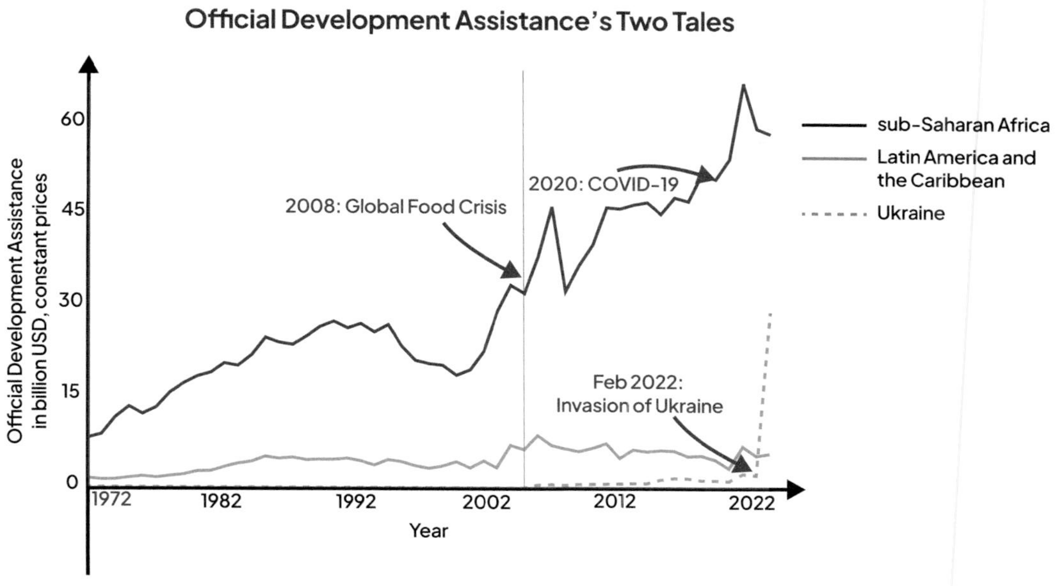

Figure 4.1 Official development assistance's two tales, OECD (2024)

for inflation), showcasing how aid amounts have not kept up with overall capital accumulation. In any year, only a handful of countries (mostly Scandinavian) meet or exceed the target of 0.7 percent of their gross national income for development assistance. South-South cooperation is increasingly important (Independent Group of Scientists Appointed by the Secretary-General 2019). Leveraging blended finance brings in private capital to supplement public investments.

Development finance institutions, mainly development banks, accounted for US$1.9 trillion in investments in 2018 – reflecting their overall importance. Multilateral development banks (MDBs), however, did not increase their net flows significantly: even though the World Bank, for example, acknowledged that we need to increase funding levels to achieve the transformative vision of the SDGs, MDB funding flows increased only from US$25 billion around 2015 to US$41 billion in 2021 largely because of the COVID-19 pandemic (Bhattacharya and Kharas 2023). On the other hand, the Asian Infrastructure Investment Bank – what James Kynge calls "Beijing's answer to the World Bank" – is stepping up its annual climate financing funding from US$2.6 billion to around US$7 to 8 billion (Kynge 2023).

The need continues for private funding to increase. Developing countries alone face an annual US$1.2 trillion funding gap to achieve sustainable development (Independent Group of Scientists Appointed by the Secretary-General 2019). Countries in both the Global South, such as Zambia (Haabazoka 2019), and the Global North, such as Italy (Johnson 2019), were subsequently keen to draw on all types of investments, including from China. Chinese investments come as part of its large Belt and Road Initiative or individual entrepreneurs (e.g., in Nigeria, Feng and Pilling 2019). The United States aimed to counter Chinese investments (e.g., in Ecuador, Sevastopulo and Long 2021). Private bankers in Geneva, London, or New York are another funding source. Moving *beyond aid* has also become the motto for many African leaders, who are eager to increase the value-added of their exports (Pilling 2024).

Private funds are magnitudes higher than the funding gaps identified. According to UNCTAD (2024), global financial assets were approaching US$500 trillion. Some of the largest investors might not be who you think they are: pension funds make up US$12.3 trillion out of US$20.7 trillion of all US institutional investors' equity and investment funds in 2023 (OECD 2025). These funds are especially adept at sustainable development, increasingly incorporating, for example, climate change risks into their portfolio. They also play a large role in influencing decisions in publicly traded companies: as one of the largest shareholders, they can force greater investments toward sustainable development (e.g., improved labor rights, reduced carbon emissions, etc.) (Independent Group of Scientists Appointed by the Secretary-General 2019). To illustrate how investments can work, we first look at an instance of *private companies drinking the SDG Kool-Aid*, followed by a deeper dive into the ability of private capital to fill funding gaps.

Private companies drinking the SDG Kool-Aid

The first time I engaged with the SDGs was in my position as a scientific advisor to a policy advocacy team. In 2014, I joined them on a trip to Charlotte, North Carolina, for the Water-Food-Energy Nexus conference. Our team was working on negotiations of the draft text of the 2030 Agenda. At this stage, my main engagement with sustainable development had been focused on the role of local NGOs, research institutes, and international organizations. I was hence surprised to find that PepsiCo was a headliner of the main panel at the conference, talking about how the company is embracing the water-food-energy nexus and was ready to contribute to sustainable development.

To illustrate PepsiCo's commitment to sustainable development, the company has developed three focal areas for their work: positive agriculture, positive value chain (including climate, water, packaging, and people), and positive choices (including expanded portfolio offerings, and brands with positive impacts). Across these three focal areas, a self-declared total of 11 SDGs are contributed to: SDG 2, SDG 3, SDG 5, SDG 6, SDG 7, SDG 8, SDG 10, SDG 12, SDG 13, SDG 15, and SDG 17. To provide a better overview of the multinational corporation's contribution, Table 4.1 offers self-declared metrics and targets as reported in PepsiCo's annual ESG report (PepsiCo 2023).

While more novel a decade ago, the call for the private sector to step up if we are to achieve sustainable transformations is omnipresent. Nearly three

Table 4.1 ESG performance metrics 2023 of PepsiCo. Source: Metrics from PepsiCo (2023)

Focal area and main SDG	Metric	2030 Target	2023	2022
Positive Agriculture [SDG 2]	Spread the adoption of regenerative agriculture practices across 7 million acres of the land used around the world to grow our crops and ingredients for our products	7 million acres	1.8 million acres	0.9 million acres
Positive Value Chain: Climate [SDG 13]	Reduce Scope 1 and 2 emissions by 75% (2015 baseline)	75%	33%	23%
Positive Value Chain: Water [SDG 6]	Replenishing back into the local watershed more than 100% of the water we use	100%	69%	45%
Positive Value Chain: Packaging [SDG 12]	Cut virgin plastic from non-renewable sources per serving across our global beverages and convenient foods portfolio by 50%	50%	1%	-2%
Positive Value Chain: People [SDG 5]	Achieve and sustain 50% women in management roles	50% (2025 target)	45%	44%
Positive Choices [SDG 3]	Reduce added sugars: 67% or more of beverage portfolio volume having less than 100 calories from added sugar per 12 oz. serving	67% (2025 target)	62%	56%

out of four of the world's largest companies are aligning their reporting at least in part to the SDGs (KPMG 2022). Out of a sample of 206 companies, SDG 13: Climate action (around 85%), SDG 8: Decent work and economic growth (75%), and SDG 12; Responsible consumption and production (around 50%) are the most frequently mentioned SDGs that companies support (GRI 2022).

There is still room to grow: only 6 percent are reporting on negative impacts on the SDGs (KPMG 2022) and only 20 percent provide evidence on their impacts (GRI 2022). Concerns are increasing around *greenhushing*, with companies speaking less about their commitments to climate change and sustainability (Mundy 2024) as they struggle to fundamentally transform their business model and try to introduce change at the margins. Mundy (2024) uses the example of an oil company executive who invests in carbon capture. Mention of *sustainability* or *emissions* on earning has more than tripled since 2019, peaked in 2021–2022, and is now stagnating or trending down. According to Mundy, the data trend captures a failure by companies to follow through on their initial pledges as they return to their focus on short-term earnings. Companies have left in droves the UN-backed Net-Zero Asset Managers climate initiative, causing the coalition to pause its activities and highlighting the limitations of voluntary commitments (Jessop and Kerber 2025). Given their profit motive, companies respond best to rules (Scheyvens, Banks, and Hughes

2016). New legislation, including sustainability disclosure rules to investors discussed below, can contribute to sustainability and emissions sticking around (Mundy 2024).

Private capital to fill the funding gap

To achieve the aspirational SDGs and their 169 targets by 2030, investments in the range from US$4 to 7 trillion (Niculescu, 2017; UNPRI, 2017) per year are needed, of which at least US$1.4 trillion per year are expected from the private sector (UNPRI, 2017). Already pre-COVID-19, the UN explicitly supported an increased role for the private sector in development, including an influx of new capital for sustainable development. With the Ukraine war diverting funds from sustainable development, the total funding gap has been estimated at up to US$135 trillion (Edgecliffe-Johnson and Neville 2022).[3] The funding gap for climate change adaptation in developing countries alone is around US$360 billion a year, with only US$21 billion currently provided through public financing (Mooney 2023).

To close the funding gap and achieve sustainable development, we need public and private companies to change their way of doing business. To adapt and grow their business, new capital needs to be mobilized.

As a first step, investors should identify what to divest from. Take the Norwegian pension fund (Norway Government Pension Fund Global), which holds US$1.3 trillion in assets. They exclude 181 companies via various negative screening strategies based on topics such as coal, nuclear, tobacco, human rights, environment, and weapons (van Zanten and Huij 2024). Pre-dating the SDGs, sustainable investment funds using positive screening focused on energy, water, and health to offer investors an additional outlet. These actively managed funds (such as iShares Global Clean Energy) include assets of up to US$6 billion (van Zanten and Huij 2024).

To further support the investment community with signals as to who is contributing the most (or least) to the achievements of the SDGs, select financial providers created SDG company ratings. These ratings offer a comparative ranking to investors and allow companies to evaluate their relative positioning among their peers. They fare better than the well-known Environmental, Social, Governance (ESG) ratings in capturing sustainability, with the SDG scores capturing positive and negative impacts as expected by investors, regulators, and climate scientists (van Zanten and Huij 2024). With no common definition of ESG used across rating agencies, operationalizing sustainable development through the 169 SDG targets and their indicators theoretically offers greater consistency. Policymakers in the EU – hoping to halt increased greenwashing (Bryan 2023) – developed a common definition for sustainable investments to regulate financial products (European Parliament 2019).

Ultimately, these rankings are only as good if they accurately and consistently reflect the respective companies' achievements. Discrepancies among different ranking systems can end up confusing sustainable investors, leading to investment inefficiencies. Worse, wrong rankings provide counterproductive financing to a sustainable transformation. Research by Bauckloh et al. (2024) found that SDG ratings from five rating providers differed substantially, with the average agreement among providers ranging from 30 to 50 percent. More contentious sectors, such as energy, and individual SDGs, such as SDG 13 on Climate Action, had higher disagreements.

VIGNETTE 1 ENVIRONMENTAL, SOCIAL, AND GOVERNANCE (ESG) POLICY

ESG policy's origin dates back to the UN calling for "better inclusion of environment, social and corporate governance factors in investment decisions" in 2004. As part of the UN Global Compact, over 20,000 companies have joined forces and committed to more principled, sustainable development and investment. ESG focus has taken off with investors since 2019, moving from less than 1 percent of earning calls to approximately 20 percent (Agnew, Klasa, and Mundy 2022). Investors have jumped on to create outlets for funding to the tune of US$2.7 trillion in assets by 2021. As progress on ESG policy is largely self-declared, fears of greenwashing become more prominent (Agnew, Klasa, and Mundy 2022). Conflict permeated around the role of weapons manufacturing: traditionally screened out of funds (via the negative screening), the war in Ukraine has raised the question if their contribution to geopolitical security should reward them with inclusion (Agnew, Klasa, and Mundy 2022).

Criticism has also risen from within the ranks of the largest investor in ESG: BlackRock.[4] Tariq Fancy, who quit as its Chief Investment Officer for Sustainable Investing, has lost faith that the market will fix market failures by embracing ESG. According to a review of Fancy by Armstrong (2021), ESG investing fails to raise the cost of capital for "bad" companies, doesn't provide higher returns and can't overcome the conflict between short-term profits and long-term horizons needed for changes to take place. Divesting from companies struggles to have a long-term impact if another buyer will purchase the (discounted) stock. In the end, leveraging an ESG lens (e.g., FTSE Russell n.d.) protects investors against social and environmental risks – rather than actively supporting companies in tackling the underlying problem (e.g., climate change).

In summary, ESG policy itself fails to systematically change the underlying dynamics that companies are accountable to their shareholders. In our current system, their fiduciary obligation remains the extraction of profits and not supporting social and environmental change (Armstrong 2021). There is hence a disconnect between how companies position themselves as the solution to global problems "without any acknowledgment of industry's complicity in *creating* at least some of these problems in the first place" (Scheyvens, Banks, and Hughes 2016, 380; *original emphasis*).

A more fundamental shift is needed whereby governance and policy-making create an enabling environment for businesses to contribute toward the SDGs. Legislation should be enacted to steer businesses toward being more socially and environmentally responsible (Scheyvens, Banks, and Hughes 2016). In Africa, illicit financial flows, such as tax evasion, are costing governments an estimated US$88 billion annually. Improving governance to curb such flows could drastically close funding gaps, as well as reversing the damage done by expropriation and illicit activities (UNCTAD 2020). Governments that lack strong institutions and transparent governance are also less likely to receive development aid as overall budget support, further fragmenting development financing (World Bank 2024a). The above reinforces how we need to activate or build capacities in governance and policymaking to supplement any transformational efforts toward more just economies and closing funding gaps.

While exact figures on funding gaps are difficult to ascertain here are two examples from sub-sectors showing the need for private capital and sector investments to step up and contribute toward sustainable development: biodiversity financing and agricultural financing. By zooming in on a sub-section, we can explore how the numbers break down, where the funding gaps are, and avenues to close them. One important note on the type of capital before diving in. In agriculture and biodiversity, *patient capital* – capital that does not demand a short turnaround for investors – is especially welcome to sustain impacts for investments that need a longer time to materialize benefits. For example, investing in agroforestry takes years to yield fruits, as trees mature. You cannot shake the tree early and you have to be patient.

THE BIODIVERSITY FINANCING GAP

As of 2019, the annual financing need for global biodiversity was estimated at US$967 billion (see Figure 4.2) (Tobin-de la Puente and Mitchell 2021). The global biodiversity financing available was US$143 billion, creating a financing gap of US$824 billion. To provide a sense of where funding for biodiversity comes from, ODA, which I think of as *the* source of financing for sustainable development, is only US$4 to 10 billion (or approximately 1 percent of the total financing needed). Most funding comes from government budgets and taxation

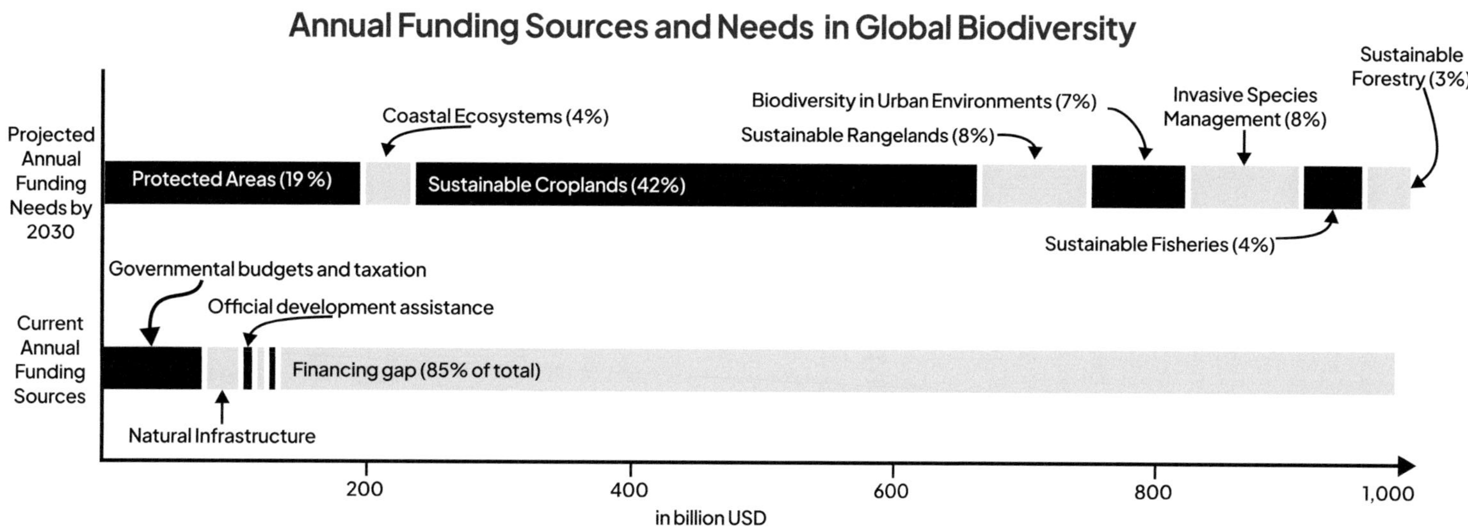

Figure 4.2
Annual funding sources and needs in global biodiversity, adapted using upper estimates from Tobin-de la Puente and Mitchell (2021)

(US$75 to 78 billion). Natural infrastructure ("networks of land and water that restore and conserve ecosystems"; Tobin-de la Puente and Mitchell 2021, 34), sustainable supply chains, and biodiversity offsets, combine for another US$39 to 44 billion in funding. Nature-based solutions and carbon markets comprise the smallest share of funding, at only US$1 billion. Less than 20 percent of all current biodiversity funding is transferred to the biodiversity-rich Global South (Tobin-de la Puente and Mitchell 2021, 39).

Figure 4.2 also captures the largest needs for annual biodiversity funding by 2030. They are sustainable croplands (ranging from US$315 to 420 billion a year), protected areas (US$149 to 192 billion a year), sustainable rangelands (US$81 billion a year), and invasive species management, sustainable fisheries, and forestry (at a combined US$78 to 168 billion a year). Biodiversity in urban environments and coastal ecosystems also has an annual funding need of around US$100 billion per year. The accounting shows the need for private sector funding to contribute toward sustainable finance.

VIGNETTE 2 HOW TO FINANCE BIODIVERSITY

Based on the UNDP and EU biodiversity finance initiative (BIOFIN) and Tobin-de la Puente and Mitchell (2021, 47), the following five funding avenues exist. For each avenue an example is presented, as cited by Tobin-de la Puente and Mitchell (2021) if you are seeking a deeper dive. These avenues are relevant well beyond biodiversity.

1. Generate revenue (e.g., payments for reduced deforestation under the REDD+ program): "Increase the number of funds deployed towards biodiversity protection through public spending, private investment, or other measures that can generate or leverage financial resources allocated to biodiversity."
2. Deliver better (e.g., biodiversity tax incentives for South Africa's protected areas): "Deliver results for biodiversity conservation through better resource management, improved efficiency, and greater alignment of incentives among actors."
3. Realign expenditures (e.g., reforming agriculture subsidies to support biodiversity in Switzerland): "Reduce investments that have negative impacts on biodiversity and redirect those financial flows towards activities that positively impact biodiversity."
4. Avoid future expenditures (e.g., Costa Rica's 3.5 percent tax on carbon emissions): "Prevent future costs through strategic investments and policy changes that protect biodiversity today and reduce the need for larger expenditures in the long term to restore or replace lost ecosystem services."

5. Catalyze (e.g., the Land Degradation Neutrality Fund): "Enhance policy, administrative or investment measures or enabling conditions that can result in new, improved or scaled-up biodiversity finance."

One of the most prominent mechanisms is payment for ecosystem services (PES). The mechanism recognizes the value gained from ecosystems (think pollination services increasing yields or watersheds purifying water). These payments are made by ecosystem users or third parties, such as governments. For example, a project I supported in the hills near Morogoro, Tanzania, was exploring if the local water processing facility is willing to pay for lower water pollution due to soil erosion control from organic farmers using terracing.

Since 2022, we have witnessed a rapid rise of "nature-positive" commitments from the private sector – with a larger push toward not just halting degradation, but regeneration of landscapes. While nature is inherently more difficult to assess than climate change, more private companies understand it pays to invest in ecosystems and biodiversity (e.g., beekeeping). Coordinated efforts are under way to channel these efforts toward a transformation at the country- and global-level (Savage 2024).

THE AGRICULTURAL FINANCING GAP

The funding gap to shift from a conventional to a regenerative agricultural system amounts to US$250–430 billion annually for the next ten years (Global Alliance for the Future of Food 2024). Around one-tenth of the funding is currently serviced. So similar to biodiversity funding, the funding gap is around 85–90 percent of the total. This funding gap ignores the hidden costs of doing business-as-usual. Using the true cost accounting method introduced in Chapter 2, Hendriks et al. (2023, 591) estimate that the annual costs from agriculture are around US$19 trillion.

Beyond regenerative agriculture, Steiner et al. (2020) identified 11 actions to transform our food system. Table 4.2 lists their actions, the targets used by the authors for their analysis, and the estimated annual cost to achieve them.

The majority of the funding is needed to reroute the system, including efforts in wetland protection, carbon payments, and training (Thornton et al. 2023). Unlocking sustainable finance, for example, has multiplier effects, with US$20 billion projected to release US$320 billion in funding (Thornton et al. 2023). Bashian et al. (2024, 19) argue that it is not a lack of funding that causes the gaps, but "missing confidence that the financing [for regenerative agriculture] will fit financiers' current risk and reward standards." Recalling our review of agricultural transformation in Chapter 2, we need to leverage all changemakers, including closing knowledge and research gaps, to support increased funding streams.

Table 4.2 Agricultural funding and focus areas, adapted from Thornton et al. (2023)

Focus area	Action	Targets	Annual costs (US$ billion)
Reroute	Ensure zero agricultural land expansion in high-carbon landscapes	Avoid conversion of 250 Mha forests and 400 Mha peatlands	753
	Enable markets and public-sector actions to incentivize climate-resilient, low emission practices	185 million farmers adopting new practices	181
	Support prosperity through mobility and rural reinvigoration	20 million rural jobs by 2030 through investing in infrastructure and youth	116
De-Risk	Secure resilient livelihoods through early warning systems and adaptive safety nets	Safety nets for 40 million rural dwellers currently dependent on humanitarian aid	5.6
	Help farmers make better choices	Connecting 200 million farmers and agribusinesses to ICT-enabled bundled advisory services	2.4
Reduce	Shift to healthy, sustainable, climate-friendly diets	Incentivize shifts to reduce beef and dairy consumption in 15 higher-income countries and the 40 largest cities	35
	Reduce food waste and loss	50 percent reduction in food waste and loss in major supply chains	12.6
Realign	Implement policy and institutional changes that enable transformation	Realign subsidies to a climate change agenda, improve ease of doing business, and improve readiness to climate change and resilience of select countries	177
	Unlock billions in sustainable finance	De-risking public investments to unlock US$320 billion in public and private capital	20
	Drive social change for more sustainable decisions	Improve education of 10 million young people on climate change and action	20
	Transform innovation systems to deliver impacts at scale	Additional funding through public investments for applied research to achieve the SDGs	15
Total			1.3 trillion

These shifts in funding also need to happen in the donor landscape. I researched this question together with Sinan Hatik (Biovision and IPES-Food 2020). We analyzed 137 agricultural research for development projects worth US$807 million from the then Bill and Melinda Gates Foundation (BMGF), the largest philanthropic donor in the world. As much as 85 percent of the foundation's investment focused on supporting industrial agriculture and increasing its efficiency. Less than 5 percent of their investments were made to focus on substitution of industrial inputs, as well as redesigning agro-ecosystems. Their funding primarily focuses on increasing the productivity of (sustainable) agriculture (related to SDG 2.3), limiting the ability to achieve a more holistic transformation needed (as reflected in SDG 2.4 and SDG 2.5).

These challenges are not unique to philanthropic organizations. Donor and recipient countries as well have the potential to contribute toward more sustained transformations. Most funding from the Swiss government, for example, supports the redesigning of agro-ecosystems, which is essential for achieving a transformation. In Kenya, a recipient country, the focus mirrors the priorities of

BMGF, with nearly three-quarters of projects supporting industrial agriculture and its efficiency (Biovision and IPES-Food 2020). Our findings re-emphasize the need to address knowledge gaps in alternatives to business-as-usual. Identifying synergies can increase the viability of the transformation itself and reduce the risk to investors.

The incremental nature of finance

Songbae Lee, former lead in agricultural finance at USAID, asked my students whether finance is incremental or transformational. I immediately had the urge to shout "transformational!" He used the increasing evidence around microfinance to argue that development finance is actually incremental. His argument goes as follows: does access to a credit card help you manage purchases across the year? Yes. Does it contribute to improving your life? Yes. Does it make you wealthy? Most likely not.

Microfinance programs facilitate steady access to capital under improved lending conditions. They help their members to smooth out unexpected shocks, as well as undertake larger purchases. However, they have no transformational effects on households by themselves. Access to finance is a tool – a lever – for change that can support transformation. But it needs to be embedded.

Microfinance interventions need to include financial education (including a focus on savings, assets, or capacity building) or the creation of economic opportunities to result in significant benefits. A failure to do so would risk having the opposite effect, as demonstrated by the indebtedness incurred from revolving debt by low-income U.S. households that lack sufficient assets (Tescher and Stone 2022).

Including a gender perspective, a systematic review of the role of microfinance programs shows they were most effective when they coincided with educational programming, entrepreneurship training, and social activities (Peters et al. 2016). Women who failed to have access to such complementary programming were at risk of increased isolation or manipulation by their household and even the larger community.

Similar challenges emerge when we move beyond the individuals to the role of finance in funding for businesses and countries.

Minimizing risks in financing development

Why is the average interest rate in lending in Kenya 12.3 percent, in Canada 2.7 percent, and in Zimbabwe 131.8 percent (World Bank 2023)? First, inflation – which devalues funds over time – and the resultant weak currency vis-à-vis the dollar explain why Zimbabwe is flying off the chart. Considering inflation and foreign currency losses, the second driver is the risk of investment losses to investors. This risk is primed higher in the Global South than in the Global North. As a result, governments borrowing at a higher interest rate (e.g., 14 percent in Kenya for a 10-year treasury bond; World Government Bonds 2025) translates into higher borrowing costs for Kenyan small and medium-sized enterprises (SMEs) from their local banks.

One instrument developed to reduce the risk of financing development is *first loss*. Actors such as USAID have offered funding to cover any initial losses to entice private investing. Take a hypothetical investment of US$100,000. Financial actors can offer to cover the first US$20,000 of losses. If an investment subsequently generates losses of US$15,000, this actor will offer to incur all the losses, ensuring that private investors can recuperate all their funding upon exit. Such blended finance interventions have become more prominent over the years in agriculture, leveraging development funding to attract risk-averse private investors or mobilize engagement from private food companies (Johnson 2020).

To illustrate how it works, next we will examine two changemakers in action: Case Study 4.1 FASA and first loss and Case Study 4.2 Impact investing.

CASE STUDY 4.1 FASA AND FIRST LOSS

An example of such a first loss funding mechanism is FASA – the *Financing for Agricultural Small and Medium Enterprises in Africa* fund – launched by the Norwegian Agency for Development Cooperation (Norad) and USAID in 2023. It aims to close the funding gaps for small and medium-sized enterprises in African agriculture, as three out of four fail to access funding from formal banks and yet are too large for microfinance. The fund offers junior investment capital – with each party's initial contributions of US$35 million – to attract additional commercial financing. Since commercial capital expects close to a market risk-adjusted rate of return, the fund offers to reduce the senior tranche's investment risk. By leveraging these funds, the goal is to support 500 small and medium-sized agricultural enterprises, creating 60,000 additional jobs and reaching 1.5 million smallholder farmers over a decade (USAID and Norad 2023).[5]

To illustrate how such financing would work, donor countries provide funds to FASA, which acts as a fund of funds. FASA, managed by a third party, subsequently invests in African agricultural funds. These funders will then make either debt or equity investments in agricultural SMEs ranging between US$200,000 to US$5 million. FASA represents the junior tranche and offers to take the first loss, making the investment more attractive for private capital to take the so-called senior tranche. Assuming a mix of funding (i.e., *blended finance*) from the private sector, development finance institutions (DFIs), and FASA, the initial investments by donors are expected to mobilize up to ten times the additional capital and significantly contribute to closing the financing gap. The donors' investments are technically grants recycled in perpetuity.

A ten times multiplier is impressive. As discussed earlier, large multipliers could close our funding gaps. My USAID colleague, however, cautioned by posing the question whether a multiplier of four or a multiplier of 100 is more desirable. If your goal is to bring in capital on the sidelines to close the funding gap, the single US dollar of funding that mobilizes 100 dollars of funding is better. Such large multipliers beg the question whether the single dollar of funding was needed in the first place for creating additionality that could not have been achieved otherwise.

Figure 4.3 shows how development funding moves from the one extreme of grant funding and philanthropy to the other extreme of leveraging commercial capital from investors. It is important to remember that these funding types are not interchangeable: money from grant funding, for example, can do entirely different work in transforming sustainable development than commercial capital, which, in its simplest form, needs to find a profitable and viable investment. Pilots of innovations or interventions to address sustainable development, for example, are usually not viable yet for commercial funding.

Impact-first investors agree to take a lower rate of financial returns in favor of generating a social and environmental impact. Blended finance ventures such as FASA are positioning themselves more toward the center of the continuums, with the primary goal of creating impact and the secondary goal of recouping all investments. These public investors (e.g., Norad) provide technical assistance and will accept some failures in their investments (given the actual risks involved); if all of them succeed, they were likely funding businesses that already would have been viable in obtaining solely commercial capital.

These ventures also offer an important learning ground for testing what type of investment mix and structure works. Centered off to the right of the continuum are so-called finance-first investors, which seek investment opportunities that create a social or environmental impact but need to ultimately provide a financial return to their investment (even if at a lower rate than commercial investors). DFIs are even larger players: the U.S. International Development Finance Corporation (DFC) (2025b) offers financing as well as feasibility studies, technical assistance and political risk insurance (e.g., political violence). The U.S. DFC primarily invests together with commercial investors that seek a risk-adjusted rate of return in developing countries. Three recent investments made by the U.S. DFC include equity investments in an Irish company for extraction of rare earth minerals in South Africa (US$50 million), a loan to facilitate construction of food storage silos in Zambia (US$10 million), or an insurance for development programming on economic reform and women's empowerment by The Asia Foundation in India, Laos, and Vietnam (US$0.2 million) (U.S. International Development Finance Corporation 2025a).

Besides managing the risk, the challenge with such investments is that to contribute to an agricultural transformation, funding opportunities in SMEs that advance sustainable development are required. If we agree on

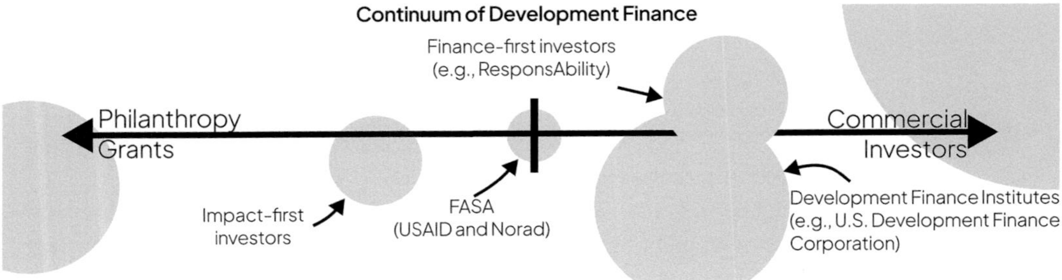

Figure 4.3 Continuum of development finance, with thanks to Songbae Lee

the basic premise that business-as-usual is no longer an option, especially in agriculture, then increasing the most profitable business lines – such as selling certified hybrid seeds that are high-input intensive – might run counter to our aim. Our concern, however, is more muted when considering the immense yield gaps in sub-Saharan Africa, as fields have been marked by the underuse of organic and synthetic inputs.

CASE STUDY 4.2 IMPACT INVESTING

A new group of actors – impact investors – has emerged. Impact investing, a term coined by the Rockefeller Foundation in 2007 (Harji and Jackson 2012), invests "into companies, organizations, and funds with the intention to generate social and environmental impact alongside a [often below market rate] financial return" (Mudaliar et al. 2017). Generating a financial return, as well as social and environmental impact is called the triple-bottom line (Miller, Ono, and Petruljeskov 2018; Barber, Morse, and Yasuda 2021). In the transformational investment space for sustainable food systems, around three out of four funds are not seeking market rate returns (Van den Eeckhout et al. 2023, 19).

Funders invest for impact with intentionality (see *impact-first* in Figure 4.3), mainly in a "mission-driven organization (for-profit, nonprofit, or hybrid) with a market-based strategy" (Rockefeller Philanthropy Advisors 2017). In contrast to conventional development grants, impact investors believe their market-based approaches increase the efficiency and effectiveness of funding.

Compared to conventional private investments, impact investors' focus on social and environmental impacts aims to strengthen the overall sustainability of interventions. To measure their impact, impact investors subscribe to the UN 2030 Agenda: 83 percent of impact investors use the SDGs to track some or all of their investments (GIIN 2020). Impact investors – and their estimated US$444 billion in invested assets as of 2018 (Global Sustainable Investment Alliance 2019) – could play a significant role in closing the SDG funding gap.

Together with colleagues, we attempted a systematic and quantitative analysis of existing impact investment deals in agriculture, and their contributions to meeting today's challenges to sustainable development (Ledermann et al. 2020). Based on our country-level analysis of over 500 impact investments, we identified that impact investing failed to reach the majority of the rural poor or smallholders. As illustrated in Figure 4.4, commercial smallholder farmers in tight value chains (i.e., farmers who own over 2 hectares of land and are producing crops for a contracted buyer) represent only 7 percent of all smallholders, yet receive 57 percent of the investments. Noncommercial smallholder farmers, who make up 60 percent of all farmers and own less than 1 hectare of land, represent only 1 percent of beneficiaries from impact investments. They lack a commercial focus in their farming and mostly produce for own consumption. They are hence not a likely outlet for a business case to development, showcasing the limitations of private capital in development.

Investments furthermore offer limited support for the ecological approaches necessary to achieve a transformation to sustainable food systems. Our research showed select additionality, as some investments were made in regions with limited financial access and investments. We concluded by arguing that impact investing currently does not provide as large a sustainable impact as would theoretically be possible.

To generate larger effects, the impact investors would need to coordinate across the four changemakers, including with government- and donor-led development efforts to build capacities for more marginalized farmers. Second, to increase the additionality of their investment, greater proximity of investors to their investment is warranted. Failure to understand local markets and conditions hinders additional capital flows. It can cause an enlarged gap between perceived versus actual risks – which perpetuates worse lending conditions, putting a strain on limited capital available. Greater proximity can also support tackling increased climate risks.[6] Third, given the existing standards of due diligence focused on financial and economic dimensions, internal capacity is warranted to better assess the environmental (and social) impacts of their investments. New regulations, such as the European Union Regulation on deforestation-free products (EUDR) that comes into effect in late 2026 (European Commission 2025), are accelerating a push in that direction (CSAF 2024).

YOUR TURN

What is the role of economy and finance in addressing your SDG? Review a business you are familiar with and investigate its commitments to the SDGs. Which SDGs are they contributing to? How do they relate to their core business? Where do you think their contribution seems more of a stretch?

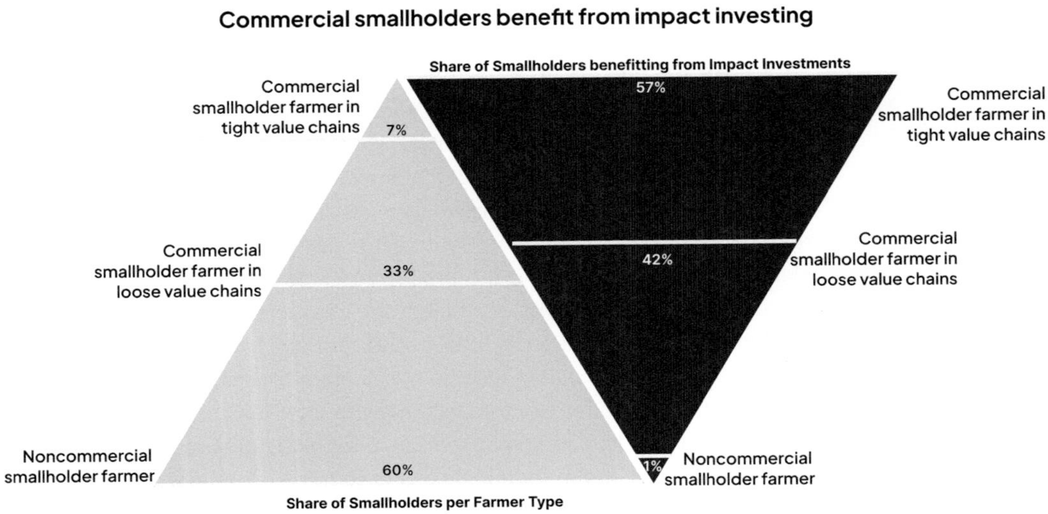

Figure 4.4 Commercial smallholders are the main beneficiaries of impact investments, adapted from Ledermann et al. (2020)

NOTES

1 More trade in sustainable technologies can speed up global adoption through technology transfer and sharing of ideas (Independent Group of Scientists Appointed by the Secretary-General 2019). High-tech examples are improved technologies on EV batteries. Low-tech examples are best practices to combat climate change.

2 The demonstration in Nairobi, Kenya, against a tax increase in June 2024 captured the younger generation rebelling against a tax increase on consumer goods without corresponding budget cuts or increased efficiencies (Apollo, Okal, and Makau 2024).

3 The UK is planning to cut development funding by US$7.5 billion to increase military spending (Gross and Pickard 2025). Even countries not known for their military prowess, such as Switzerland, have opted to cut development aid by US$282 million to boost its own military (Keystone-SDA 2024).

4 By 2025, BlackRock quit climate groups and stepped away from diversity targets, as it was facing lawsuits and backlash in a shifting US domestic environment (Masters and Temple-West 2025; Pitcher 2025).

5 The FASA fund complements the successful launch of Aceli in Africa, "a market incentive facility working to mobilize $1.5 billion in private sector lending to small and medium enterprises (SMEs)" (Aceli Africa 2024a). Aceli stands out from its peers by taking gender into account in its lending decisions (Aceli Africa 2024b).

6 Prominent funders, such as the Council on Smallholder Agricultural Finance (CSAF) and its 19 members, finance transformations to regenerative agriculture or agroforestry. They also invest in renewable energy as off-grid solar power offers to lower costs and improve energy supply for rural SMEs (CSAF 2024).

REFERENCES

Aceli Africa. 2024a. "Aceli Stakeholder Convening." Entebbe, Uganda: Aceli Africa.

Aceli Africa. 2024b. "Gender Inclusive Lending for Agriculture in Africa." Available at: https://aceliafrica.ams3.digitaloceanspaces.com/wp-content/uploads/2024/09/12142815/AceliAfrica_GenderInclusion_LearningBrief_2024.pdf.

Agnew, H., A. Klasa, and S. Mundy. 2022. "How ESG Investing Came to a Reckoning." *Financial Times*, June 6. https://www.ft.com/content/5ec1dfcf-eea3-42af-aea2-19d739ef8a55.

Apollo, R., J. Okal, and J. Makau. 2024. "Youth Uprising: How Gen-Z Protests Could Shift Kenya's Power Structures." Available at: https://www.african-cities.org/youth-uprising-how-gen-z-protests-could-shift-kenyas-power-structures/.

Armstrong, R. 2021. "The ESG Investing Industry Is Dangerous." *Financial Times*, August 24. https://www.ft.com/content/ec02fd5d-e8bd-45bd-b015-a5799ae820cf.

Barber, B., A. Morse, and A. Yasuda. 2021. "Impact Investing." *Journal of Financial Economics* 139 (1): 162–185. https://doi.org/10.1016/j.jfineco.2020.07.008.

Bashian, A., A. Mishra, D. Haynes, D. Bennell, K. Rudzinski, R. Raimond, S. Farley, S. Lang, T. Davids, and T. Crosby. 2024. "Financing for Regenerative Agriculture". Pollination Group. Available at: https://www.rockefellerfoundation.org/wp-content/uploads/2024/06/Financing-for-Regenerative-Agriculture-Final.pdf.

Bauckloh, T., J. Dobrick, A. Höck, S. Utz, and M. Wagner. 2024. "In Partnership for the Goals? The Level of Agreement between SDG Ratings." *Journal of Economic Behavior & Organization* 217: 664–678. https://doi.org/10.1016/j.jebo.2023.11.014.

Bhattacharya, A., and H. Kharas. 2023. "How to Make MDBs Fit for Purpose to Support SDG Implementation." In Center for Sustainable Development at Brookings (Ed.), *The SDG Second Half: Ideas for Doing Things Differently*. Washington, DC: Center for Sustainable Development at Brookings Institute.

Biovision, and IPES-Food. 2020. "Money Flows: What Is Holding Back Investment in Agroecological Research for Africa?" Biovision Foundation & International Panel of Experts on Sustainable Food System. Available at: https://www.ipes-food.org/_img/upload/files/Money%20Flows_Full%20report.pdf.

Bryan, K. 2023. "EU Regulators Flag Rising Greenwashing Practices by Banks." June 1. Available at: https://www.ft.com/content/5d236244-e073-412d-b981-0d2757f60b4b.

CSAF. 2024. "State of the Sector 2024." Available at: https://csaf.org/wp-content/uploads/2024/07/CSAF_State_of_Sector_2024.pdf.

Edgecliffe-Johnson, A., and S. Neville. 2022. "Bill Gates Warns Ukraine War Is Sapping Europe's Foreign Aid Budgets." *Financial Times*, September 13.

European Commission. 2025. "Regulation on Deforestation-free Products." Available at: https://environment.ec.europa.eu/topics/forests/deforestation/regulation-deforestation-free-products_en (accessed February 14, 2025).

European Parliament. 2019. Regulation (EU) 2019/2088. Available at: Brussels@ European Parliament.

Feng, E., and D. Pilling. 2019. "The Other Side of Chinese Investment in Africa." *Financial Times*, March 27. https://www.ft.com/content/9f5736d8-14e1-11e9-a581-4ff78404524e.

FTSE Russell. n.d. "Sustainability and ESG." Available at: https://www.lseg.com/en/ftse-russell/sustainability-and-esg-data (accessed February 14, 2024).

GIIN. 2020. "Understanding Impact Performance: Agriculture Investments." Available at: https://thegiin.org/assets/Understanding%20Impact%20Performance_Agriculture%20Investments_webfile.pdf.

Global Alliance for the Future of Food. 2024. "Cultivating Change: A Collaborative Philanthropic Initiative to Accelerate and Scale Agroecology and Regenerative Approaches." Available at: https://futureoffood.org/wp-content/uploads/2024/05/GA_CultivatingChange_Report_052124.pdf.

Global Sustainable Investment Alliance. 2019. "Global Sustainable Investment Review 2018." Available at: http://www.gsi-alliance.org/wp-content/uploads/2019/03/GSIR_Review2018.3.28.pdf.

Goldman, L,. M. Tsan, R. Dogandjieva, C. Colina, S. Daga, and V. Woolworth. 2018. "Inflection Point: Unlocking Growth in the Era of Farmer Finance." USAID, The MasterCard Foundation, Global Development Incubator, Dalberg. Available at: https://mastercardfdn.org/wp-content/uploads/2018/06/Inflection-Point_April-20160-accessible.pdf.

GRI. 2022. "State of Progress: Business Contributions to the SDGs." Available at: https://www.globalreporting.org/media/ab5lun0h/stg-gri-report-final.pdf.

Gross, A., and J. Pickard. 2025. "Development Minister Resigns in Protest at Keir Starmer's Aid Cuts." *Financial Times*, February 28. https://www.ft.com/content/d2e361c1-b291-47a2-aa57-ada9a2b41f50.

Haabazoka, L. 2019. "Developing Nations' Investment Drive Does Not Harm Sovereignty." *Financial Times*, January 30. https://www.ft.com/content/80fbb606-248a-11e9-b329-c7e6ceb5ffdf.

Harji, K., and E. T. Jackson. 2012. "Accelerating Impact: Achievements, Challenges and What s Next in Building the Impact Investing Industry." New York: The Rockefeller Foundation.

Hendriks, S., A. de Groot Ruiz, M. Herrero Acosta, H. Baumers, P. Galgani, D. Mason-D'Croz, C. Godde, K. Waha, D. Kanidou, J. von Braun, M. Benitez, J. Blanke, P. Caron, J. Fanzo, F. Greb, L. Haddad, A. Herforth, D. Jordaan, W. Masters, C. Sadoff, … M. Watkins. 2023. "The True Cost of Food: A Preliminary Assessment." In J. von Braun, K. Afsana, L. O. Fresco and M. H. A. Hassan (Eds.), *Science and Innovations for Food Systems Transformation*. Cham: Springer International Publishing, pp. 581–601.

Independent Group of Scientists Appointed by the Secretary-General. 2019. *Global Sustainable Development Report 2019: The Future is Now – Science for Achieving Sustainable Development*. New York: United Nations. Available at: https://sdgs.un.org/gsdr/gsdr2019.

Jessop, S., and R. Kerber. 2025. "Investor Climate Group Suspends Activities after BlackRock Exit." Reuters. https://www.reuters.com/sustainability/sustainable-finance-reporting/investor-climate-group-suspends-activities-after-blackrock-exit-2025-01-13/.

Johnson, M. 2019. "Italy Endorses China's Belt and Road Initiative." *Financial Times*, March 23.

Johnson, S. 2020. "Blended Finance Eases Burden for Farmers in Poor Countries." *Financial Times*, September 23. https://www.ft.com/content/805cb168-2fb8-4705-8d97-53d06a59a236.

Keystone-SDA. 2024. "Swiss Parliament Cuts Foreign Aid Budget." *Swissinfo.ch*, December 4. Available at: https://www.swissinfo.ch/eng/swiss-politics/swiss-parliament-cuts-foreign-aid-budget/88466732.

KPMG. 2022. "Reporting on the UN Sustainable Development Goals." Available at: https://kpmg.com/xx/en/home/insights/2022/09/survey-of-sustainability-reporting-2022/sdg.html.

Kynge, J. 2023. "'China's World Bank' Plans to Triple Climate Change Lending by 2030." *Financial Times*, September 24. https://www.ft.com/content/127e92aa-b247-4e6d-9dff-3e8dd642ae62.

Ledermann, S., A. Muller, S. Le Clec'h, S. Lee, and D. N. Bresch. 2020. "Agricultural Impact Investments for Sustainable Development: Current Potential and Shortcomings." Unpublished Working Paper. Washington DC and Zürich.

Masters, B., and P. Temple-West. 2025. "BlackRock Quits Climate Change Group in Latest Green Climbdown." *Financial Times*, January 9. https://www.ft.com/content/f0fb9841-db1d-442e-a757-1a1327497fb1.

Miller, C., T. Ono, and M. Petruljeskov. 2018. *Agricultural Investment Funds for Development*. Rome: FAO.

Mooney, A. 2023. "World Way Off Track to Meet Paris Climate Goals, Says UN Report." *Financial Times*, September 8.

Mudaliar, A., A. Pineiro, R. Bass, and H. Dithrich. 2017. "The State of Impact Measurement and Management Practice." Available at: https://thegiin.org/assets/2017_GIIN_IMM Survey_Web_Final.pdf.

Mundy, S. 2024. "Have CEOs Really Gone Quiet on Climate Change?" *Financial Times*, September 30. https://www.ft.com/content/df7310c8-3e17-4eaf-bcf2-207865f45c01.

Niculescu, M. 2017. "Impact Investment to Close the SDGs." Available at: https://www.undp.org/content/undp/en/home/blog/2017/7/13/What-kind-of-blender-do-we-need-to-finance-the-SDGs-.html.

OECD. 2025. OECD Data Explorer: Institutional Investors' Assets and Liabilities. Available at: https://data-explorer.oecd.org

PepsiCo. 2023. "ESG Performance Metrics". Available at: https://www.pepsico.com/docs/default-source/sustainability-and-esg-topics/2023-esg-summary/2023-esg-performance-metrics.pdf.

Peters, M. D. J., C. Lockwood, Z. Munn, S. Moola, and R. K. Mishra. 2016. "People's Views and Experiences of Participating in Microfinance Programmes: A Systematic Review of Qualitative Evidence." Available at: www.gov.uk/research-for-development-outputs/

Pilling, D. 2024. "Could T-Shirts Be the Way to Industrialise an African Nation?" *Financial Times*, August 29. https://www.ft.com/content/92b4f84d-229d-405b-85ec-7869529b7633.

Pitcher, J. 2025. "BlackRock's 'Woke' Era Is Over." *The Wall Street Journal*, March 2. https://www.wsj.com/business/blackrocks-woke-era-is-over-ebdbd6e7.

Rockefeller Philanthropy Advisors. 2017. "Impact Investing: An Introduction." Available at: http://www.rockpa.org/wp-content/uploads/2017/10/RPA_PRM_Impact_Investing_Intro_WEB.pdf.

Savage, S. 2024. "The New Corporate Green Goal: Being 'Nature Positive'." *Financial Times*, October 20. https://www.ft.com/content/4d12f8d1-c0df-4ab6-b374-741e9517448b.

Scheyvens, R., G. Banks, and E. Hughes. 2016. "The Private Sector and the SDGs: The Need to Move Beyond 'Business as Usual'." *Sustainable Development*. 24: 371–382. 10.1002/sd.1623

Sevastopulo, D., and G. Long. 2021. "US Development Bank Strikes Deal to Help Ecuador Pay China Loans." *Financial Times*, January 14. https://www.ft.com/content/affcc432-03c4-459d-a6b8-922ca8346c14.

Steiner, A., G. Aguilar, K. Bomba, J. P. Bonilla, A. Campbell, R. Echeverria, R. Gandhi, C. Hedegaard, D. Holdorf, and N. Ishii. 2020. *Actions to Transform Food Systems Under Climate Change*. Wageningen, The Netherlands: CGIAR Research Program on Climate Change.

Susskind, D. 2024. *Growth: A History and a Reckoning*. Cambridge, MA: The Belknap Press of Harvard University Press.

Tescher, J., and C. Stone. 2022. "Revolving Debt's Challenge to Financial Health and One Way to Help Consumers Pay It Off." Brookings Institute. Available at: https://www.brookings.edu/articles/revolving-debts-challenge-to-financial-health-and-one-way-to-help-consumers-pay-it-off/ (accessed April 3, 2024).

Thornton, P., Y. Chang, A. M. Loboguerrero, and B. Campbell. 2023. "Perspective: What Might It Cost to Reconfigure Food Systems?" *Global Food Security* 36. https://doi.org/10.1016/j.gfs.2022.100669.

Tobin-de la Puente, J., and A. W. Mitchell (Eds.) 2021. *The Little Book of Investing in Nature*. Oxford: Global Canopy.

U.S. International Development Finance Corporation. 2025a. "Active Projects." Available at: https://www.dfc.gov/what-we-do/active-projects.

U.S. International Development Finance Corporation. 2025b. "Our Products." Available at: https://www.dfc.gov/what-we-offer/our-products.

UN Global Compact. "Sustainable Finance." Available at: https://unglobalcompact.org/what-is-gc/our-work/financial (accessed February 14, 2025).

UNCTAD. 2020. "Tackling Illicit Financial Flows for Sustainable Development in Africa." Available at:. https://unctad.org/system/files/official-document/aldcafrica2020_en.pdf.

UNCTAD. 2024. "Trillion-Dollar Shift Urgently Needed to Align Global Finance with Climate and Development Goals." Available at: https://unctad.org/news/trillion-dollar-shift-urgently-needed-align-global-finance-climate-and-development-goals.

UNPRI. 2017. "Impact Investing Market Map: White Paper Document For Consultation." Available at: https://www.unpri.org/about/pri-teams/investment-practices/impact-investing-market-map.

USAID. 2023. "Finance: Unlocking Capital Flows." Available at: www.usaidgov.

USAID, and Norad. 2023. "Bridging the Gap: Financing Africa's Agricultural Growth." Available at: https://www.norad.no/contentassets/78e1bef38ef143349449581a16d73a37/fasa-fund.pdf.

Van den Eeckhout, H., I. Mandela, S. M. Akite, and R. Raimond. 2023. "Food Systems Investing in East Africa" (Transformational Investing in Food Systems (TIFS)). Available at: https://www.tifsinitiative.org/wp-content/uploads/TIFS_Investing-in-East-Africa-Food-Systems-Aug2023-Final.pdf.

van Zanten, J. A., and J. Huij. 2024. "ESG to SDG: Do Sustainable Investing Ratings Align with the Sustainability Preferences of Investors, Regulators, and Scientists?" SSRN Electronic Journal. Available at: www.researchgate.net/publication/362682111_ESG...

World Bank. 2023. "World Development Indicators." Available at: https://datatopics.worldbank.org/world-development-indicators/.

World Bank, 2024a. *Aid Circumvention: The Elusive Dream of Putting Countries in the Driver's Seat*. Washington, DC: The World Bank Group; International Development Association; Development Finance.

World Bank. 2024b. "Remittances Slowed in 2023, Expected to Grow Faster in 2024." Available at: https://www.worldbank.org/en/news/press-release/2024/06/26/remittances-slowed-in-2023-expected-to-grow-faster-in-2024.

World Government Bonds. 2025. "10 Years Government Bond Yield." Available at: https://www.worldgovernmentbonds.com (accessed January 15, 2025).

5 Individual and collective action

THE CASE FOR INDIVIDUAL AND COLLECTIVE ACTION IN ACHIEVING THE SDGs

A common theme of our book is that you cannot (successfully) impose sustained change from the top down. Change occurs at an individual or collective level. As a result, our third changemaker – individual and collective action – plays a crucial role in any attempts to achieve sustainable development. We will dive deeper into how change happens, including bottom-up development, in Chapter 13.

Building on our discussion on governance and policymaking, an active civil society and its empowered individuals play a large role in achieving an ambitious goal-based agenda, such as the SDGs. I remember well encountering activist and scholar Naomi Klein's first work – *No Logo* – over two decades ago. It captured the overarching trend of the increased power of corporations vis-à-vis the government in an era of consumer capitalism. Timely for a Western audience, she captured the role of (youth) movements in pushing back against corporate exploitation. By creating global alliances, Klein (2002) was convinced that "global citizens will eventually find sustainable solutions for this sold planet."

Dan Hancox (2019) revisited the book together with Naomi Klein 20 years after its publication:

> [T]he thrust of the book, and the global justice movement in general, was not about making more tasteful shopping decisions, but understanding how corporations shape our lives and culture and using that knowledge to try to, at least, stop them from running riot.

In my classroom discussions at that time, the collective actions by citizens were connected to the rise of the Fair Trade or anti-globalization movements, fears of greenwashing, or, in my particular case, the increase in certified organic agriculture. What was missing in my discussions were accounts of activism for sustained change from the Global South.

For sustainable development to succeed, citizens across the Global North and the Global South need to be active participants in setting their own development priorities. Citizen engagement is crucial to hold policymakers accountable. An effective example is a participatory budgeting process, as seen in Belo Horizonte. Accountability and resultant buy-in are further strengthened through participatory monitoring by citizens, including peer evaluations (also

DOI:10.4324/9781003400028-8

featured in Chapter 15). An involved and engaged citizenry will have stronger impacts on sustainable development (Independent Group of Scientists Appointed by the Secretary-General 2019). The involvement of civil society in the negotiations of the SDGs and the Sustainable Development Solutions Network (SDSN), both discussed later, go a long way toward ensuring stronger impacts.

The changemaker of *individual and collective action* is closely linked to increasing human capabilities: empowerment, self-determination, and active participation as the foundation for human well-being (Independent Group of Scientists Appointed by the Secretary-General 2019). They reinforce each other as changemaker and transformation hotspot. Addressing poverty (SDG 1) or inequality (SDG 10) will not only reduce marginalization in society, but also will allow people to raise their voices. Trade unions or other organizations that support collective action among marginalized groups are needed to advocate for policies that reduce inequality (Coolin, Sapienza, and Sumner 2024). Collective action is also a powerful force in addressing injustices on water and sanitation access (SDG 6). Meeting the basic right and necessities of water will reduce your labor burden and enable your self-determination. Collective action – and the resultant increased sense of agency and hope – have also been linked to improved mental health (Schwartz et al. 2022).

Progress through individual and collective action in one SDG can unlock co-benefits to achieve others: if you are sick (SDG 3), you cannot actively participate, take advantage of improved educational outcomes (SDG 4), or contribute to innovations (SDG 9). Nazanin Zaghari-Ratcliffe, imprisoned in Iran for six years, captured the relationship between the individual and the collective: "Freedom is not an island; it is in connections and conversations, in everyday human things" (Zaghari-Ratcliffe 2024).

Empowering women is a proven strategy to make sure that decisions on the distribution of public goods and the impacts of policymaking do not further exacerbate gender inequalities. Increased representation of women in policymaking has been linked to more career aspirations and improved educational outcomes for adolescent girls, as well as reductions in conflicts and a more durable peace (Independent Group of Scientists Appointed by the Secretary-General 2019).

Women and children are 14 times more likely than their male counterparts to die during a natural disaster. Raising their voices is long overdue. Women and youth need to be at the decision-making table as they offer insights and solutions to address climate change (Independent Group of Scientists Appointed by the Secretary-General 2019). Women are also at the forefront of recent lawsuits across the globe, arguing that a lack of adequate climate actions by their countries violates their rights (Bryan and Hancock 2024). One successful case – nine years in the making – was the European Court of Human Rights' ruling "that the Klimaseniorinnen [climate seniors] association of Swiss retirees had their right to 'respect for private and family life' violated by [the Swiss government's] failure to protect them from the impact of heatwaves caused by climate change" (Bryan and Hancock 2024). Climate litigation is also reaching record numbers across the Global South (with the largest share in Latin America), with nearly 70 percent of them successful (Tigre 2024). Our actions matter and make a difference. As stated by Professor Celeste Saulo, Secretary-General of the World Meteorological Organization (WMO):

> We have to do something. It's not about sitting at home and looking at this [climate change] as if it were a movie. This is not a movie. This is real life. You cannot turn off the TV and pretend nothing is happening. You are in the middle of the story.
>
> (cited in Mooney 2024)

The importance of continued learning

Life is a lifelong learning journey. I need to stop and remind myself to be open to learn. I am regularly inspired by the insights my students bring into the classroom and beyond. On a research trip to Central Malawi that I led, a group of graduate students joined us who were getting their Master's degree in International Education. Our main mandate for our research was to investigate the effects of climate change and gender on sustainable agriculture. These students, however, identified the critical function of primary and secondary education in fostering a mindset that can support stronger adoption of sustainable agricultural practices in the future. Efforts include closing knowledge gaps around climate change, with large deficits among young people in lower-income countries (UNICEF and Gallup n.d.). Beyond climate change, "the provision of supportive and stimulating environmental conditions in early childhood, formal and informal high-quality education, and lifelong learning" can improve individual competencies to make decisions that can accelerate a sustainable transformation (Independent Group of Scientists Appointed by the Secretary-General 2019).

Investing in early childhood education (SDG 4.2) is widely recognized as one of the best established avenues for transformation (Sachs et al. 2019). Education can also counter stigmas and discrimination early on, leading to improved well-being. Enriching young students with more holistic, sustainability-focused education that is rooted in their communities and addressing real-life questions they face are expected to pay multiple dividends later in life (Pramling Samuelsson and Kaga 2008).

More inclusive knowledge creation offers a shift away from the dominant paradigm of top-down social, technological, and institutional innovation. Especially in the Global South, sustainability transitions need to empower grassroots movements and strengthen the visibility of indigenous, informal knowledge to contribute to addressing our challenges. To further examine the role of knowledge in stimulating action, we will focus next on its role in rural development.

KNOWLEDGE'S ROLE IN STIMULATING INDIVIDUAL AND COLLECTIVE ACTION

Development and the subsequent adoption of agricultural technologies and practices are enabled by three main factors: labor, capital, and land (Hayami and Ruttan 1971). Dominant agricultural technologies, such as tractors, reduce constraints on labor (i.e., labor-saving), but demand a significant capital investment (i.e., capital-intensive). In a sub-Saharan African context, where labor constraints are comparatively less limited, capital-intensive investments traditionally focused on increasing the adoption of hybrid seeds (land-saving), synthetic fertilizer (land-saving), and pesticides (labor-saving). What they all have in common is their capital-intensive nature. Prominent scholarship on the diffusion of agricultural practices in Africa subsequently mirrors such capital-intensive technologies (e.g., on improved crop varieties; Abdulai 2023).

The paradigm of top-down diffusion of technologies to stimulate an increase in productivity and income (Moseley 2017) is reaching its ecological and economic limits (e.g., see Dawson, Martin, and Sikor, 2016 for Rwanda). Policymakers and scholars are turning toward more sustainable, bottom-up agricultural practices (e.g., in Malawi, see Bezner Kerr et al. 2019). Accelerated after the 2007–2008 food crisis, recent research looks to improve the assessment of sustainable agricultural practices and their adoption (Kassie et al. 2013; Kassie et al. 2015; Manda et al. 2016; Mponela et al. 2022). These diverse practices (such as intercropping, composting, and agroforestry) share a common trait: these practices are not defined by being capital-intensive. Their defining characteristic is the missing fourth factor: knowledge.

In contrast to capital-intensive practices (e.g., hybrid seeds), I argue that knowledge of agricultural practices (e.g., intercropping) could be more widely shared. As referenced in our book's title image, knowledge can travel and spread throughout a village, whereas a proprietary seed that is costly cannot. With investments in human capital, such as education, barriers for adoption are expected to be lower as knowledge-intensive practices can reach the poorest percentile along our income distribution. With agriculture's pro-poor growth properties (Figure 2.11 in Chapter 2), the positive effect on income growth is theoretically greatest in agriculture for the poorest farmers.

There are still social barriers in diffusion. Sharing can be limited to one's own social strata or network. Knowledge, however, can also be passed on from generation to generation. As most rural youth in sub-Saharan Africa continue to gain an income source from agriculture, the ability to share, learn, and adapt is of the utmost importance. As discussed in Chapter 6, some relevant knowledge is currently latent. It lies dormant in the form of traditional knowledge that is at risk of being lost. Attempts to revive such knowledge are laudable, as they can create an additional source to draw upon for our transformation. Given its complementary nature, knowledge is adaptable to one's own local farming contexts. Farmers are innovators, especially when they are able to take and manage the risks that come with engaging in new trials and the resultant errors. Resource-poor farmers impacted by shocks, however, struggle to adopt new practices (Amare et al. 2021; Dercon and Christiaensen 2011; Hellmuth et al. 2009; Shiferaw et al. 2015). Reducing the barriers to adoption through knowledge-intensive approaches to agricultural development could improve their ability to reduce their own vulnerabilities.

The rise of ICT promises more cost-efficient pathways for spreading agricultural technologies – and knowledge in particular. The drawback of knowledge is that it is frequently not patentable. As a common good, it demands public investments that are often in short supply. I co-founded a social enterprise, *farmbetter* (featured in Chapter 9), to contribute to filling this gap. Our aim was to empower smallholder farmers to improve their resilience. I elaborate more on this knowledge-intensive, bottom-up approach to development in Chapter 13. We will also cover the diversity of knowledge in our next changemaker – *Science and Technology*.

Significant development efforts have focused on strengthening civil societies both in the Global North and the Global South (e.g., SDC 2025), with the understanding that "[i]ncreasing the civic space for people to organize and participate in public dialogue and decision-making increases the likelihood of arriving at representative outcomes" (Independent Group of Scientists Appointed by the Secretary-General 2019, 36). Greater civic engagement and action can emerge from awareness campaigns supported by laws and regulations, financial incentives, taxes or fines that leads to behavioral changes (Thaler and Sunstein 2021).

THE ROLE OF CIVIL SOCIETY IN NEGOTIATING THE SDGs

> "We the peoples" are the celebrated opening words of the Charter of the United Nations. It is "we the peoples" who are embarking today on the road to 2030. Our journey will involve Governments as well as parliaments, the United Nations system and other international institutions, local authorities, indigenous peoples, civil society, business and the private sector, the scientific and academic community – and all people. Millions have already engaged with, and will own, this Agenda. It is an Agenda of the people, by the people and for the people – and this, we believe, will ensure its success.
>
> (Paragraph 52 of *2030 Sustainable Development Agenda*, United Nations 2015)

The negotiations in the run-up to the SDGs and the 2030 Agenda were lauded as the most inclusive yet of civil society. NGO representation increased from 250 during the first global summit on Human Environment in Stockholm in 1972 to nearly 10,000 accredited representatives at the Rio+20 conference held in 2012 that gave birth to the SDGs (Bäckstrand 2013). The foundation I worked for held a high-level side-event at Rio+20, offering an informal space for governmental and non-governmental actors to come together, build alliances and strengthen knowledge exchange. My colleagues regularly shared draft language with political representatives to revise wording through the stock-taking and negotiation sessions held.

Civil society could contribute to the Open Working Group (OWG) established in 2013 that led the 2030 Agenda process. According to the UN, "more than 1.4 million participants from governments, civil society, the private sector, academia, and research institutions" took part through global, national, and regional consultations (Kamau et al., 2018; cited in Sénit 2019, 699). Beyond civil society, individuals also were consulted: nine million citizens responded to a *MyWorld* survey (deployed online, via text message, and print) (Sénit 2019).

According to Sénit (2020), more marginal actors – while present at the negotiating table – failed to have their views effectively reflected. More radical proposals, such as addressing inequality (SDG 10) through an elimination of both extreme poverty and extreme wealth, were unsuccessful (Sénit 2019).

Through a discourse analysis of over 100 negotiation texts, Sénit shows that the dominant theme at the SDG negotiations centered on progressive sustainability (70 percent). Progressive sustainability reflects a state-centric view on equity and responsibilities in sustainable development that fails to "include broader understandings of equity towards the marginalized and physically absent such as the global poor, let alone future generations and non-humans" (Sénit 2020, 427). A more radical perspective on sustainability that offers to move us away from growth-based economics towards a focus on equity and human rights was only represented in 6 percent of the negotiations (Sénit 2020, 418).

Greater representation of leaders from the Global South, such as Paula Caballero from Colombia, ensured that the SDGs were not just an extension of the Millennium Development Goals (MDGs) (Fukuda-Parr and Muchhala 2020). In the end, the SDGs represented a stark shift away from the MDGs' focus on basic needs toward greater inclusiveness, environmental sustainability, and economic equality. While countries from the Global South were best represented, Sénit argues that the process fell short of bringing civil society to the center of the table:

> While the negotiations involved more than ten million civil society voices and resulted in an agreement that claims to "transform our world" and "leave no-one behind," the understandings of transformation and inclusiveness conveyed in the SDGs negotiations were eventually biased toward the interests of the most represented actors in the [Open Working Group (OWG)]: developing states. This comes with no surprise, as, in the end, developing countries recorded more seats and speaking slots in the OWG than the other actors, while in intergovernmental negotiations states are the final

> decision-makers of the exact framing of global agreements, with civil society playing a minor role.
>
> (Sénit 2020, 426)

As recounted by a civil society actor attempting to reframe a focus on growth toward one on environmental limits to growth, including planetary boundaries: "[I]nfluential delegations said: 'we're not going to participate in that discussion, we need to develop, if I can't talk to my citizens about improving and developing, I'm out of office'" (Sénit 2019, 702).

As a whole, civil society actors were successful in refining the goals and targets, even if less ambitious than hoped. Based on the 2030 Agenda negotiations, the UN has embraced a more participatory path post-2015. Civil society organizations play a crucial role in translating the SDGs to the larger population, furthering awareness and bottom-up change (Bizikova and Reilly-King 2017). Citizen science data can support or lead monitoring and evaluating sustainable development efforts (Espinosa and Rangel 2022), which is the focus of Chapter 6.

VIGNETTE 1 THE SUSTAINABLE DEVELOPMENT SOLUTIONS NETWORK (SDSN)

One of the most prominent civil society groups to advance the SDGs is the Sustainable Development Solutions Network (SDSN). Founded in 2012 under then UN Secretary-General Ban Ki-Moon, the network has over 1,800 members from over 100 countries with the goal to implement the UN 2030 Agenda and the Paris Agreement on Climate Change (SDSN 2024).

To showcase their work, let's examine the Swiss SDSN chapter that was founded by work colleagues. According to their own website, the Swiss chapter "links science, business, civil society, politics and administration to advance the implementation of the 17 SDGs in Switzerland and beyond" (SDSN Switzerland 2024). Using a snapshot from their electronic newsletter in July 2024, the network was active in highlighting to citizens and policymakers how Switzerland is losing ground:

- Switzerland is higher ranked in the global football rankings, than its sustainable development ranking (where it dropped from 15th to 22nd spot).
- As extreme weather events ravaged the southern parts of Switzerland, the network notes Switzerland is failing on SDG 13 (Climate Action). Considering greenhouse gases emissions from imports abroad and domestic emissions from fossil fuels and cement production, the country ranks 163rd out of 167 in the world.

Besides spurring additional momentum to close gaps, the network regularly hosts high-level events featuring SDSN-funded research on a regenerative economy (Unter et al. 2024) or food system (SDSN Switzerland 2022). The network aims to engage across the political spectrum, with its cross-party steering group including even members from the right political landscape. Given Switzerland's direct democracy, including the right to promote citizen-led initiatives or referendums, SDSN also led the establishment of a citizens' council (in German, *Bürgerinnenrat*) that identified 126 recommendations shared with Switzerland's executive and legislative branch for consideration. Recommendations included calls for greater food justice, such as labels on the working conditions in supply chains, lower prices for sustainable food, and providing food vouchers for lower income households (Schmid et al. 2024).

VIGNETTE 2 VOLUNTARY LOCAL REVIEWS

Local governments have joined efforts to measure their progress on SDGs, using *Voluntary Local Reviews* (McArthur 2023). With New York City pioneering it in 2018, by 2024, nearly 200 such reviews have been completed (United Nations n.d.). These reviews are meaningful as they complement voluntary national reviews.[1] In Chapter 16 we revisit why these local reviews are a reason for not giving up.

Bizikova and Reilly-King (2017) argue that civil society organizations "act as catalysts for critical global and national agendas, and assist in bringing citizens' voices to national debates and the development of national strategies." Together with Indigenous peoples, local communities, and academia, they hold a critical role in these voluntary local and national reviews. They already have effectively pressurised governments to implement the SDGs or contributed to monitoring efforts of progress (Bizikova and Reilly-King 2017). After the first Trump administration's withdrawal from the Paris Agreement in 2017, for example, over 300 local governments, 400 higher education groups, 2,500 businesses, 13 tribal nations, 800 faith groups and others declared "we are all in" (America is all in 2025).

Beyond the SDSN network introduced earlier, prominent platforms include *Southern Voices*, which brings together 70 think-tanks from the Global South to generate evidence and create a space to hold governments accountable to the SDGs. *CIVICUS* (2025), an alliance of more than 15,000 members across 175 countries, aims to "strengthen citizen action and civil society throughout the world." They advocate for civil society having a seat at the table of UN reforms. Their flagship *State of Civil Society Report: 2024* draws on over 200 interviews and covers over 100 countries, noting an increase in repression of civil society and shrinking of civic space as activists fight for climate change or gender rights (CIVICUS 2024).

Local or sub-national governments and sub-national actors are a driving force in providing push-back to national policy changes. With the arrival of the second Trump administration in January 2025 came a domestic policy push to repeal the IRA. In two years of the IRA, over 300,000 jobs were created. Over 200 energy projects are located in Republican districts, which is nearly twice as many as in Democratic districts. For this reason, even Republican office holders are speaking up, hoping to avoid partisan politics and ideology affecting their local economies (DeConcini, Rennicks, and Hyman 2024). Their actions stand in contrast to the dismantling of USAID, which lacked sub-national or local constituencies outside of the Washington, DC, area to lobby for USAID's continuation.

INTERSECTIONALITY AND OUR ACTIONS

As we engage in individual and collective action, it is important to recognize how we are all affected differently by sustainable development challenges and opportunities. Drawing on the concept of intersectionality, our life circumstances affect how we engage with each other. Our identities – including gender, race, class, age, or migration status (UNPRPD and UN Women 2022) – are heavily informed by our geographic location and can explain power differences and resultant marginalization. Where ageism and sexism dominate, for example, an older woman in her sixties struggling to find a job is less likely to succeed than a younger, male colleague with equal or lower qualifications. Gender inequalities are particularly pronounced where poverty and marginalization intersect (Kabeer 2015). Using an intersectionality lens, we recognize that these inequalities are not static, but co-exist in a dynamic relationship (UNPRPD and UN Women 2022). The dynamics include shifting power relationships that shape social inequalities (Larson et al. 2016).

Intersectionality has made forays into international human rights law (UNPRPD and UN Women 2022). The Convention on the Rights of Persons with Disabilities (CRPD, signed in 2006) recognizes how multiple forms of discrimination intersect and asks signatory states to address them.[2] A better understanding of the multiple forms of discrimination through intersectionality and existing human rights instruments can improve our progress toward achieving the "*Leave no one behind* mantra" (UNPRPD and UN Women 2022). In the Global North, for example, forced drugging in US nursing homes of older women and men who have disabilities violates their rights to "physical integrity, health care, freedom of dignity, and free and informed consent" (UNPRPD and UN Women 2022, 10). In the Global South, the case of sexual violence of girls in the Philippines based on their gender, disability, age, and socioeconomic status violates their rights to "equality before the law, including access to justice, physical integrity, safety of the person, access to education, and freedom of expression" (UNPRPD and UN Women 2022, 10). You can also apply an intersectionality lens to your own individual or collective practice (see Table A.I in Appendix A).

Understand who you are and the role of privilege

As a self-reflexive development practitioner, it is helpful to understand who we are, our background, and how we are perceived. By understanding the power dynamics and the role of privilege, we can achieve better policy outcomes. As argued by Thomas (2022) on environmental injustices: "No, the environment and climate don't discriminate, but people do, and therefore so do environmental policies."

Reviewing the impact of COVID-19 on inequalities, Perry, Aronson, and Pescosolido (2021, 3) posited that "[d]isasters are exogenous shocks to social systems that reveal enduring failings and inequities." *New York Times* columnist David Brooks (2005) captured how "[f]loods wash away the surface of society, the settled way things have been done. They expose the underlying power structures, the injustices, the patterns of corruption and the unacknowledged inequalities." If you can see the underlying role of politics and power, as it relates to the environment and our interactions with it, you will be better positioned to identify enduring levers for change.

As you engage in discovery and self-discovery, it will feed into your individual and collective practice. When we educate others, we also need to keep our positionality in mind, to avoid "shaming those who are faced with systemic barriers to living low-waste and sustainably" (Thomas 2022). Ambika Rajyagor, advocate for the Special Needs and Disabled community and cofounder of Disabled & Outdoors highlights how ableism intersects with environmental causes: "[e]coableism creates a divide that 'others' disabled environmentalists ... Allies to the disabled environmentalist community should help advocate for accessible sustainable practices and materials that are considerate of disabled needs" (cited in Thomas 2022).

On accounting for and embracing diversity

When developing interventions, we cannot lose sight of the diversity across people. Accounting for diversity and heterogeneity is expected to improve the results of interventions (UNPRPD and UN Women 2022). Using the example of people with disabilities, women with disabilities are twice as likely to be the victims of gender-based violence, and are facing greater challenges in getting access to justice (UNPRPD 2020). A three-year project by the UN CRPD, including nine UN agencies, strengthened the integration of rights of persons with disabilities into governance and policymaking by advancing the ratification of the

CRPD by Timor-Leste (Saleem 2022). It funded capacity building for service providers to prevent and respond to gender-based violence and aimed to improve access for women with disabilities to legal services (UNPRPD 2020). The project also worked toward increasing the availability of data, as persons with disabilities are notoriously under-counted. By accounting for diversity, the project can contribute to reducing inequality (SDGs 5 and 10), improving women's well-being (SDG 3), and unlocking their potential to more effectively improve their livelihoods (SDGs 1 and 4).

Diversity is to be cherished and built upon, not to be diminished or ignored. In Brazil, Indigenous peoples' struggle for cultural diversity continues. According to Krenak (2020, 36), resistance included "refusing to accept the idea that we're all the same. There are still upwards of 250 ethnicities in Brazil, and they speak over 150 languages and dialects."

Intersectional environmentalism and collective action

The SDGs are lopsidedly tilted toward the environment (with 11 SDGs versus 1 MDG focused on it). Individual and collective action has historically played an oversized role in achieving environmental change. In the USA, the Earth Day movement started in 1970 and led to the creation of the Environmental Protection Agency (EPA) and the passage of the Clean Air Act (CAA) later that year. Here, I want to introduce an application of intersectionality to environmental activism and movements by drawing on Leah Thomas' book, *Intersectional Environmentalism* (Thomas 2022). She argues that:

> [while] primarily white communities in the U.S. saw a 70 percent reduction in air pollution after the CCA was passed, low-income and Black, Indigenous, Asian, Pacific Islander, and Latinx communities were seemingly left out of the parameters of protection outlined in these laws.
>
> (Thomas 2022)

According to Diandra Market, intersectional environmentalism represents the lens we apply to achieve the overall goal of environmental justice (Thomas 2022). On these quests for environmental justice, Thomas places people of color who frequently have been less visible at the center of environmental advocacy and education. She argues that in the USA, concerns of people of color were still marginal, as they "had to perform additional emotional labor to advocate for correcting environmental injustice impacting their communities" (Thomas 2022).

The case of Warren County in North Carolina, where toxic waste was dumped in a predominantly African American neighborhood from the late 1970s to the early 1980s, illustrates well the importance of individual and collective action. Having exhausted the court systems, civil disobedience campaigns and Dr. Bullard's groundbreaking book, *Dumping in Dixie: Race, Class, and Environmental Quality* (2018, originally published in 1990) brought national attention to the cause. While the protests failed to stop the dumping, it served as the birthplace of the environmental justice movements (NRDC 2023). Two decades later, the site was ultimately decontaminated.

Dr. Bullard's work laid bare how Black communities in the southern USA experienced discrimination around toxic waste. Race continues to be strongly correlated with proximity to toxic waste facilities (Thomas 2022). Predominantly Black and migrant communities in the US Gulf Coast region bordering Texas and Louisiana have lived near petrochemical factories for decades (Smyth and McCormick 2023). With the boom in exports of liquified natural gas, new infrastructure investments are extending pollution in what activists term *sacrifice zones* – with pollution and negative health effects increasing (Smyth and McCormick 2023).

By now, intersectional environmentalism movements span the globe. For example, the Indigenous Landback movement, active in North America and Australia, aims to reclaim "everything stolen from the original peoples: land, language, ceremony, food, education, housing, healthcare, governance, medicines, kinship" (Landback 2024). Their work pushes us to move beyond land acknowledgments toward restorative and transformative justice (Fisk et al. 2021).

THE ROLE OF JUSTICE: THE EXAMPLE OF FOOD

Access to food and sovereignty over its production and consumption are at the heart of activists' efforts. We will dive deeper into food sovereignty through Case Study 5.1 on La Via Campesina – a popular transnational movement. Before diving into their work, I want to review two concepts that are at the heart of activism: food justice and food apartheid.

Drawing on definitions offered by Dara Cooper, a national organizer of the National Black Food and Justice alliance, Thomas (2022) articulates food justice as "a framework [that goes] beyond access [to food] to ensure that our communities have not only the right but the ability to have community control of our food, including the means of production and distribution."

Her framing captures the struggles experienced by African American farmers in the USA. Over the last decade, their share of farming dropped from 14 percent (around a million of Black-run farms) to 2 percent (or less than 35,000), with landholdings declining by 90 percent compared to just 2 percent by white farmers (Tabuchi and Popovich 2023).

Tabuchi and Popovich (2023) attribute this decline to "decades of racial violence and unfair lending and land ownership policies." Beyond farmers, food justice efforts include integrating refugees and newcomers into the USA to improve well-being and nutrition (International Rescue Committee 2024). *Food apartheid* moves beyond *food justice* by calling out the exploitative political economy and uneven power relationships within food systems, including a lack of self-determination and access to healthy food (Thomas 2022).

CASE STUDY 5.1 LA VIA CAMPESINA

In agrarian development[3], civil society is playing an increasingly more visible role in influencing policymaking. *La Via Campesina* is the world's largest and most prominent transnational peasants movement (Altieri and Toledo 2011), combining 180 organizations and home to over 200 million peasants spread across 81 countries (La Via Campesina 2024). It advocates for pro-smallholder policies across the Global North and the Global South, including the USA, Canada, Italy, India, or Kenya. One of its hallmark efforts is broadening our understanding around food security. My own research focuses on the availability, accessibility, and utilization of food (Pray and Ledermann 2016).

La Via Campesina heralded the term *food sovereignty* to emphasize the socio-cultural importance of food production and consumption as well its environmental dimension. As captured in the Declaration of Nyéléni in 2007 (nyeleni.org), where over 500 representatives from over 80 countries came together, food sovereignty is "the right of peoples to healthy and culturally appropriate food produced through ecologically sound and sustainable methods, and their right to define their own food and agricultural system."

Through its campaigning, La Via Campesina positions itself as a central pillar of a transformation to sustainable, local food systems. Leveraging individual and collective action to change local production and influence governance and policymaking, their movement is closely aligned with the rise of agroecology. The Food and Agriculture Organization of the United Nations (FAO) defines agroecology as:

> a holistic and integrated approach that ... seeks to optimize the interactions between plants, animals, humans and the environment while also addressing the need for socially equitable food systems within which people can exercise choice over what they eat and how and where it is produced.
>
> (FAO n.d.)

Agroecology's three pillars are the science and research of ecological production, its practice by farmers across the globe, and a social movement. While research has historically been under-funded (cf. Hatik and Ledermann 2020), and the practice evolving (e.g., FAO 2018), La Via Campesina is one of the main actors spearheading its movement.

With the global food crisis from 2007 to 2008, calls for self-sufficiency increased as countries were struggling to quell unrest where access to food was limited. Both La Via Campesina and agroecology gained momentum in its aftermath (Altieri and Toledo 2011), culminating in the *First International*

Symposium on Agroecology for Food and Nutrition Security (FAO 2014), which I attended at the FAO headquarters in Rome. Then Director General, José Graziano da Silva, closed the symposium by stating that the stakeholders "opened a window introducing agroecology in the cathedral of the Green Revolution [the FAO]" (FAO 2018).

The UN also declared 2014 the Year of Family Farms, highlighting the importance of their contributions to global food production (e.g., my research with FAO colleagues; Graeub et al. 2015). In 2018, another milestone was reached through La Via Campesina's active engagement within the civil society framework at the UN. The UN General Assembly adopted the *United Nations Declaration on the Rights of Peasants and Other People Working in Rural Areas* (La Via Campesina 2018). An illustrated version of the declaration, authored by La Via Campesina, captures the essential nature of peasants and rural workers not just as recipients of human rights, but as active participants in struggles to improve their livelihoods (La Via Campesina 2020). The next step is to put the declaration into action, as the declaration is:

> [a] tool to mobilise communities and organise political training. It is essential to use the Declaration in legal procedures in defence of peasants and to call on society to develop regional and national strategies aimed at implementing this UN Declaration. This tool allows us to pressure our governments and governance institutions at all levels, to fulfil their obligation of ensuring the dignity and justice for those who produce to guarantee people's food sovereignty.
>
> (La Via Campesina 2020, 6)

La Via Campesina has received its share of criticism over the decades. Its focus on local knowledge and practices runs counter to the dominant agricultural development paradigm focusing on modernization through intensification of agricultural production. Criticism includes the social movement's rejection of the Green Revolution, as it maximized production and ignored social and environmental costs. The movement also argues that the oversized role of foundations in the policymaking of developing countries, such as the Gates Foundation, undermines the democratic process of a more citizen-led, informed policy development through civil society (Genetic Literacy Project 2019). Critics argue that their stance against agricultural cooperations, including frequent activism against genetically modified crops, risks slowing down progress in improving food security (Aerni 2011). In 2021, La Via Campesina decided to boycott the high-level UN Food Systems Summit, which included the active participation of the private sector and foundations. Their decision highlights the continued tensions in taking a principled stand for its members (La Via Campesina 2021) while remaining a strong voice for marginalized peasants at the premier stages of policymaking.

YOUR TURN

What is the role of individual and collective action in addressing your SDG? How does your analysis of your SDG change when you use an intersectionality lens?

NOTES

1 The USA did not submit its first national review until July 2023.

2 The UN Committee on the Rights of Persons with Disabilities spelled out the following example (cited in UNPRPD and UN Women 2022):

> while denial of access to general health-related information due to inaccessible format affects all persons on the basis of disability, the denial to a blind woman of access to family planning services restricts her rights based on the intersection of her gender and disability ....

3 With thanks to Olyvia Tremper, who researched food sovereignty with me.

REFERENCES

Abdulai, A. 2023. "Information Acquisition and the Adoption of Improved Crop Varieties." *American Journal of Agricultural Economics* 105 (4): 1049–1062. https://doi.org/10.1111/ajae.12419.

Aerni, P. 2011. "Food Sovereignty and its Discontents." *ATDF Journal* 8 (1/2): 23–39.

Altieri, M. A., and V. M. Toledo. 2011. "The Agroecological Revolution in Latin America: Rescuing Nature, Ensuring Food Sovereignty and Empowering Peasants." *Journal of Peasant Studies* 38 (3): 587–612. https://doi.org/10.1080/03066150.2011.582947.

Amare, M., B. Shiferaw, H. Takeshima, and G. Mavrotas. 2021. "Variability in Agricultural Productivity and Rural Household Consumption Inequality: Evidence from Nigeria and Uganda." *Agricultural Economics* 52: 19–36. https://doi.org/10.1111/agec.12604.

America is all in. "America is all in to Achieve the Goals of the Paris Agreement." Available at: https://www.americaisallin.com. (accessed February 14, 2025).

Bäckstrand, K. 2013. "Civil Society Participation in Sustainable Development Diplomacy. Towards Global Stakeholder Democracy?" Paper presented at the 8th Pan-European Conference on International Relations, Warsaw, 18–21 September 2013.

Bezner Kerr, R., J. Kangmennaang, L. Dakishoni, H. Nyantakyi-Frimpong, E. Lupafya, L. Shumba, R. Msachi, G. O. Boateng, S. S. Snapp, A. Chitaya, E. Maona, T. Gondwe, P. Nkhonjera, and I. Luginaah. 2019. "Participatory Agroecological Research on Climate Change Adaptation Improves Smallholder Farmer Household Food Security and Dietary Diversity in Malawi." *Agriculture, Ecosystems & Environment* 279: 109–121.

Bizikova, L., and F. Reilly-King. 2017. "*Do We Need to Engage Civil Society Organizations in Implementing SDGs?*" New York: IISD.

Brooks, D. 2005. "The Storm After the Storm." *The New York Times*, September 1. https://www.nytimes.com/2005/09/01/opinion/the-storm-after-the-storm.html.

Bryan, K., and A. Hancock. 2024. "The Pensioners and Babies Behind a New Era of Climate Lawsuits." *Financial Times*, April 12. https://www.ft.com/content/5be99340-E847-488c-b901-b6e260faff8e.

Bullard, R. D. 2018. *Dumping in Dixie: Race, Class, and Environmental Quality*, 3rd edn. New York: Routledge.

CIVICUS. 2024. "State of Civil Society Report 2024." Available at: https://www.civicus.org/index.php/state-of-civil-society-report-2024.

CIVICUS. 2025. "Who We Are." CIVICUS. Available at: https://www.civicus.org/index.php/who-we-are (accessed February 14, 2025).

Coolin, G., E. Sapienza, and A. Sumner. 2024. *The Politics of Inequality*.Nairobi: UNDP.

Dawson, N., A. Martin, and T. Sikor. 2016. "Green Revolution in Sub-Saharan Africa: Implications of Imposed Innovation for the Wellbeing of Rural Smallholders." *World Development* 78: 204–218. https://doi.org/10.1016/j.worlddev.2015.10.008.

DeConcini, C., J. Rennicks, and G. Hyman. 2024. "Trump May Thwart Federal Climate Action, but Opportunities for Progress Remain." Available at: https://www.wri.org/insights/trump-climate-action-setbacks-opportunities-us (accessed February 14, 2025).

Dercon, S., and L. Christiaensen. 2011. "Consumption Risk, Technology Adoption and Poverty Traps: Evidence from Ethiopia." *Journal of Development Economics* 96: 159–173. https://doi.org/10.1016/j.jdeveco.2010.08.003.

Espinosa, C., and G. Rangel. 2022. "What Roles Do Civil Society Organizations Play in Monitoring and Reviewing the Sustainable Development Goals? An Exploration of Cases from Ecuador, Colombia, and Argentina." *Tapuya: Latin American Science, Technology and Society* 5 (1). https://doi.org/10.1080/25729861.2022.2143669.

FAO. 2014. "International Symposium on Agroecology for Food Security and Nutrition." Available at: https://www.fao.org/about/meetings/afns/en/.

FAO. 2018. "FAO's Work on Agroecology: A Pathway to Achieving the SDGs." Available at: https://openknowledge.fao.org/server/api/core/bitstreams/9ebe6217-1bdb-4b25-b9ae-4c93a73aab21/content.

FAO. n.d. "Agroecology Knowledge Hub." Available at: https://www.fao.org/agroecology/overview/en/.

Fisk, J. J., L. A. Jacobs, B. U. K. Russo, E. Meier, A. Nakachi, K. K. P. Spencer, K. Kaulukukui-Narikawa, A. W. Datta, and K. Quiocho. 2021. "Cultivating Sovereignty in Parks and Protected Areas: Sowing the Seeds of Restorative and Transformative Justice Through the #LANDBACK Movement." *Parks Stewardship Forum* 37 (3). https://doi.org/10.5070/p537354734.

Fukuda-Parr, S., and B. Muchhala. 2020. "The Southern Origins of Sustainable Development Goals: Ideas, Actors, Aspirations." *World Development* 126. https://doi.org/10.1016/j.worlddev.2019.104706.

Genetic Literacy Project. 2019. "La Via Campesina: Peasants Cite 'Food Sovereignty' To Denounce Science-Based Farming Improvements." Available at: https://geneticliteracyproject.org/glp-facts/la-via-campesina-peasants-cite-food-sovereignty-denounce-science-based-farming-improvements/ (accessed March 23, 2025).

Graeub, B. E., M. J. Chappell, H. Wittman, S. Ledermann, R. Bezner Kerr, and B. Gemmill-Herren. 2015. "The State of Family Farms in the World." *World Development* 87: 1–15,

Hancox, D. 2019. "No Logo at 20: Have We Lost the Battle Against the Total Branding of Our Lives?" *The Guardian*, August 11. https://www.theguardian.com/books/2019/aug/11/no-logo-naomi-klein-20-years-on-interview.

Hatik, S., and S. T. Ledermann. 2020. "Bill and Melinda Gates Foundation's Funding for Agricultural Research." Available at: https://www.ipes-food.org/_img/upload/files/Money%20Flows_Full%20report.pdf.

Hayami, Y., and V. W. Ruttan. 1971. *Agricultural Development: An International Perspective*. Baltimore, MD: The Johns Hopkins University Press.

Hellmuth, M., D. Osgood, U. Hess, A. Moorhead, and H. E. Bhojwan. 2009. "Index Insurance and Climate Risk: Prospects for Development and Disaster Management." New York: International Research Institute for Climate and Society. https://www.preventionweb.net/files/10755_ClimateandSocietyIssueNumber21.pdf.

Independent Group of Scientists Appointed by the Secretary-General. 2019. *Global Sustainable Development Report 2019: The Future is Now – Science for Achieving Sustainable Development*. New York: United Nations. Available at: https://sdgs.un.org/gsdr/gsdr2019.

International Rescue Committee. 2024. *Cultivating Youth Food Justice Programs*.New York: International Rescue Committee.

Kabeer, N. 2015. "Gender, Poverty, and Inequality: A Brief History of Feminist Contributions in the Field of International Development." *Gender and Development* 23: 189–205. https://doi.org/10.1080/13552074.2015.1062300.

Kassie, M., M. Jaleta, B. Shiferaw, F Mmbando, and M. Mekuria. 2013. "Adoption of Interrelated Sustainable Agricultural Practices in Smallholder Systems: Evidence from rRural Tanzania." *Technological Forecasting and Social Change* 80 (3): 525–540. https://doi.org/10.1016/j.techfore.2012.08.007.

Kassie, M., H. Teklewold, M. Jaleta, P. Marenya, and O. Erenstein. 2015. "Understanding the Adoption of a Portfolio of Sustainable Intensification Practices in Eastern and Southern Africa." *Land Use Policy* 42: 400–411. https://doi.org/10.1016/j.landusepol.2014.08.016.

Klein, N. 2002. *No Logo*. New York: Picador.

Krenak, A. 2020. *Ideas to Postpone the End of the World*. Trans. A. Doyle. Toronto: Anansi International.

La Via Campesina. 2018. "Finally, UN General Assembly Adopts Peasant Rights Declaration! Now Focus Is on Its Implementation." Available at: https://viacampesina.org/en/finally-un-general-assembly-adopts-peasant-rights-declaration-now-focus-is-on-its-implementation/.

La Via Campesina. 2020. "United Nations Declaration on the Rights of Peasants and Other People Working in Rural Areas." Available at:. https://viacampesina.org/wp-content/uploads/2020/04/UNDROP-Book-of-Illustrations-l-EN-l-Web.pdf.

La Via Campesina. 2021. "The UN Food Systems Summit Is Hogwash. It Is a Threat to Peoples' Food Sovereignty." Available at: https://viacampesina.org/en/the-un-food-systems-summit-is-hogwash-it-is-a-threat-to-peoples-food-sovereignty/.

La Via Campesina. 2024. "Home." Available at: https://viacampesina.org/en/.

Landback. 2024. "Manifesto." Available at: https://landback.org/manifesto/ (accessed October 30 2024).

Larson, E., A. George, R. Morgan, and T. Poteat. 2016. "10 Best Resources On... Intersectionality with an Emphasis on Low- and Middle-Income Countries." *Health Policy Plan* 31 (8): 964–969. https://doi.org/10.1093/heapol/czw020.

Manda, J., A. D. Alene, C. Gardebroek, M. Kassie, and G. Tembo. 2016. "Adoption and Impacts of Sustainable Agricultural Practices on Maize Yields and Incomes: Evidence from Rural Zambia." *Journal of Agricultural Economics* 67 (1): 130–153. https://doi.org/10.1111/1477-9552.12127.

McArthur, J. W. 2023. "The 'Second Half' of the SDG Era: Ideas for Doing Things Differently." In Center for Sustainable Development at Brookings (Ed.), *The SDG Second Half: Ideas For Doing Things Differently*. Washington, DC: Center for Sustainable Development at Brookings.

Mooney, A. 2024. "Meteorologist Celeste Saulo: 'Climate Change Is Not a Movie. This Is Real Life'." *Financial Times*, August 23. https://www.ft.com/content/4e4f9747-9bc8-4242-a43d-0463c0620607.

Moseley, W. G. 2017. "One Step Forward, Two Steps Back in Farmer Knowledge Exchange: 'Scaling-Up' as Fordist Replication in Drag." In J. Sumberg (Ed.), *Agronomy for Development: The Politics of Knowledge in Agricultural Research*. London: Routledge, pp. 79–90.

Mponela, P., J. Manda, M. Kinyua, and J. Kihara. 2022. "Participatory Action Research, Social Networks, and Gender Influence Soil Fertility Management in Tanzania." *Systemic Practice and Action Research* 36 (1): 141–163. https://doi.org/10.1007/s11213-022-09601-3.

NRDC. 2023. "The Environmental Justice Movement." Available at: https://www.nrdc.org/stories/environmental-justice-movement.

Perry, B. L., B. Aronson, and B. A. Pescosolido. 2021. "Pandemic Precarity: COVID-19 Is Exposing and Exacerbating Inequalities in the American Heartland." *Proceedings of the National Academy of Sciences* 118. https://doi.org/10.1073/pnas.2020685118.

Pramling Samuelsson, I., and Y. Kaga. 2008. *The Contribution of Early Childhood Education to a Sustainable Society*. Paris: UNESCO.

Pray, C., and S. Ledermann. 2016. "Genetically Engineered Crops and Certified Organic Agriculture for Improving Nutrition Security in Africa and South Asia." In *Hidden Hunger: Malnutrition and the First 1,000 Days of Life: Causes, Consequences and Solutions*. Basel: S. Karger AG. https://karger.com/books/book/157/Hidden-Hunger Malnutrition-and-the-First-1-000-Days.

Sachs, J. D., G. Schmidt-Traub, M. Mazzucato, D. Messner, N. Nakicenovic, and J. Rockström. 2019. "Six Transformations to Achieve the Sustainable Development Goals." *Nature Sustainability*. Sustainable Development Solutions Network. Available at: http://www.jstor.org/stable/resrep25838

Saleem, A. 2022. "Timor-Leste Ratifies CRPD." Available at: https://timorleste.un.org/en/203983-timor-leste-ratifies-crpd.

Schmid, P., L. Léa, C. Michael, and S. Bieri. 2024. "Creating Pathways to Just and Sustainable Food Systems with Citizen Assemblies." *Innovation: The European Journal of Social Science Research* 37 (3): 832–850. https://doi.org/10.1080/13511610.2024.2309173.

Schwartz, S. E. O., L. Benoit, S. Clayton, M. F. Parnes, L. Swenson, and S. R. Lowe. 2022. "Climate Change Anxiety and Mental Health: Environmental Activism as Buffer." *Current Psychology* 1–14. https://doi.org/10.1007/s12144-022-02735-6.

SDC. 2025. "Good Governance." Available at: https://www.eda.admin.ch/deza/en/home/themes-sdc/state-economic-reforms.html.

SDSN Switzerland. 2022. "Summary and Explanation to Pathways to the Future of Food." Available at: https://sdsn.ch/wp-content/uploads/2023/05/SDSN_Summary_guidelines_food_future_screen.pdf.

SDSN Switzerland. 2024. "SDSN Switzerland – From Knowledge to Action for the 2030 Agenda." Available at: https://sdsn.ch/en/ (accessed February 14, 2025).

SDSN, UN. n.d. "Sustainable Development Solutions Network." Available at: https://www.unsdsn.org.

Sénit, C.-A. 2019. "Leaving No One Behind? The Influence of Civil Society Participation on the Sustainable Development Goals." *Environment and Planning C: Politics and Space* 38 (4): 693–712. https://doi.org/10.1177/2399654419884330.

Sénit, C.-A. 2020. "Transforming Our World? Discursive Representation in the Negotiations on the Sustainable Development Goals." *International Environmental Agreements: Politics, Law and Economics* 20 (3): 411–429. https://doi.org/10.1007/s10784-020-09489-1.

Shiferaw, B., T. Kebede, M. Kassie, and M. Fisher. 2015. "Market Imperfections, Access to Information and Technology Adoption in Uganda: Challenges of Overcoming Multiple Constraints." *Agricultural Economics* 46: 475–488.

Smyth, J., and M. McCormick. 2023. "'Sacrifice Zone': Booming US LNG Sector Leaves Its Mark on the Gulf." *Financial Times*, November 27. https://www.ft.com/content/5c051556-6efd-4db3-9152-508d7e0de566.

Tabuchi, H., and N. Popovich. 2023. "Two Biden Priorities, Climate and Inequality, Meet on Black-Owned Farms." *The New York Times*, January 31. https://www.nytimes.com/2021/01/31/climate/black-farmers-discrimination-agriculture.html.

Thaler, R. H., and C. R. Sunstein. 2021. *Nudge: The Final Edition*. New Haven, CT: Yale University Press.

Thomas, L. 2022. *The Intersectional Environmentalist*. New York: Voracious.

Tigre, M. A. 2024. "*Climate Ligitation in the Global South: Mapping Report*." New York: Columbia Law School Sabin Center for Climate Change Law. Available at: https://scholarship.law.columbia.edu/cgi/viewcontent.cgi?article=1231&context=sabin_climate_change.

UNICEF, and Gallup. n.d. "The Changing Childhood Project." Available at: https://changingchildhood.unicef.org/stories/the-other-climate-crisis.

United Nations. 2015. *Transforming Our World: The 2030 Agenda for Sustainable Development*. New York: United Nations.

United Nations. n.d. "Voluntary Local Reviews." Available at: https://sdgs.un.org/topics/voluntary-local-reviews.

UNPRPD. 2020. "Empower for Change – Reducing Violence and Discrimination against Women and Children with Disabilities in Timor-Leste." Available at: https://www.un.org.tl/en/publications/send/8-human-rights/21-empower-for-change-reducing-violence-and-discrimination-against-women-and-children-with-disabilities-in-timor-leste.

UNPRPD, and UN Women. 2022. "Intersectionality Resource Guide and Toolkit." Available at: https://www.unwomen.org/sites/default/files/2022-01/Intersectionality-resource-guide-and-toolkit-en.pdf.

Unter, K. L. Vogel, J. Walls, C. Küng, and J. Tamayo. 2024. "Towards Defining a Regenerative Economy." Available at: https://sdsn.ch/wp-content/uploads/2024/06/Towards-Defining-a-Regenerative-Economy-HSG-IWOe-SDSN.pdf.

Zaghari-Ratcliffe, N. 2024. "The Feeling of Freedom." *Financial Times*, December 6.

6 Science and technology

THE IMPORTANCE OF CLOSING RESEARCH AND DEVELOPMENT GAPS

Innovations lie at the core of making progress on many sustainable development challenges. Straightforward examples include recycling to deal with increased waste, or the development of solar power to increase green energy production. More complex examples are scientific advances in forecasting extreme weather events (Mooney 2024) and understanding how they are linked to climate change (e.g., the work by World Weather Attribution) (Hodgson 2021; Hodgson and Heal 2021). Such scientific advances are needed to solve remaining puzzles, such as the growth in methane emissions from wetlands in the tropics to adapt our mitigation strategies (Hook and Campbell 2022).

To move beyond business-as-usual, the creation, diffusion and exchange of knowledge and technology are of the utmost importance (Independent Group of Scientists Appointed by the Secretary-General 2019, 36). SDG 9 specifically captures *industry, innovation and infrastructure*. Where innovations already exist, we need to scale them up or out. Where innovation gaps exist, research and development (R&D) investments play a crucial role.

R&D investments, however, are highly uneven, with 80 percent of the global sum of US$1.7 trillion per year spent on R&D accounted for by ten countries alone (UNESCO Institute for Statistics n.d.). The United States spends nearly US$500 billion on R&D alone each year (Independent Group of Scientists Appointed by the Secretary-General 2019, 37). Around three-quarters of that spending in the USA and the UK comes from the private sector (Murray 2023).

These inequalities in R&D threaten to widen the gaps as the least developed countries lack access to locally relevant innovations (Independent Group of Scientists Appointed by the Secretary-General 2019, 37). Inequality in access to technologies is omnipresent: global gendered inequalities in mobile internet use fell to 23 percent, but remained 58 percent in South Asia and 41 percent in sub-Saharan Africa (Independent Group of Scientists Appointed by the Secretary-General 2019, 37). Inequalities have consequences: in agriculture, for example, closing the gender gap in uneven access to resources would increase yields by 20–30 percent and reduce hunger by 12–17 percent (FAO 2011, 5).

To close gaps in R&D, we should focus more on boosting research capacity in the Global South (Independent Group of Scientists Appointed by the Secretary-General 2019, 124). Table 6.1 captures the stark regional differences

DOI:10.4324/9781003400028-9

Table 6.1 Inequalities in research (UNESCO Institute for Statistics 2024)

Region	R&D workers per 1 million	Compared to global average (%)
World	1353	–
North America and Western Europe	4746	+251
Central and Eastern Europe	2332	+72
East Asia and the Pacific	1851	+37
Central Asia	662	-52
Arab States	630	-53
Latin America and the Caribbean	625	-54
South Asia and West Asia	311	-77
sub-Saharan Africa	94	-94

using an indicator for SDG Target 9.5 (UNESCO Institute for Statistics 2024) – number of research and development workers per 1 million. With the global average of 1,353 R&D workers per 1 million people, the United States, Canada, and Western Europe have 2.5 times as many R&D workers. At the other extreme, South Asia and West Asia have only one-quarter of workers, with sub-Saharan Africa only having one-twentieth of workers. Put differently, for every R&D worker in sub-Saharan Africa, you have over 50 R&D workers in North America and Western Europe.

SCALING INNOVATIONS AND ADAPTATION

Closing these gaps is also important, given the context-dependence of innovations, their adaptation processes and capacity building needs. You likely already have encountered a myriad of technological solutions to sustainable development. We covered in Part II Overview the limitations of solely relying on technologies to transform development, without activating other changemakers. We also need to address the necessity to adapt technologies to each user and their geographical context. Even the least controversial ones, such as solar power, need to be adapted to local contexts or include capacity-building efforts.

Let us use a thought experiment: if we were to import a German or Taiwanese solar panel into Malawi, we would need, at minimum, to build the capacity on how to maintain that solar panel installation locally. Beyond maintenance, the solar panel itself might need to be adapted. Constructed for an average Swiss household's energy consumption (of around 5,000 kilowatt (kW) hours per year in my father's case), the 8,000 kW hours produced from the 32 solar panels on my father's roof might be excessive. They are even wasteful, when surplus solar energy cannot be fed into the wider electricity grid or battery-stored locally. In rural Malawi, electricity is usually limited to power lights or recharging cell phones. A micro-solar solution is needed. One such provider, VITALITE (2024), has adapted solar technology to offer a small bundle that includes a single 6 W or 10 W solar panel with the capacity to provide up to eight hours of lighting and phone charging. Other development efforts have adapted the household-based concept described above to create micro-solar grids that connect entire rural villages (Self Help Africa 2024). Similar efforts have been done by my colleague in building a mini-hydroelectric plant in the Ludewa district in neighboring Tanzania. The project, implemented over six years, invested heavily in capacity building, yielding 300 kW of renewable energy that successfully supplies energy to nearby villages (ACRA n.d.).

Scaling relies on the successful diffusion of an innovation (see also Chapter 13). Rogers (2003) wrote the seminal work on the topic. A potential adopter assesses five key attributes: (1) What is the relative advantage gained from adoption?; (2) How compatible is the innovation with my current livelihood?; (3) How complex is the innovation to master?; (4) Can I try it out first?; and (5) Can I observe the benefits myself?. Combined with other factors, such as existing social norms, social capital, and marketing efforts, the answer to these five questions determines their rate of adoption. Innovators and early adopters are the first to integrate an innovation into their livelihoods, because of their ability to take risks, higher social status and levels of education. Laggards are the last to adopt, as they lack access to financial and social capital (Rogers 2003).

While diffusion of innovations is at the heart of achieving sustainable development, any benefits accrued might not be evenly shared across society. In its early stages, diffusion efforts can enlarge the wealth gap. In agriculture, early adopters often are wealthier, better educated farmers with larger landholdings (Rogers 2003). Assuming that these early adopters gain windfall profits, innovations also tend to first spread across a social stratum, rather than between them (failing to reach lower social strata) (Ledermann, Pray, and Anderson 2024). Building upon the concept of intersectionality introduced in Chapter 5, individuals who experience multiple dimensions of marginalization (e.g., gender combined with caste and education) are likely the last to benefit from an innovation.

Leveraging the changemakers of governance and policymaking, as well as economy and finance, can reduce barriers to adoption. With government funding in retreat, a more fundamental transformation might be needed. The next section investigates the role of science in the political and social spheres and discusses how to draw on indigenous, local knowledge.

THE ROLE OF SCIENCE IN THE POLITICAL AND SOCIAL SPHERES

Science is crucial to bring meaning to the challenge of sustainable development, as well as identifying solutions. Because of the pressures of delivering results both in science and development, we are incentivized to simplify our research or interventions and possibly stay away from tackling more challenging problems, such as energy transitions (Independent Group of Scientists Appointed by the Secretary-General 2019, 112). To address the SDGs, research can be guided by the 2030 Agenda by "[shifting its focus] from understanding phenomena (e.g., social inequality) to identifying and detailing ways of improving them (e.g., policies of redistribution, most inclusive economic models)" (Independent Group of Scientists Appointed by the Secretary-General 2019, 113).

For more complex problems, research can be co-designed and by co-producing knowledge to address the wicked problems of sustainable development in accordance with the 2030 Agenda (Independent Group of Scientists Appointed by the Secretary-General 2019, 113). Scientists can also address gaps within the 2030 Agenda. One example is a lack of attention to spiritual values attached to natural resources (Independent Group of Scientists Appointed by the Secretary-General 2019, 117). Another example is culture which "received insufficient attention as an intrinsic component of sustainable development and must be translated and embedded in national and local development" (Independent Group of Scientists Appointed by the Secretary-General 2019, 117).

ADVANCES IN INTEGRATING INDIGENOUS KNOWLEDGE

In the context of limited funding and wicked problems, the role of indigenous knowledge should play a larger role: "indigenous knowledge has guided societies and supported sustainable management of resources, especially in regions where practices have been known for hundreds of years" (Independent Group of Scientists Appointed by the Secretary-General 2019, 120). In contrast to the positivist focus of Western knowledge, the strong embeddedness at the individual, communal, and societal levels of indigenous knowledge has produced highly relevant knowledge around "megatrends such as biodiversity, climate change adaptation and land conversation" (Independent Group of Scientists Appointed by the Secretary-General 2019, 121).

One such example is sustainable land management methods documented by practitioners around the globe by the World Overview of Conservation Approaches and Technologies (WOCAT) platform. Regional centers work across ecosystems, such as the International Institute of Tropical Agriculture (IITA) across the African continent or the International Centre for Integrated Mountain Development (ICIMOD) in the Himalayan region. As the value of traditional and indigenous knowledge becomes more prominent, organizations, like ICIMOD, play a dual role in building capacity for their local knowledge not to be exploited by for-profit companies who patent non-original inventions (such as turmeric used for wound healing or neem for treating malaria) (Prasad Oli and Devi Dhakal 2008). Having established the complementary value of traditional knowledge, let us review calls for greater diversity in Case Study 6.1.

CASE STUDY 6.1 CALLS FOR KNOWLEDGE DIVERSITY

> [W]e need to replace the present-day "monoculture" of Western scientific thinking, which attempts to homogenize "valid" knowledge, with an "ecology of knowledge" that "opens up the dominant canon" and affords "an equality of opportunities to different kinds of knowledge.
>
> (De Sousa Santos 2007a; cited in Krenak 2020, 29)

Indigenous activists and actors from the Global South have long called for a greater diversity of knowledge. Building on our case of La Via Campesina in Chapter 5, I want to introduce two voices for our case: Ailton Krenak, a Brazilian Indigenous leader, scholar, and activist (Krenak 2020, 29), and Boaventura De Sousa Santos, a Portuguese academic and a leading figure in the World Social Forum that shaped a new lens for seeing the world (De Sousa Santos 2006).

As captured in the opening quote from De Sousa Santos, calls for a diversity of knowledge are captured in the concept of ecology of knowledge. Besides increasing the diversity, Krenak (2020, 29) emphasizes how knowledge should move us: "knowledge ecology should be an integral part of our everyday experience, inspiring our choices about where we want to live and the experience we want as a community."

Any attempts to diversify our knowledge include the decision on what *counts* as valid knowledge. For Krenak (2020), it includes a shift in our relationship between nature and ourselves, stripping the environment of its cultural meaning: "When we say that the mountain is telling us it is going to rain and that today will be a prosperous day, they say: 'Mountains are just mountains. They don't tell us anything'" (Krenak 2020, 50). He continues by arguing that:

> [w]hen we depersonalize the river, the mountain, when we strip them of their meaning – an attribute we hold to be the preserve of the human being – we relegate these places to the level of mere resources for industry and extractivism.
>
> (Krenak 2020, 50)

We will review further the limits to growth and regenerating nature in Chapter 12.

Reintegrating the social and cultural into our scientific domain is central to such an epistemological shift. An epistemology from the Global South "contrary to hegemonic epistemologies in the [W]est, does not grant a priori supremacy to scientific knowledge (heavily produced in the North), thus allowing for a more just relationship among different kinds of

> knowledge" (De Sousa Santos 2007b, 19). A dialogue that cuts across cultures and translates different knowledges – as captured in the idea of the ecology of knowledge – has the potential to address shared challenges and move us closer toward achieving social justice goals (De Sousa Santos 2007b).

Krenak's and De Sousa Santos' writings showcase the limits of undertaking sustainable development within a Western-centric value system. Localization efforts, previously led by the USA and its development agency – USAID – to devolve more power to implementation by local actors are noteworthy. However, the evaluation of such efforts continues to be dominated by indicators of growth and measurable, positivist changes. Localization efforts alone struggle in supporting a more radical transformation of how Indigenous actors and diverging value positions can be entertained.

We also need to account for how grant funding in development comes with both responsibilities and accountabilities. In the case of the US government, accountabilities rest with its Congress which holds the power of the purse. With the oversight provided, there's limited wiggle room as to what are acceptable metrics to use – especially when considering that donors are most interested in aggregated, cumulative impact from their funding.[1]

Where would one find donor funding that does not come with these restrictions? It is more likely in places where accountability does not loop back to governments or political institutions, but has more philanthropic origins. As all aid is self-aid to varying degrees (Sachs 2010), we should not romanticize the alternatives. Foundations established based on the values of an individual or group's endowment are better positioned to support and reflect transformative action as part of their funding calls. An example is the Arcadia Fund, which allows unsolicited applications and has awarded around US$1 billion to "work to preserve endangered cultural heritage, protect endangered ecosystems, and promote access to knowledge" (Arcadia n.d.).

When it comes to funding science and research, there is also a need to move beyond the existing power dynamics inherent in research collaborations. Case Study 6.2 expands on holistic development (Kassie et al. 2020), covered in the Part II Overview. Clapperton Chakanetsa Mavhunga's research (Mavhunga 2018) allows us to dive deeper into how local knowledge production takes place, and discover the interplay between local and Western epistemologies.

In my own work, I aim to move away from developing a proposal first, and identifying a partner in the Global South second (in order to fulfill a precondition for funding). I rather engage in a dialogue with my colleagues in the Global South regarding what research we should jointly engage in or what ideas they had for which they had trouble securing funding. Beyond academic research,

CASE STUDY 6.2 SHIFTING KNOWLEDGE PRODUCTION IN SUSTAINABLE DEVELOPMENT

In his book, *The Mobile Workshop: The Tsetse Fly and African Knowledge Production*, Dr. Mavhunga (2018) provides a new temporal account of the practice of science and technology in a Southern African context. The book centers on the Tsetse fly, which we briefly encountered in Part II overview. Due to bites on livestock and humans, the fly is causing estimated economic losses in the billions of US dollars each year. Mavhunga captures the story of how local Zimbabweans and later their colonizers built knowledge on how to co-exist and manage the tsetse fly. He writes extensively in *Chidzimbahwe* for which I have included the English translation in parentheses as needed. His study is groundbreaking, as in the history of science in Africa, Africans are rarely considered intellectual agents.

Mavhunga establishes for the reader that local Zimbabweans developed their own methods to stave off the fly, including using repellants of herbs or smoke. They also adapted their livelihoods to co-exist with the Tsetse fly, including their patterns for herding livestock to minimize the risk of the Tsetse fly. In this process, their knowledge production relied on the scientific method, including experimenting, observing, and tracking patterns.

> Vanhu vatema (black people) created numerous stratagems to manage and coexist with mhesvi (Tsetse) that vachena (whites) later borrowed and deployed to control the insect. These included moto (fire), specifically late burning of forests; manhuwe (repellents); miteyo (traps); kugarisika kwevanhu (human settlement); rusosa/ruzhowa (fences); the use of muchetura (poisons) or mishonga yezvipukanana (pesticides); and so on.
>
> (Mavhunga 2018, 11–12)

With the arrival of the colonizers, their mode of larger-scale livestock holding demanded a shift away from co-existence to eradication of the Tsetse fly.

> Of course, these vachena (whites) were newcomers to mhesvi (tsetse), and though they had political power, designed activities, and directed what, how and where information was to be gathered, they could not personally live with mhesvi for sustained periods without falling victim to the bite of another chipukanana (small insect): hutunga (the one that gores), which vachena called mosquito. Only African men required as "flyboys" (fly catchers) could make the regular fly rounds or inspections of the targeted areas. They produced the information that the white entomologists compiled into "scientific reports", the object of which was to know the pestiferous mobilities of the insect.
>
> (Mavhunga 2018, 9)

Mavhunga's work is a rare example. Too often, we are forced to see *Africa* through the eyes of the colonizers, without giving credit to or showcasing the local knowledge contributions.

The Zimbabwean research assistants were rendered invisible in the scholarly accounts published in Western journals. So were the hunters or spraymen who were exposed to harmful DDT. So were forced resettlements where black people were pushed to the unhealthy margins in proximity to Tsetse habitats to serve as an early-warning system.

Mavhunga uses the term *thingification* to denote how locals were mere instruments or objects (Mavhunga 2018, 13). Black people's knowledge served as "mere raw material for white people, who produced true knowledge ... The natives became samples, specimens, data, and at best, informants, rather than intellectual agents in their own right" (Mavhunga 2018, 14).

Based on his findings, he argues for a move toward self-reintellectualization, whereby black people at "home and abroad [are] researching, reinstalling, reasserting, and revaluing themselves not just as eaters of other people's knowledge, but producers of their own" (Mavhunga 2018, 19). His work shapes our understanding of the foundational role played by black people, and how it was appropriated and translated into the Global North's scientific journals. It pushes us to advance frameworks within which knowledge can be co-produced effectively. It serves as a powerful counterpoint: even though formally trained scientists are rare per the R&D gaps outlined earlier, it does not mean that science is not being produced.

Mavhunga's work is a powerful testament of shaking the stereotype that African innovations or science are not relevant when addressing sustainable development issues. The work on holistic development, featured in the Part II Overview, captured my African colleagues' achievement in deploying cutting-edge science that is built on interactions with rural Africans (Kassie et al. 2020). The Tsetse research, for example, involved identifying which animal scents were most effective at either attracting or repelling the fly, drawing heavily on local herders' experiences. These cutting-edge African researchers from *icipe* (International Centre for Insect Physiology and Ecology) are at the forefront of developing integrated solutions that address complex problems that colleagues around the globe are struggling with. As reflected in our work in Tolay, Ethiopia, they are leading experts in identifying co-benefits and trade-offs across silos by leveraging new approaches. An example is the integrated (programming) approach of One Health, which aims to "improve health and well-being through prevention of risks and mitigation of diseases that originate at the interface between humans, animals and their natural environments" (Independent Group of Scientists Appointed by the Secretary-General 2019, 124). A cornerstone of their work is to train the next generation of

African scientists: *icipe* effectively engages in capacity building, having trained 237 fellows and 33 postgraduate scholars in 2023 alone (*icipe* 2023).

To further this shift in knowledge production, such excellence in scientific production needs to be supported through a reallocation of power in grant-making decisions and research collaborations. As noted by Nordling (2015, 25), "the role of partners in Africa and other developing regions was 'often still primarily limited to collecting data', whereas partners from the developed world played a leading role in analysing and publishing the findings." As funders, similar to localization efforts, are slowly shifting the power dynamics, researchers in the Global North hold considerable agency in how they approach the crucial funding process.

the concept of transboundary research partnerships captures how to set an agenda together, promote mutual learning, and include diverse stakeholders (Stöckli, Wiesmann, and Lys 2018).

It is only by increasing the diversity of knowledge that counts, shifting the knowledge production, working across communities, and embracing scientific collaborations and more equal partnerships that the most wicked and complex problems of sustainable development can be addressed. This step represents the focus of our next section.

SCIENTIFIC PARTNERSHIPS AND COLLABORATIONS TO ADDRESS COMPLEX AND WICKED DEVELOPMENT CHALLENGES

"It takes a village" – or so the famous adage goes. Popularized by Hillary Clinton's (1996) book titled the same, the proverb with contested African origins emphasizes the roles of communities in raising the next generation (Goldberg 2016). The village, as an entry point for development in the 21st century, reached its zenith with the *Millennium Villages Project*. Powered by Dr. Jeffrey Sachs and Columbia University's The Earth Institute, the project located across 15 East and West African sites aimed to achieve the MDGs through targeted interventions between 2005 and 2015 (Millennium Promise Alliance n.d.).

Working myself in nearby project sites in Kenya, what struck me as unique was the village-level setup whereby beneficiaries were reached across entire sites rather than just single households. A more systematic implementation of development was meant to result in larger, community-wide impacts. In the end, the project was marred by controversy (*The Economist* 2012), including around the impacts observed. Because of a lack of control sites for assessing the effectiveness of the interventions (Duff-Brown 2018), it was difficult to attribute the progress made to the millions of aid spent.

While Millennium Villages are recorded in the annals of development efforts, the SDGs carry forward the spirit of looking at the importance of communities and scalable transformations. SDG 17: *Partnerships for the Goals* emphasizes how achieving development needs to draw on all types of knowledge and solutions, moving beyond the traditional Global North/Global South pathways. As captured by Targets 17.6 and 17.7, sharing of science, innovation and technology should occur across borders:

> 17.6. Enhance North-South, South-South and triangular regional and international cooperation on and access to science, technology and innovation and enhance knowledge sharing on mutually agreed terms ....

As an example of South-South cooperation, Brazil, for example, has emerged as an agricultural powerhouse that has inspired efforts to replicate its agricultural productivity across the Global South, such as in Nigeria (Schipani 2020).

> 17.7. Promote the development, transfer, dissemination and diffusion of environmentally sound technologies to developing countries on favorable terms ...

The United States and India, for example, agreed in September 2024 on a roadmap to "expand bilateral technical, financial, and policy support to expand complementary U.S. and Indian manufacturing capacity for clean energy technologies and components" (The White House 2024).

Efforts to improve knowledge transfer are needed to address the set of wicked and complex sustainable development challenges. Drawing on the work of Peter Messerli and Sabin Bieri, Table 6.2 breaks down development challenges into the following five helpful categories.

Table 6.2 The wicked and complex nature of sustainable development. Source: Independent Group of Scientists Appointed by the Secretary-General (2019), 111–112

Type of Sustainable development challenge	**Description**
Simple	Largely uncontested scientific evidence forms the basis for decision-making and planning, such as recycling.
Complex	Evidence is not contested, but there are many gaps in knowledge. The way forward can be illuminated by increasing the understanding of coupled social and ecological systems, such as more environmentally friendly farming practices that both local and transnational companies find economical to adopt.
Complicated	Sufficient evidence is available, but implementation requires societal consensus. For example, policies of modest carbon taxation and income redistribution. Those challenges require communications efforts to raise awareness, mobilize responses, spur negotiations, circumnavigate vested interests and create adequate societal demand for action.

(*Continued*)

Table 6.2 (Continued)

Type of Sustainable development challenge	Description
Wicked	Not wicked in the sense of evil, but rather wickedly difficult. Here low factual certainty is combined with low societal support. Purely fact-based decisions no longer seem possible, which can make those challenges appear insurmountable. They include decarbonization, for example, or ways of creating sustainable food systems.
Chaos	The issues are unknowable and non-negotiable. Those include the turning of religious fundamentalism to terrorist violence, for example, or the full harm of crossing ecological tipping points.

LEVERAGING A DIVERSITY OF KNOWLEDGE AND SHIFTING KNOWLEDGE PRODUCTION

We can leverage a transdisciplinary approach that can account for a greater diversity of knowledge to address challenges where lower factual certainty exists, irrespective of the level of sociopolitical agreement. Where societal consensus is lacking yet evidence appears well-established (e.g. complicated challenges such as wealth redistribution), integrating more diverse ecologies of knowledge might yield significant benefit.

For complex challenges, knowledge gaps can be closed by leveraging a more inclusive form of knowledge production. Their example used of more environmentally sustainable farming is appropriate, as it reflects a case where local knowledge production is paramount to the success of an intervention.

For wicked challenges, where knowledge gaps are contested and change needs to take place beyond the household or local level, ecological and sociocultural dimensions of knowledge production become more important. The example of creating sustainable food systems, the focus of our first transformation in Chapter 7, represents one such challenge. Co-production of knowledge in alignment with societal buy-in is expected to yield a productive avenue for overcoming these challenges.

Ahuja (2022) interviewed Dr. Kris de Meyer, director of the UCL Climate Action Unit, who captures well how knowledge itself will not solve it all: "Many climate scientists think that if only people were made aware of the problem, even made fearful enough, then they will act, but fear doesn't drive action in predictable ways." We will revisit how knowledge and awareness translate into action in Part IV.

ARTIFICIAL INTELLIGENCE AND ITS EQUITY PROBLEM

The news is dominated by the rise of artificial intelligence (AI) and its promise (and some of its pitfalls). As an observer of the trends, the hype surrounding AI acts as an extension of the Internet-of-Things (IoT) gospel (Rifkin 2014) and the relevance of ICT as our third (Rifkin 2011) and fourth Industrial Revolution (Schwab 2017). Claims are made that AI, IoT, and ICT will bring an end to capitalism as we know it (Rifkin 2014), offer to level the global playing field (Bourguignon and Scott-Railton 2015), and create new avenues for achieving sustainability (Kaplinsky 2022). These narratives have acquired additional credence through the Global Pandemic and resultant lockdowns and social distancing between 2020 and 2022.

By 2025, AI appears ready to usurp the narrative, threatening to take down prior commitments to sustainable energy transformation with it (Crownhart 2024). President Trump pronounced during his first week back in office how he expected to meet AI's energy demand: "Nothing can destroy coal. Not the weather, not a bomb—nothing" (Gruver and Brown 2025). The lifespan of a new coal power plant built today is more than 40 years (Independent Group of Scientists Appointed by the Secretary-General 2019), so today's decisions on how to power AI have consequences for generations to come.

AI euphoria is masking the deep inequities upon which technological advances are built, and that they threaten to exacerbate (see also our discussion on COVID-19 and inequality in Chapter 2). Reversing progress made on environmental justice lays bare these inequities for everyone to see. Before jumping too far into the future, let us investigate the trends and how they play out on the ground now.

Reversing downward trends with the internet

While China dominated the news cycle around COVID-19, some of the most severe restrictions occurred in a country you likely did not have on your radar: Uganda. Holding the global record for the length that schools were kept closed during the pandemic, half of all students stopped learning even as remote options were deployed (Blanshe and Latif Dahir 2021). With most of the population depending on agriculture for their livelihoods (Atamanov et al. 2021), agricultural advisory services initially came to a halt.

Pivoting to remote options proved a significant challenge given pre-pandemic trends: research and development funding were on a continuous downward trend in agriculture. Accordingly, fewer farmers received advice from agricultural extension officers. Women and more marginalized farmers were even less likely to receive advisory services. Given these inequalities, shifting toward digital modes of service delivery, such as using smartphones that are traditionally owned by wealthier farmers, would further perpetuate inequities. While digital extension services offer viable future avenues, especially if led via the private sector, renewed investments in the public sector and their offerings would be essential to better support marginalized farmers (Ledermann, Pray, and Anderson 2024).

"Can you hear me now?"

The above situation becomes more complicated because agriculture's defining property is context. My soil and microclimate might be different than the one of a farmer a mile down the road. As a result, leapfrogging brings risks in adopting technologies that were not developed for the relevant socio-cultural, ecological, and economic environments. Prominent examples include the time-consuming efforts in East Africa to identify effective drought-tolerant seeds using genetic research successfully deployed in the USA (Schnurr and Dowd-Uribe 2021). Outside agriculture, the use of cellphone towers to leapfrog landline connections represents a success story: only 2 percent of households in sub-Saharan Africa had a working landline phone connection (Pew Research Center 2015) as smartphone adoption reached over 50 percent by 2022 (GSMA 2023).

One of the most prominent cases playing out is the use of AI. As a co-founder of a social enterprise aiming to empower smallholder farmers to improve their resilience via digital advice, I am intimately aware of the power and promises that come from deploying AI in a chatbot or to improve matching algorithms for more effective service delivery. Besides the persistent gap in internet penetration, with only 23 percent of the population in sub-Saharan Africa connected to mobile internet (GSMA 2023), AI's Achilles' heel is its dependency on the English language. Leapfrogging development by leveraging the power of AI risks further exacerbating inequalities between those who have (with access to smartphones, the internet, and versed in English) and those who lack (without access to one of the three).

For example, English is the official language used in Malawian schools. In reality, the other eight official languages, such as Chewa, dominate rural areas. With limited investments made in these language models, the gaps between English and non-English speakers will be further stretched. Co-production of knowledge, hence, is unlikely to occur unless significant investments are made into documenting such knowledge and the supporting language. It showcases the importance of the levers of governance in supporting science and technology. The German Ministry for Economic Cooperation and Development and its development branch (Deutsche Gesellschaft für Internationale Zusammenarbeit; GIZ) and their project *Artificial Intelligence for All – FAIR Forward* represent one donor that stepped up its investments to improve access to AI for local innovations, including support for the use of local language models (GIZ n.d.).

Such efforts echo the calls for greater localization in development. The assumption is that the *closer to the issue, the closer to the solution.* We discussed the importance of the village and the community. However, it is important not to fall into the trap of romanticizing local knowledge. As noted by the late renowned participatory development scholar and practitioner, Robert Chambers (1983, 2017), there are limits to indigenous, traditional knowledge and where it will benefit from drawing upon external scientific insights (e.g., for diseases).

GETTING THE RIGHT DATA TO CREATE TRANSFORMATIVE ACTION

To stimulate action across the changemakers of governance and policymaking, economy and finance, and individual and collective action, the changemaker of science and technology also needs to elevate whose knowledge is visible and then build upon it. To get us started, we need access to data and insights that accurately capture realities on the ground (Center for Sustainable Development 2023). Rulers used censuses to get a top-down understanding of the subjects they governed, often to extend state control (e.g., Bourdelais 2004, for French censuses). A lack of data hampers decision-making by policymakers. But wrong data that fails to make the invisible visible will result in worse decision-making. Neglecting to account for the complexity and diversity are at the core of failures of top-down interventions (Scott 1998).

At the positivist end, the Center for Sustainable Development (2023) identified opportunities, for example, such as "harnessing breakthroughs in ocean measurement technologies to improve ocean management and decision making." Such new technologies improve our climate modeling of complex issues, such as ecosystem and carbon cycling impacts of deep seabed mining (Center for Sustainable Development 2023). My former colleagues organized multi-stakeholder workshops in the Global South that leveraged advanced system dynamic modeling to achieve sustainable development. At a side-event I moderated to highlight the value of these efforts, the colleague in charge of modeling bluntly stated: "Garbage in, garbage out." That was not the pitch the audience was expecting. But they captured the main challenge: the more complex our methods or tools (including AI), the greater the damage done by using incomplete data. The main drive of the programming was to build capacity within national governments to take the lead. In the end, limited funding and poor data weakened the intended impact of taking complexity into account when making budgeting decisions for sustainable development.

Beyond the government, a push for more data and greater transparency is needed to engage our changemaker in economy and finance. The Center for Sustainable Development (2023) uses the case of the negative impacts from the clothing industry's overproduction: about 30 percent of over 100 billion garments produced are never sold. To make a greater push toward a circular economy and address externalities, new industry standards are called for, informed by such data (Center for Sustainable Development 2023) (see also our fast fashion Case Study 8.1 in Chapter 8). With deregulation and trade barriers rising to the top of policymakers' agenda in 2025, hard-earned achievements made on sustainability reporting in the EU, however, are unraveling (Hancock 2025), illustrating the importance of incentives and popular support.

Recalling the knowledge ecology, we need a new generation of scholars who can speak to the different epistemologies. These concerns are more elevated as we face the rise of disinformation that upends order (Tett 2025). To get the data right, this includes training on how to integrate divergent or silent voices. Leveraging participatory methods is usually at the top of the list. From a pragmatic perspective, we also need rigorous mixed methods approaches whose research outputs can enter the dominant discourse and showcase the cracks within their canon. Even as we push for AI to advance our understanding of complex challenges, such as weather forecasting, researchers have proven the value of integrating local forecasting knowledge to increase the accuracy of modeling efforts (Gbangou et al. 2021).

BRINGING IT ALL HOME: CHANGEMAKERS IN ACTION

Roberts (2015), an advisor for the US government in Asia, captured a story that showcases well how changemakers work together. Having observed how the use of development funds "often fails to move the needle" (Roberts 2015), he was surprised how leveraging low-cost technology combined with science communication and public diplomacy can translate into collective action and produce change that practitioners like him thought unimaginable.

His story starts with a new Twitter account from the US Embassy in Beijing called @BeijingAir. With high pollution levels in Beijing, the embassy's air-quality monitor was set up in 2008 to tweet on an hourly interval its readings on outdoor air quality. It offered previously unavailable data to the public on $PM_{2.5}$ airborne particle levels, which cause serious health damage (Roberts 2015). Using the US air pollution standards set by its Environmental Protection Agency (EPA), tweets that recorded record levels of $PM_{2.5}$ caused a stir among Beijing residents. Just as my daughter's school in Washington, DC, prevented her from playing outside during days of heavy air pollution, parents across Beijing took this new data point into account in their daily household activities.

Given the disconnect between the official Chinese data and the public diplomacy tweetbot, residents started to "put pressure on Chinese officials to acknowledge the scale of the problem and start taking proactive measures to tackle it" (Roberts 2015). With effect: the Chinese government not only set up its own $PM_{2.5}$ stations (to the tune of 500 stations across 70 cities), it also "pledged hundreds of billions of dollars for cleaning the air and began to implement pollution reduction targets for major cities" (Roberts 2015).

Spin-off efforts were launched across Asian countries, all captured through the original website (aqicn.org) established by the *Air Quality Index* team at the US Embassy. Click on it: I found a station less than 200 meters from my home. By now, the site covers more than 250,000 air quality monitoring stations in 2000 cities across more than 130 countries (see Figure 6.1).

The map in Figure 6.1 also shows that air quality monitors are lacking in still too many places. In sub-Saharan Africa, except for South Africa, most air quality monitors are hosted by US embassies. No monitors exist for around a dozen African countries. Keep this technological divide in the front of your mind as we move toward our transformations. Remember the privilege you

On a normal September day - Air pollution across the globe

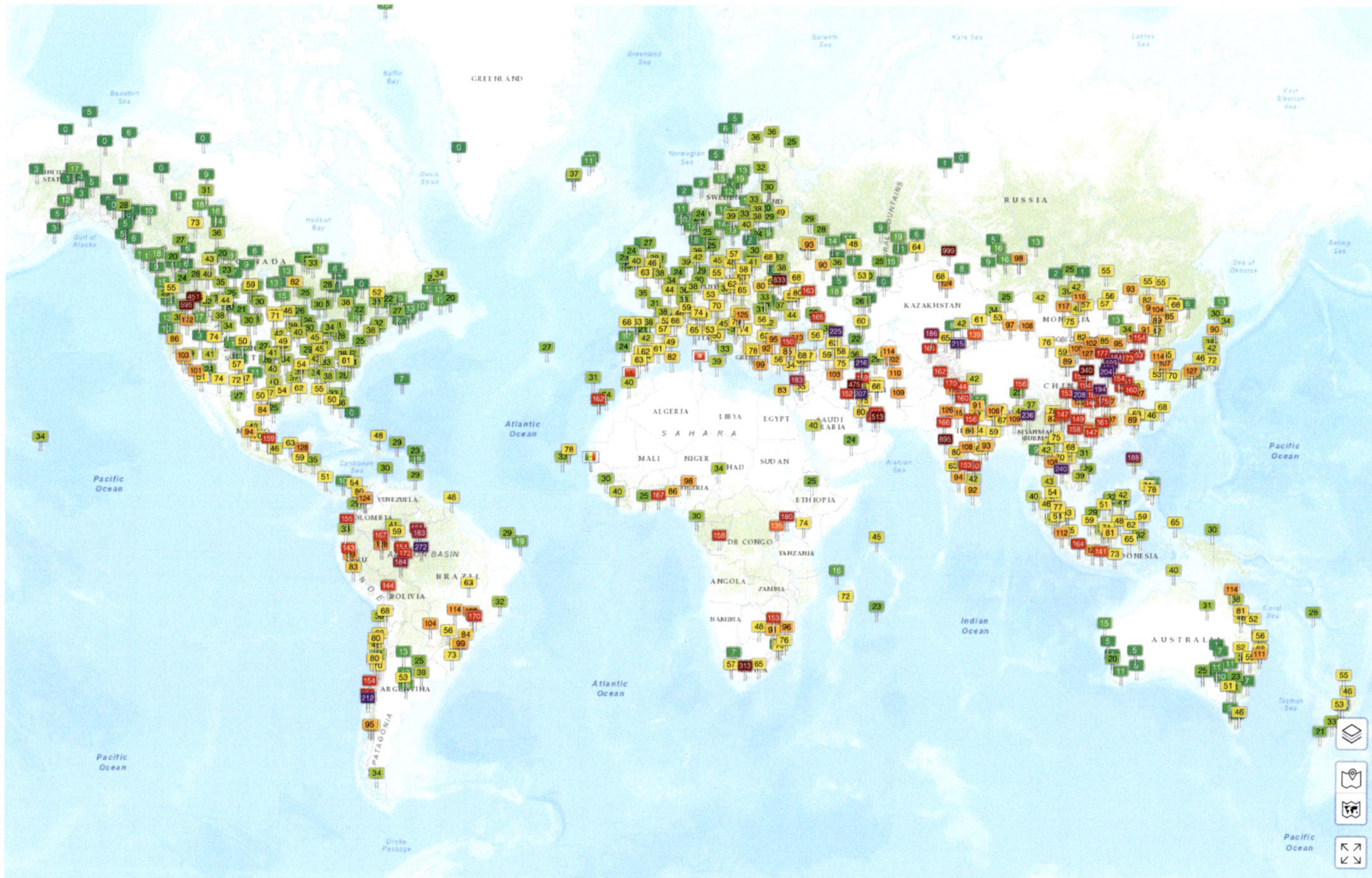

Figure 6.1 On a normal September day – air pollution across the globe, Air Quality Index team, September 25, 9.35 p.m. EST (www.aqicn.org)

might enjoy. And the aspirations we all share for a better life. As the 2030 Agenda is focused on leaving no one behind, let us focus on how to transform sustainable development for all.

YOUR TURN

- What is the role of science and technology in addressing your SDG?
- What changes if using an intersectionality lens?

NOTE

1 The unilateral dismantling of USAID in 2025 by the executive branch put the balance of power into question.

REFERENCES

ACRA. n.d. "Hydroelectric Power for 20 Isolated Rural Villages in Ludewa District – Tanzania." Available at: https://www.acra.it/cosa-facciamo/aree-di-intervento/tanzania/programma-energia.

Ahuja, A. 2022. "'Magic Numbers' Are Clouding the Climate Debate." *Financial Times*, September 20. https://www.ft.com/content/7e047a60-c2eb-47ed-b6bf-fff4260079c9.

Arcadia. n.d. "How We Work." Available at: https://arcadiafund.org.uk/how-we-work. (accessed February 14, 2025).

Atamanov, A., F. Cochinard, J. Ilukor, and G. Ponzini. 2021. "A Year into the COVID-19 Pandemic: What Rounds 4 and 5 of Uganda High-Frequency Phone Survey Tell." World Bank Blogs (blog). Available at: https://blogs.worldbank.org/opendata/year-covid-19-pandemic-what-rounds-4-and-5-uganda-high-frequency-phone-survey-tell.

Blanshe, M., and A. L. Dahir. 2021. "Uganda Reopens Schools After World's Longest Covid Shutdown." Available at: https://www.nytimes.com/2022/01/10/world/africa/uganda-schools-reopen.html.

Bourdelais, P. 2004. "The French Population Censuses: Purposes and Uses During the 17th, 18th and 19th Centuries." *The History of the Family* 9 (1): 97–113. https://doi.org/10.1016/j.hisfam.2001.10.005.

Bourguignon, F., and T. Scott-Railton. 2015. *The Globalization of Inequality*. Princeton, NJ: Princeton University Press.

Center for Sustainable Development at Brookings. 2023. *Forging Next Steps into the SDG Second Half*. Washington, DC: Brookings Institute.

Chambers, R. 1983. "Whose Knowledge?" In R. Chambers, *Rural Development: Putting the Last First*. London: Routledge.

Chambers, R. 2017. "Can We Know Better?" In R. Chambers, *Can We Know Better?* New York: Practical Action Publishing.

Clinton, H. R. 1996. *It Takes a Village: And Other Lessons Children Teach Us*. New York: Simon & Schuster.

Crownhart, C. 2024. "AI Is an Energy Hog. This Is What It Means for Climate Change." Available at: https://www.technologyreview.com/2024/05/23/1092777/ai-is-an-energy-hog-this-is-what-it-means-for-climate-change/.

De Sousa Santos, B. 2006. *The Rise of the Global Left: The World Social Forum and Beyond*. London: Bloomsbury Publishing.

De Sousa Santos, B. 2007a. *Another Knowledge Is Possible: Beyond Northern Epistemologies.Reinventing Social Emancipation*. London: Verso.

De Sousa Santos, B. 2007b. "The World Social Forum and the Global Left." Available at: https://ces.uc.pt/publicacoes/oficina/ficheiros/286.pdf.

Duff-Brown, B. 2018. "The Fog of Development: Evaluating the Millennium Villages Project." Available at: https://fsi.stanford.edu/news/fog-development-evaluating-millennium-villages-project.

FAO. 2011. *The State of Food and Agriculture: Women in Agriculture, Closing the Gender Gap for Development*. Rome: FAO.

Gbangou, T., E. Van Slobbe, F. Ludwig, G. Kranjac-Berisavljevic, and S. Paparrizos. 2021. "Harnessing Local Forecasting Knowledge on Weather and Climate in Ghana: Documentation, Skills, and Integration with Scientific Forecasting Knowledge." *Weather, Climate, and Society* 13 (1): 23–37. https://doi.org/10.1175/wcas-d-20-0012.1.

GIZ. n.d. "FAIR Forward – Artificial Intelligence for All." Available at: https://www.giz.de/expertise/html/61982.html.

Goldberg, J. 2016. "It Takes a Village to Determine the Origins of an African Proverb." Available at: https://www.npr.org/sections/goatsandsoda/2016/07/30/487925796/it-takes-a-village-to-determine-the-origins-of-an-african-proverb.

Gruver, M., and M. Brown. 2025. "Trump Pumps Coal as Answer to AI Power Needs But Any Boost Could Be Short-Lived." *AP*, 2025.

GSMA. 2023. *The Mobile Economy 2023*. Available at: https://www.gsma.com/mobile economy/wp-content/uploads/2023/03/270223-The-Mobile-Economy-2023.pdf.

Hancock, A. 2025. "Brussels Under Pressure to Curb Green Agenda in Response to Trump." *Financial Times*, January 26. https://www.ft.com/content/da348979-0261-4468-ba93-d6164fb1865b.

Hodgson, C. 2021. "Scientists Warn of Increased Frequency of Extreme Weather Events." *Financial Times*, August 23. https://www.ft.com/content/20327e57-282d-421d-90da-725db3b1e31b.

Hodgson, C., and A. Heal. 2021. "Fires and Floods: Can Science Link Extreme Weather to Climate Change?" *Financial Times*, August 4. https://www.ft.com/content/fe4e658e-0473-4f98-b995-4606aefa90bc.

Hook, L., and C. Campbell. 2022. "Methane Hunters: What Explains the Surge in the Potent Greenhouse Gas?" *Financial Times*, August 22. https://www.ft.com/content/9ef195d6-dcc3-4378-bb35-2721981d6416.

icipe. 2023. "Annual Report." Available at: https://www.icipe.org/sites/files_force/icipe-Annual-Report2023.pdf.

Independent Group of Scientists Appointed by the Secretary-General. 2019. *Global Sustainable Development Report 2019: The Future Is Now – Science for Achieving Sustainable Development*. New York: United Nations. Available at: https://sdgs.un.org/gsdr/gsdr2019.

Kaplinsky, R. 2022. *Sustainable Futures: An Agenda for Action*. Cambridge: Polity Press.

Kassie, M., Z. Abro, T. Wossen, S. T. Ledermann, G. Diiro, S. Ballo, and L. Belayhun. 2020. "Integrated Health Interventions for Improved Livelihoods: A Case Study in Ethiopia." *Sustainability* 12 (6). https://doi.org/10.3390/su12062284.

Krenak, A. 2020. *Ideas to Postpone the End of the World*. Trans. A. Doyle. Toronto: Anansi International.

Ledermann, S., C. Pray, and J. R. Anderson. 2024. "Observations on Status and Trends of Agricultural Extension and Inequality in Uganda." *World Development – Sustainability* 4 (June). https://doi.org/10.1016/j.wds.2024.100147.

Mavhunga, C. C. 2018. *The Mobile Workshop: The Tsetse Fly and African Knowledge Production*. Cambridge, MA: MIT Press.

Millennium Promise Alliance. n.d. "The Millennium Villages Project." Available at: https://www.millenniumpromise.org/millenniumvillages.

Mooney, A. 2024. "Meteorologist Celeste Saulo: 'Climate Change Is Not a Movie. This Is Real Life'." *Financial Times*, August 23. https://www.ft.com/content/4e4f9747-9bc8-4242-a43d-0463c0620607.

Murray, A. H. 2023. "Billionaires Want to Save the World. What's So Wrong with That?" *The Financial Times*, June 29. https://www.ft.com/content/f5fa7da8-f3de-4930-a460-631e353ef683.

Nordling, L. 2015. "Research: Africa's Fight for Equality." *Nature* 521: 24–25. https://doi.org/10.1038/521024a.

Pew Research Center. 2015. "Cell Phones in Africa: Communication Lifeline." Available at: https://www.pewresearch.org/global/2015/04/15/cell-phones-in-africa-communication-lifeline/.

Prasad Oli, K., and T. Devi Dhakal. 2008. "Traditional Knowledge in the Himalayan Region." Available at: https://www.icimod.org/resources/359..

Rifkin, J. 2011. *The Third Industrial Revolution: How Lateral Power Is Transforming Energy, the Economy, and the World.* New York: Palgrave Macmillan.

Rifkin, J. 2014. *The Zero Marginal Cost Society: The Internet Of Things, the Collaborative Commons, and the Eclipse of Capitalism.* New York: Palgrave Macmillan.

Roberts, D. 2015. "Opinion: How the US Embassy Tweeted to Clear Beijing's Air." *Wired.*

Rogers, E. M. 2003. *Diffusion of Innovations.* 5th edn. New York: Free Press.

Sachs, W. 2010. *The Development Dictionary: A Guide to Knowledge as Power.* 2nd edn. London: Zed Books.

Schipani, A. 2020. "Nigerian Businessman Targets Brazil-Style Farming Revolution at Home." *Financial Times*, October 21. https://www.ft.com/content/1bc033a6-4647-44d1-a7ca-047f64f84643.

Schnurr, M. A., and B. Dowd-Uribe. 2021. "Anticipating Farmer Outcomes of Three Genetically Modified Staple Crops in Sub-Saharan Africa: Insights from Farming Systems Research." *Journal of Rural Studies* 88: 377–387. https://doi.org/10.1016/j.jrurstud.2021.08.001.

Schwab, K. 2017. *The Fourth Industrial Revolution.* New York: Crown Business.

Scott, J. 1998. *Seeing Like a State.* New Haven, CT: Yale University Press.

Self Help Africa. 2024. "Lighting the Path to Progress: Bringing Solar-Power to Rural Malawi." Available at: https://selfhelpafrica.org/us/bringing-solar-power-to-rural-malawi/.

Stöckli, B., U. Wiesmann, and J.-A. Lys. 2018. *A Guide for Transboundary Research Partnerships: 11 Principles*, 3rd edn. Bern, Switzerland: Swiss Commission for Research Partnerships with Developing Countries (KFPE). Available at: https://portal-cdn.scnat.ch/asset/4d8372ee-27d7-558a-b9db-77bf409ceef4/KFPE-11P-7Q-3d2018.pdf?b=370a2788-58f3-581e-85d7-2418b87c662e&v=8ef687bc-7b14-5a4f-ad9e-bf494cddc1d7_0&s=gLwVbtPON190F6auadVe-DPd_IOknFA_pssGzeBGB68h8nfbZlcnUaztPkFbUOKFef8V2w2-5NzkK61J2WEdN4JXXheHl2gX1f6cLTTro8WFmDkhlKvHQjqRQDZdPmPo3whl0-P3PWWEWWinhvJ5hWLGC5hYwVTAPC0kLEC-F7g.

Tett, G. 2025. "Five Ways to Fix the Knowledge Crisis in the Age of Trump." *Financial Times*, January 24. https://www.ft.com/content/b4de95a2-c0da-4de8-8ad4-097f0d045633.

The Economist. 2012. "Millennium Bugs." *The Economist (London).* May 14. https://www.economist.com/feast-and-famine/2012/05/14/millennium-bugs.

The White House. 2024, "Roadmap for U.S.-India Initiative to Build Safe and Secure Global Clean Energy Supply Chains." Available at: https://www.whitehouse.gov/briefing-room/statements-releases/2024/09/21/roadmap-for-u-s-india-initiative-to-build-safe-and-secure-global-clean-energy-supply-chains/.

UNESCO Institute for Statistics. 2024. "Science,Technology and Innovation: 9.5.2 Researchers (in Full-Time Equivalent) Per Million Inhabitants." Available at: https://uis.unesco.org

UNESCO Institute for Statistics. n.d. "How Much Does Your Country Invest in R&D?". Available at: https://uis.unesco.org/apps/visualisations/research-and-development-spending/.
VITALITE. 2024. "Empowering Malawi: VITALITE's Journey to Sustainable Energy." Available at: https://vitalitemw.com/f/empowering-malawi-vitalites-journey-to-sustainable-energy.

Part III

Transformation hotspots

Part III Overview

OUR MAP

Part III is organized as follows: Our map (see Figure III.1) builds upon the framework from the Independent Group of Scientists Appointed by the Secretary-General (2019). The map allows us to navigate to where the greatest transformative opportunities exist and build our expertise on how we can leverage the different changemakers to achieve and sustain them.

Each transformation hotspot is represented by a commissioned drawing from South African artist Marizanne Linde. I use her drawings, as well as the corresponding SDGs as entry points for a synopsis at the beginning of each chapter that captures the challenges we face (United Nations 2024) and why addressing the SDGs through these hotspots matter.

To tackle the complexity inherent within transformations and sustain momentum, we need to harness co-benefits and navigate trade-offs across the interventions proposed. Based on our introduction of the framework by Nilsson, Griggs, and Visback (2016) in *Part II Overview*, Chapters 7–12 will conclude with an opportunity for you to review synergies and trade-offs. Additional exercises are available online via our companion website: sustainabledevelopment.me.

DOI:10.4324/9781003400028-11

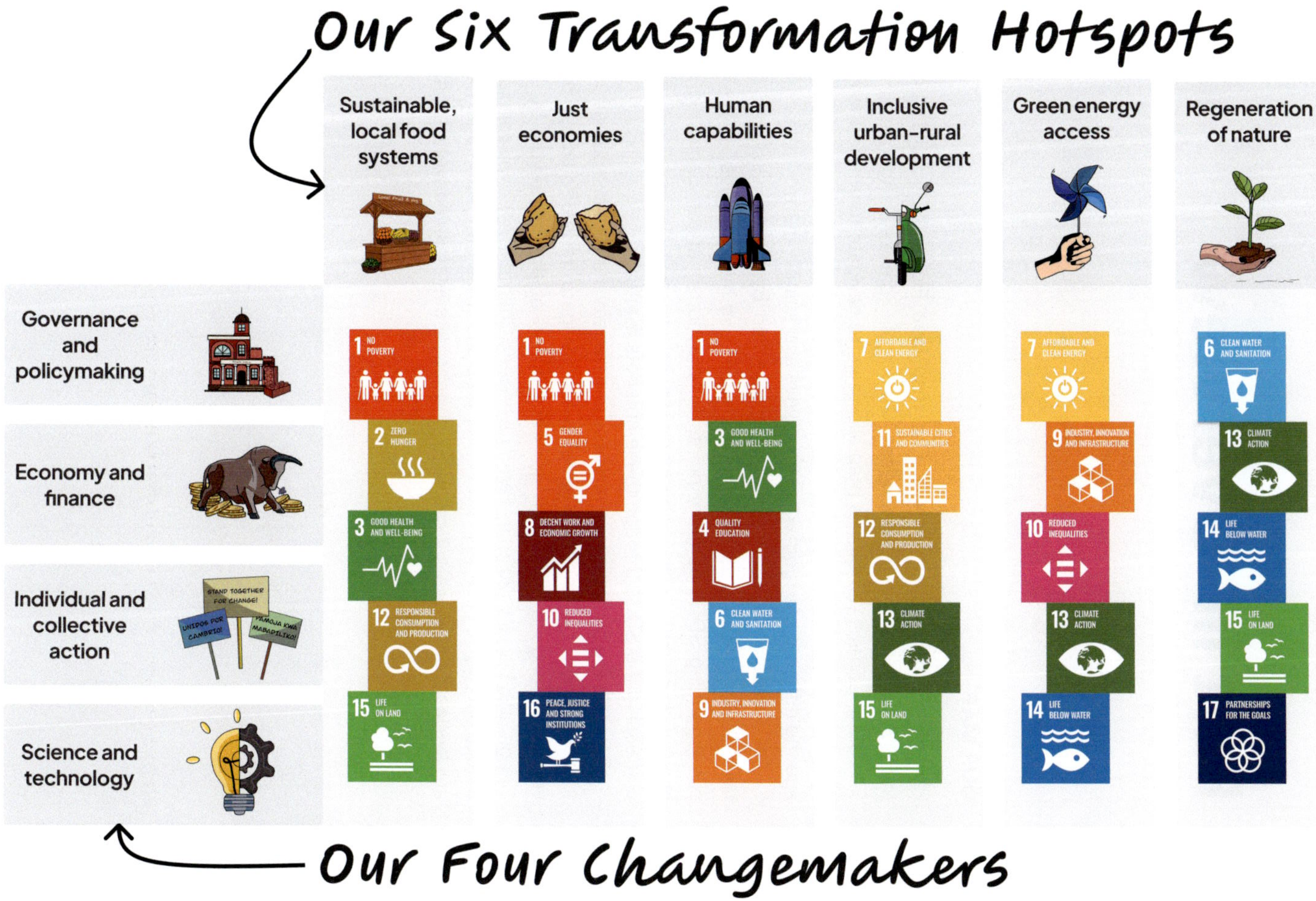

Figure III.1

Mapping our transformation hotspots, adapted from Independent Group of Scientists Appointed by the Secretary-General (2019)

REFERENCES

Independent Group of Scientists Appointed by the Secretary-General. 2019. *Global Sustainable Development Report 2019: The Future is Now – Science for Achieving Sustainable Development*. New York: United Nations. Available at: https://sdgs.un.org/gsdr/gsdr2019.

Nilsson, M., D. Griggs, and M. Visback. 2016. "Map the Interactions between Sustainable Development Goals." *Nature* 534: 320–322. https://doi.org/10.1038/534320a.

United Nations. 2024. *The Sustainable Development Goals Report 2024*. New York: United Nations.

7 Sustainable, local food systems

RECAP OF CHALLENGES AND WHY THIS HOTSPOT MATTERS

We covered in Chapter 2 why "business-as-usual is no longer an option" in agriculture based on the long-term trends. Using the latest data available (United Nations 2024), this transformation hotspot can address the following five challenges.

SDG 1: No Poverty

As COVID-19 increased extreme poverty for the first time in decades, wealthier governments provided financial support to their population (United Nations 2024, 8). In Latin America, extreme poverty has increased to levels last seen two decades ago, increasing calls to strengthen the welfare state (Stott 2021). By 2022, extreme poverty retreated to its pre-pandemic levels, with one major exception: in low-income countries, 23 million additional people are living in extreme poverty (United Nations 2024, 8).

Sustainable food systems have a proven track record of lifting up the poorest of the poor.

DOI:10.4324/9781003400028-12

SDG 2: Zero Hunger

While governments slashed agriculture budgets during COVID-19, many countries are now reinvesting in agriculture and making significant progress toward achieving productive and sustainable agriculture. For example, countries in Europe and North America are close to achieving SDG 2 (scoring 4.1 out of 5.0). In sub-Saharan Africa, funding as a share of overall government expenditures and progress toward achieving productive and sustainable agriculture is lagging (2.7 out of 5.0) (United Nations 2024, 10–11).

Transforming sustainable, local food systems directly supports achieving more productive and sustainable agriculture. Mobilizing financial support domestically and from abroad can accelerate progress made.

SDG 3: Good Health and Well-Being

The good news is that child mortality rates of under-5-year-olds hit a global historic low in 2022. The bad news is that the rate of decline has slowed down. Three out of four countries not on track are located in sub-Saharan Africa. Malnutrition poses a global problem, with trends reflecting that one out of five children under the age of 5 is expected to be stunted by 2030 (United Nations 2024, 10, 13). More lives are lost to poor nutrition (three million to high intake of sodium, three million to low intake of whole grains, and two million to low intake of fruits) than to smoking each year (Afshin et al. 2019).

A more sustainable, local food system promises not only to improve food availability but also improve the nutritional diversity available.

SDG 12: Responsible Consumption and Production

Food losses continue to pile up. Some 13 percent of food is lost after harvest, with an additional 19 percent wasted at the food retailer or during consumption. While nearly 800 million people go hungry, about one billion meals are wasted every day (United Nations 2024, 32).

Reducing waste post-harvest and during consumption in a transformed food system is one of the simplest win-wins: increase food availability, reduce greenhouse gas emissions and pressure on agricultural land.

SDG 15: Life on Land

Each year, 100 million hectares of productive land are degraded. We frequently focus on deforestation and increased soil erosion in lowland areas in agriculture. These land cover changes are, however, also impacting mountain ecosystems. In Europe and North America, for example, approximately 6 percent of all mountain land is degraded, increasing biodiversity loss through unsustainable modes of production (United Nations 2024, 39).

Moving toward more sustainable agricultural production and consumption (Tilman and Clark 2014; Willett et al. 2019) holds the key not only to reversing biodiversity loss but also to improving food security. Investing in mountain ecosystems not only improves agricultural production, but improves livelihoods downstream (IUCN n.d.).

AN AGRICULTURAL TRANSFORMATION

Each of our transformation hotspots features a shift from a current, unsustainable state toward a more sustainable state. For our first hotspot – Sustainable, local food systems – I want to take the time and illustrate how such a transformation can be visualized. Figure 7.1 builds upon an adapted framing from the International Assessment of Agricultural Knowledge, Science and Technology for Development (IISTAD) (McIntyre et al. 2009). The figure represents a paradigm shift where we are moving from a current conventional system that is high in productivity yet unsustainable to a more sustainable system (see also Brescia 2017).

Using the example of organic agriculture (cf. Ledermann 2012), Figure 7.1 captures the initial reduction in productivity over the transition period (in organic agriculture referred to as the 'conversion' period from non-organic to certified organic) before recovering to a similar or higher state. A farmer who stops using chemical pesticides and fertilizers will undergo a period whereby yields are lower since their whole farming system is adapting to new techniques (e.g., substituting synthetic with organic fertilizers). In organic agriculture, transition support comes from the state or higher transition prices from buyers. Especially in low-input environments, such a transformation from conventional to organic

Figure 7.1 Sustainable food production systems, adapted from McIntyre et al. (2009) and Brescia (2017)

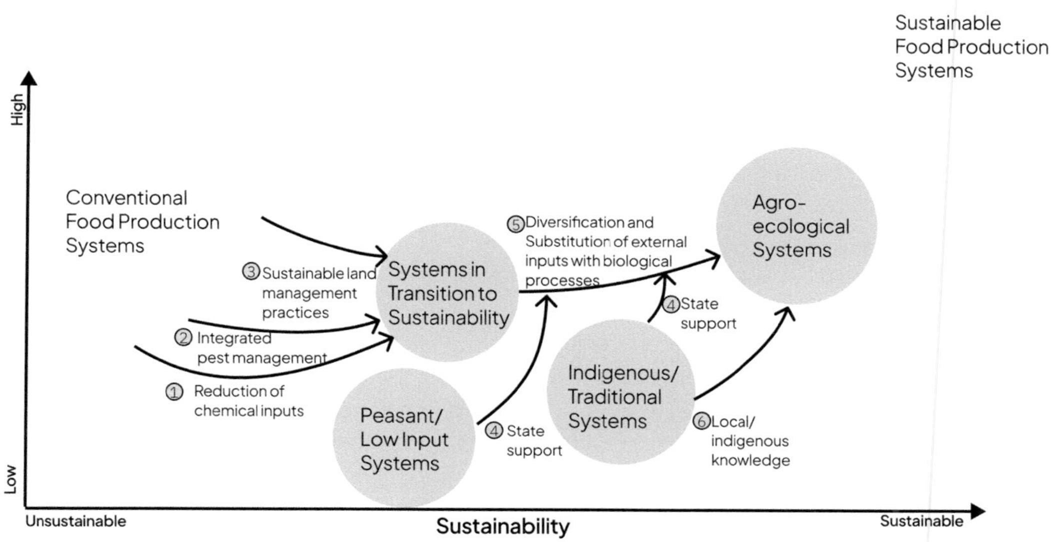

agriculture does not have to end up with a lower overall yield that is solely compensated with an organic price premium (Schader et al. 2021). The organic yield debate, ongoing for two decades (e.g., Avery 2007; Badgley et al. 2007), shows that organic and conventional yields ultimately are comparable (e.g., Seufert and Ramankutty 2017).

The transition challenge is captured in points 1–3 and 5: in the first step, a move from a conventional system involves a reduction of chemical inputs (1), including a focus on greater adoption of integrated pest management (2) and sustainable land management practices (3). Once reaching the state of an agricultural system in transition to sustainability, increased diversification and substitution of external inputs with biological processes (5) are the next steps. These changes take place not only at the plot-level, but include moving toward a more holistic, farm- or landscape-level approach that leads to agro-ecological systems.

Having followed the dominant path of transition from conventional to sustainable systems, the *peasant/low input systems* and the *indigenous/traditional systems* are marked by lower levels of productivity yet marginally higher levels of sustainability. A transformation in agriculture will need to draw on state support (4), as they have been under-funded or ignored, given their status as non-commercial smallholders.

Figure 7.1 covers the agricultural transition from a production and sustainability perspective. Achieving a transformation to local, sustainable food systems, however, moves well beyond production. We have already discussed the amount of food being wasted – in consumption, during transportation, and in production. Figure 7.2 zooms out and provides a greater perspective on how food systems operate and need to be transformed.

Beyond sustainable food production, more sustainable, local food systems cover five additional circular steps: food processing, distribution, access, consumption, and resource and waste recovery before feeding back into food production (see Kopainsky, Tribaldos, and Ledermann 2017, for a system dynamics perspective). Cross-cutting all these steps is the endeavor to push toward the local to strengthen resilience and self-sufficiency.

Food processing

At the processing stage, waste reduces the amount of food produced that is translated into actual calories consumed. This stage furthermore captures the challenge of limited capacities in certain regions to process the food. An example is processing cocoa into chocolate bars (e.g., in Ghana; Pilling 2021) or dairy milk into yoghurts. These steps create value addition that can increase demand for products, lift prices for producers, and with exports, improve the terms-of-trade. A frequently referenced case to improve terms-of-trade is processing raw

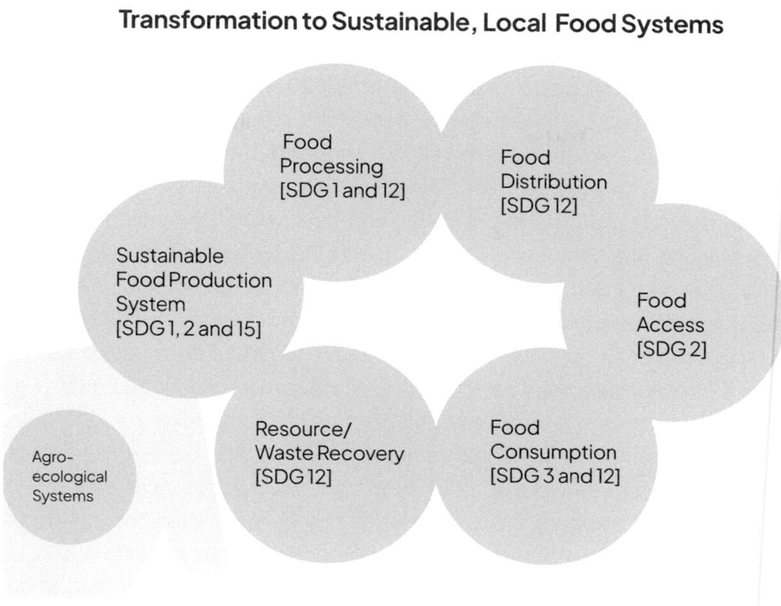

Figure 7.2 Transformation to sustainable, local food systems, adapted from https://cdn3.locable.com

materials, such as cotton, into clothing to further economic growth (e.g., Pilling 2024). Processing depends also upon access to additional imported ingredients or material. Take the yoghurts: you need access to machinery or, even more basic, you need access to sanitized cups.

Human ingenuity, however, is strong: I recall being in Central Kenya on a monitoring trip visiting a project implemented by KALRO – the Kenyan Agriculture and Livestock Research Organization. Working with smallholder farmers, the project supported a transformation of their livelihoods, introducing new ecological technologies, including Push-Pull (see Case Study 7.1) and biogas. The slogan at the time – apt for Kenya's entrepreneurial spirit – was *Agriculture as a business!*. As farmers increased their dairy productivity through integrating improved fodder crops for their livestock, they also were eager to increase their value-addition (which was part of KALRO's mandate). Farmers shared how they made additional money selling their milk in local markets. But they also learnt how to process the milk into popular flavored yoghurt drinks. While in the process of obtaining the national food inspection certificates, women farmers proudly presented their yoghurts to me in simple, unbranded plastic bags. Their drive was based on teamwork. Farmers organized themselves into cooperatives, as investments otherwise into necessary equipment would have been beyond their reach. And together, they could dream of entering the marketplace with value-added goods.

Food distribution

A cornerstone of sustainable, local food systems is creating more resilient, shortened supply chains. The global pandemic from 2020 to 2022 laid bare the risks inherent in depending on long-distance trade for essential goods. Food prices rose during COVID-19, putting a strain on net importers (Terazono 2021). Climate change is keeping the prices high (Savage 2024). In the USA, the pandemic resulted in backed-up ports, freight trains, and overworked long-distance truckers, accentuating strains in supply chains (Lynch 2021). The lack of cold storage and collapse in demand from their customers in restaurants, schools, and hotels resulted in a spike of waste. At the early stages of the pandemic, *The New York Times* reported that dairy farmers in the USA were forced to dump up to 3.7 million gallons of milk every day. They cite a single chicken processor that destroyed three-quarter of a million unhatched eggs each week (Yaffe-Bellany and Corkery 2020). These immense food wastes rattled consumers and policymakers alike, given that grocery stores were short of supply and food prices increased.

More local food systems are expected to not only increase the resilience during crises but also contribute to reducing greenhouse gas emissions by cutting down transportation. Latest accounting for carbon emissions finds that so-called food miles amount to around 3.0 $GtCO_2e$ (or 3.5–7.5 times higher than the previous estimates), or one-fifth of total food-system emissions (Li et al. 2022). Except for airfreight, localness alone is not a determinant for lower greenhouse emissions (Stein and Santini 2022, 79): container ships or trains can have lower emissions than small trucks due to economies of scale.

Matching local supply with demand is at the core of food distribution. In sustainable development, the stress frequently is placed on limited local supply of a set commodity. Interventions hence focus on how to increase the productivity without negatively impacting the environment (recall transition points 1–3 in Figure 7.1). By now, development interventions have found a myriad of avenues to do just that. Popular interventions from the menu of development practitioners include: leveraging more complex biological interactions (e.g., see the Push-Pull Case Study 7.1), composting or integrated soil fertility management (ISFM), bio pesticides or integrated pest management (IPM), soil erosion measures or agroforestry. While the impacts of projects might differ – some targeting hunger (SDG 2), with others focusing on improving life on land (SDG 15) – they usually share the long-term objective to increase productivity in a rural context. The same goes for efforts to transform more sustainable, local food systems, as we are looking to support a more self-sustaining, holistic landscape of agricultural production.

Development projects risk overlooking a fundamental challenge: solely increasing local supply of a single commodity might be detrimental to the local economy and result in weakening the intervention's effects on poverty (and health as discussed next in '*Food access*'). In rural areas with poor transportation and limited market access, increasing yields can lead to lower prices.

For instance, where maize is a staple crop in sub-Saharan Africa, more supply without trade depresses prices, causing inefficiencies and reducing the cost-effectiveness of interventions.

As a result, when transforming rural or peri-urban landscapes into sustainable, local food systems, we need to focus on increasing local supply and ensuring that avenues exist for farmers to trade their excess produce beyond their existing markets. Failure to do so would not only have detrimental effects on their ability to gain additional income for their families, but also increase unnecessary (post-)harvest losses. More profoundly, it reinforces that if we throw all our eggs into one basket, we often fall short of creating sustainable, holistic development. Increased diversification, as opposed to maximizing production for a single commodity, can counter the above and create added benefit of increasing resilience to shocks. Research and development efforts are underway in Africa to improve food security and nutrition using historically neglected, underutilized species (so-called orphan crops) (FAO 2024a, 2024b).

Food access

Following up on our discussion above, producing more maize without having an export outlet, would make it ultimately more affordable to local households. Maize represents a staple food for around one billion people globally and contributes up to 50 percent of their daily diet (Poole, Donovan, and Erenstein 2021). Globally, three crops (maize, rice, and wheat) combine to make up approximately 60 percent of all calories consumed (Willett et al. 2019).

Increased access will be welcome in areas where food insecurities dominate. For a balanced diet, people need access to non-cereals, including roots, vegetables, fruits, legumes, soybeans, and nuts or seeds (Willett et al. 2019). If waste is reduced (see also *Waste recovery*) and we shift to a healthier diet, maize production is projected to decrease as opposed to increasing by approximately 70 percent by 2050 (see Figure 7.3). Legume production, in response, would increase the most, over 200 percent, by 2050. Therefore, a transformation to more local, sustainable food systems involves also empowering farmers to diversify their production.

One avenue to achieve a transformation through improved food access is the establishment of fruit tree portfolios. As advanced by the World Agroforestry Center (ICRAF) in projects in Western and Eastern Kenya, integrating edible fruits, seeds, nuts, and oils can address challenges in seasonable availability and improve nutrition, including micronutrient gaps, throughout the year (McMullin et al. 2019). The portfolios were identified based on three factors: food production diversity and seasonality; monthly breakdown of food insecurity and harvest months for food; and existing dietary gaps.

The resultant portfolios aim to fill any harvest gaps and address micronutrient shortcomings through a more diversified portfolio. In the example of

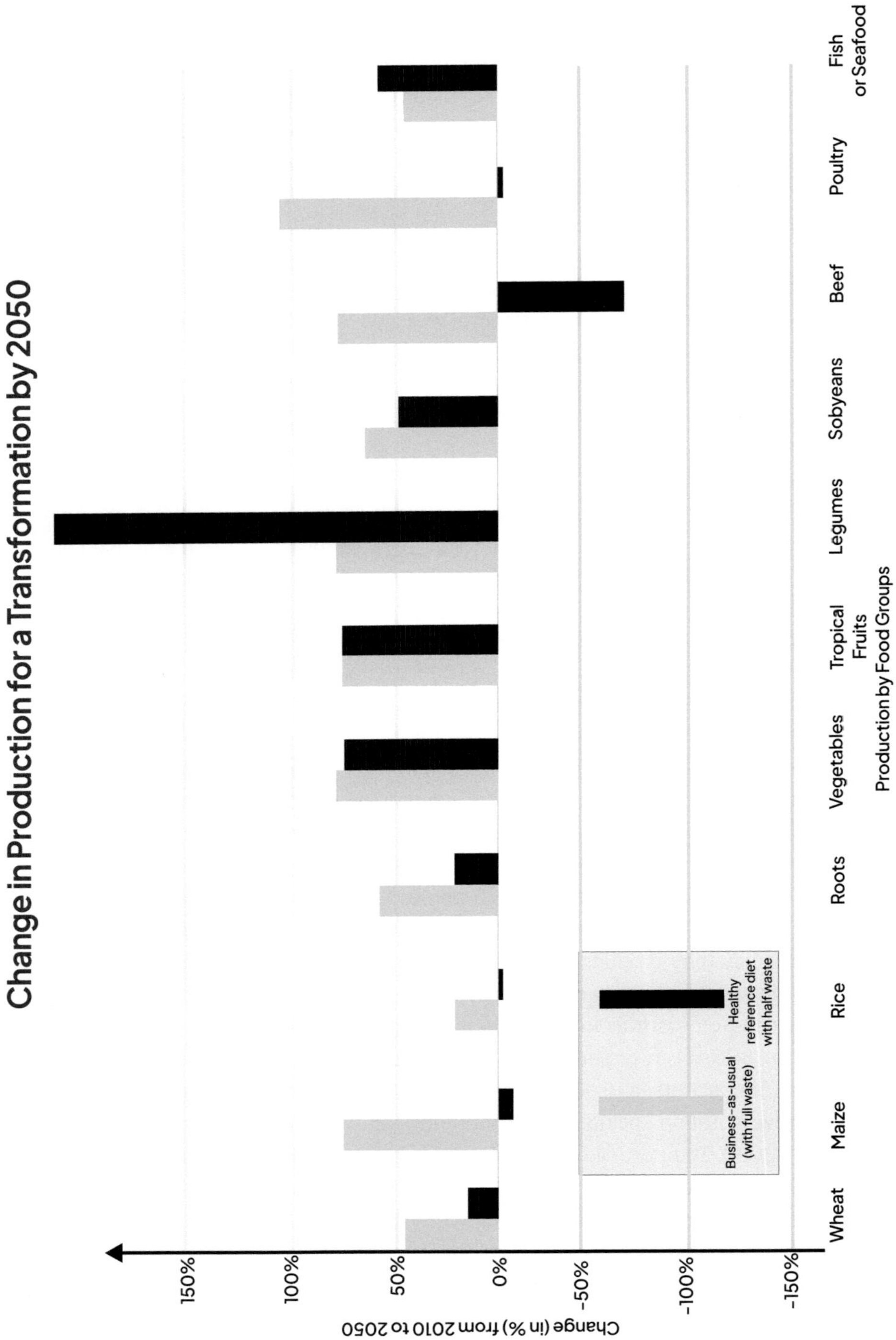

Figure 7.3
Change in production for a transformation by 2050, adapted from Willett et al. (2019)

Ngobit in Laikipia County, Kenya (World Agroforestry et al. n.d.), the average farm size is 1 hectare, with six family members and only three food trees. Some 92 percent of households experienced food insecurity in a 12-month period. The average farm has limited-to-low diversity, with one food tree species, two pulses, two staples, and three vegetables grown on average. As illustrated in Figure 7.4, the three months with the highest food insecurity are December, January, and February. It is during those periods specifically that farmers identified passion fruits, bananas, potato, peas, and beans as priorities to invest in. Additional high iron, vitamin A or C options include vegetables (such as the stinging nettle or pumpkin leaves, as well as tree tomato/tamarillo, custard apple or lemon for fruits, and cowpeas for pulses) (World Agroforestry et al. n.d.).

Beyond production, farm and off-farm income and improved processing and storage determine food access (cf. Pray and Ledermann 2016). As noted by Tilman and Clark:

> Unless the nutrition transition that is under way is changed, diabetes, chronic heart disease and other diet-related chronic non-communicable diseases will become the dominant global disease burden, often affecting even the poorer members of poorer nations for whom appropriate health care is unavailable.
>
> (2014, 521)

For a transformation to occur, food access in local systems needs to ensure that the dietary choices are accessible. Mediterranean, Pescatarian, and Vegetarian diets have already been shown to significantly improve human health (SDG 3),[1] reduce agricultural emissions (SDG 13), and reduce land pressures (SDG 15) (Tilman and Clark 2014). The challenge, of course, is to identify and elevate healthy dietary choices in a world of divergent tastes and incomes.

Food consumption

Consumption in local food systems is determined by limiting factors, such as the ability to process and store (Kansanga et al. 2025). Given these limits, a local food system is pushed toward offering consumers more limited choices that depend on seasonal patterns of food availability.

The move toward more local production and consumption already gained traction pre-COVID with campaigns to buy locally and seasonally to reduce food miles. Vegetable and fruit consumption – driven by consumer demand for non-seasonal products – contribute around one-third of total transport-emissions in the global food system (Li et al. 2022). More local production and consumption also got a boost by antitrust action taken during the pandemic against price gauging by a highly concentrated food supply chains (Foroohar 2023), with 80 percent of the US beef market, for example, controlled by four companies.

An example from my previous coworkers is their CLEVER project, a traveling exhibit to increase education and awareness for sustainable consumption

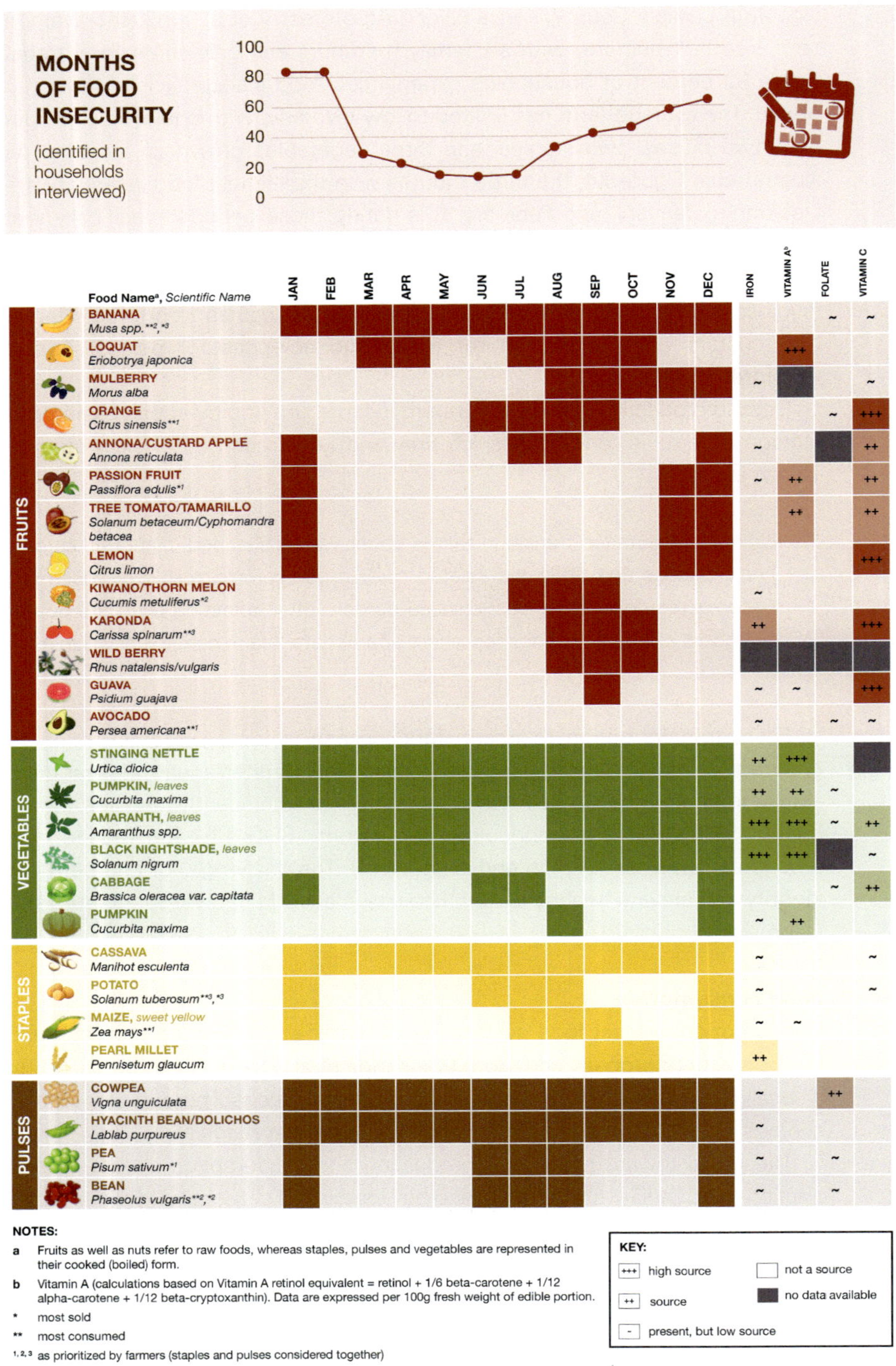

Figure 7.4
Nutrient portfolio in Ngobit, Kenya, World Agroforestry Centre. 2019. Nutritious Food Portfolios for Targeting Year-Round Food Harvest and Nutrient Gaps – Rumuruti. Nairobi: World Agroforestry Centre.CC BY 4.0

Figure 7.5 A mobile supermarket for educational purposes, Peter Lüthi/Biovision Foundation

(Biovision Foundation 2024a). In a movable shipping container (see Figure 7.5), you have a model of a small Swiss supermarket, outfitted with educational material and non-perishable replicas of produce and products. The pedagogical experience is gamified, by asking shoppers to fill their shopping basket with a set number of ingredients based on a recipe as their mission. Their shopping decisions – e.g., an organic greenhouse tomato from Spain versus a conventional one grown in Switzerland – are then subsequently scored across six dimensions: climate, pollution, resource use, biodiversity, livelihoods, and social compatibility and animal cruelty.

To increase the draw, local celebrities are used in the competition, with shoppers receiving a receipt of their tally and whether they bested the celebrity in their decision making. Staff is on hand to offer additional guidance and allow for a deeper dive. The program is marketed specifically to school classes, as it can be integrated into an existing teaching curriculum via a field trip. A latest version also offers the experience via a webshop (clever-konsumieren.ch) and teaching materials to develop specific school curricula developed in collaboration with the National Competence and Service Centre for Education for Sustainable Development. It also features tips on how to increase sustainable consumption, including addressing food waste.

The project captures well the interplay between our changemakers in policymaking, finance, individual and collective action, and science and technology: the mobile supermarket is creating more engaged citizens by demystifying the effects of their individual actions through consumption based on a scientific analysis of each decision. The creation of increased awareness and sensibilities feeds into complementary initiatives around *Food Future Switzerland*, which stimulates citizen-led discussions creating holistic policies on local, sustainable food systems, and larger global initiatives through the Swiss SDSN chapter (Biovision Foundation 2024b).

Zooming out from Switzerland, any local food system needs to compete against processed food distributed by large multinationals. Andrew Jacobs and Matt Richtel wrote a rich exposé on the rise of Nestlé in Latin America, and the contribution of processed foods to obesity and malnutrition (Jacobs and Ritchel 2017). According to Nestlé's internal assessment of their food and drinks by revenue, more than half failed to meet thresholds set for healthy food (Evans 2021). Professor Monteiro from the University of São Paulo, Brazil, illustrates the challenges around improving nutrition and public health:

> What we have is a war between two food systems, a traditional diet of real food once produced by the farmers around you and the producers of ultra-processed food designed to be over-consumed and which in some cases are addictive ... but one food system has disproportionately more power than the other.
>
> (Jacobs and Ritchel 2017)

The quote captures well that a transformation won't just produce winners. If business-as-usual is no longer an option, it inevitably entails a shift in power. To consumers, to farmers, or to innovators that translate waste into use.

Waste recovery

A first step in food waste recovery post-consumption includes composting. Initially established as a small startup from North Carolina, USA, in 2011, CompostNow (2025) has expanded to divert over 91 million pounds of food waste in the region, generating 30 million pounds of compost and nearly preventing 10 million pounds of methane emissions since its founding. Their service is emblematic of how a transformation can loop back to sustainable production, as their compost is allocated to over 100 local gardens and farms (CompostNow 2025). Additional efforts at recovery include donation of food or converting food into animal feed. Leveraging data can help: a grocery store in a wealthy Kansas City suburb in the USA reduced food waste going to a landfill by over 1,000 percent in a single year, as well as increasing its donations by 1,300 percent, using zero waste scorecards (EPA 2024).

Avocados offer an innovative example of pre-consumption waste reduction. Notorious for their high water demand (and as a hip breakfast when served on toast), avocados that did not meet market standards are disposed of in the field or discarded. The *Chosen Foods* company has built a business-model out of transforming the waste into processed avocado oil, generating extra income for farmers and meeting increased consumer demand for alternative cooking oils (Masters 2024).

Efforts at waste recovery can also be linked to the circular economy movement. The cornerstone of the circular economy centers the practice of reusing, refining, or recycling materials to minimize waste, raw materials consumption, subsequent pollution, and climate emissions (Independent Group of Scientists Appointed by the Secretary-General 2019, 62). Recycling, waste minimization, waste management, and generating energy from waste present a US$1 trillion investment opportunity that can have a significant impact on SDGs 13, 14, and 15 (Channell et al. 2022). It represents the largest out of all investment opportunities identified (Channell et al. 2022), demonstrating the transformation's (untapped) potential to address holistic development and viability of the finance and economy changemaker.

WHAT MATTERS MOST

Willett et al. (2019) identify five strategies to achieve large-scale food transformations. Activating our changemakers, they include securing commitments for shifts to healthy diets, to intensifying production of healthier food, improved governance of natural resources, and reducing food loss and waste (Willett et al. 2019, 478–483).

Just as agriculture is context-dependent, the effects of more local food systems on resilience are also highly context-dependent (Stein and Santini 2022, 81). Out of 40 papers reviewed by Stein and Santini (2022), the results in Table 7.1 capture the amount that more local systems can contribute to sustainability.

Not a single study reviewed by Stein and Santini found that a local food system increases food security throughout; it always depends on the context. The current conventional production paradigm is poorly set up to transform fully into a local food system tomorrow. Less than a third of the world could satisfy its food demand locally (Stein and Santini 2022, 81). A classic limitation is animal feed, which is rarely sufficient in Europe to support current livestock levels.

Begging for more radical or integrated transformations, let us turn our attention next to the Case Study 7.1 of Push-Pull, which offers its greatest value to crop-livestock farmers.

Table 7.1 Local food systems as solutions? adapted from Stein and Santini (2022)

Local food systems' contribution to...	**Yes**	**Some**	**Uncertain**	**No**
GHG emissions	1 [4]	0	19 [68]	8 [29]
Food security	0	2 [13]	3 [20]	10 [67]
Economic sustainability	6 [40]	1 [7]	8 [53]	0

Note: Percentages in square brackets.

CASE STUDY 7.1 PUSH-PULL

Push-Pull is an ecological agricultural practice to improve both productivity and resilience of smallholder farmers in sub-Saharan Africa, which is explained in detail later. Professor Zeyaur Khan and his team of scientists developed the practice at the turn of the century at their *icipe* research station in Mbita Point, Western Kenya. Currently, over 200,000 farmers in Eastern Africa and beyond have adopted the practice. With its entry point of sustainably increasing maize and sorghum productivity, the practice, for example, has spread to Ethiopian smallholder farmers as depicted in the Tolay region in the Part II Overview.[2]

Plant science and indigenous technology for ecological development

A picture says a thousand words. In our case, look at Figure 7.6.

Most first-time observers of Push-Pull might mistake the field as being completely weed-infested, when in reality it is highly productive and doubling yields for its owner.[3] Push-Pull was originally invented with an attempt to reduce losses from the maize stem borer, a type of moth. Stem borer damage can amount to 50 percent of yield losses on a smallholder farm. Economic losses across he continent are estimated at more than US$1 billion a year. The stem borer is notoriously difficult to control, given its damage is done by nourishing itself inside of the stem. It is attracted to the maize by its volatiles (colloquially, plant perfumes). The *icipe* scientists identified a leguminous companion crop, desmodium, that exudes volatiles that end up making the plot no longer attractive when intercropped. Intercropping desmodium *pushes* the moths away before they lay eggs (see Figure 7.7). To stop moths from moving to the neighboring field, a complementary *pull* effect was devised to draw them away from the maize. Trap crops, such as Napier grass, exude the right plant volatiles. The plants – when feasted upon by the hatched stem borer larvae – release a sticky substance which traps the larvae and results in a natural control for around 80–90 percent of the pest population.

Serendipitously, *icipe* scientists identified that the chemicals released from the roots of the desmodium plant suppress *Striga* weed – also known as a witchweed – from attaching itself to the roots of the maize plant. Twenty-four percent of fields across sub-Saharan Africa are infested with Striga. Farmers who have a large infestation can face almost complete harvest loss, as the parasitic weed attaches itself to the maize plant and competes for plant nutrients. Thousands of these seeds rest dormant in the field until they find a suitable host plant. Planting desmodium reduces the Striga seed bank, causing suicide germination as the plant cannot find a host.

Push-Pull combines innovative ecological science to identify the invisible plant volatiles and long-used, traditional knowledge of intercropping to establish a sustainable solution. The adoption benefits from using perennial crops to establish the intercropping and the border cropping. The plants stay healthy for at least five years, so farmers do not have to reinvest in obtaining the (novel) seeds each year. The practice is furthermore enhanced by being technology-agnostic: a farmer can use any type of maize they prefer, from open-pollinated varieties (OPVs) to hybrid seeds that are optimized for higher yields. Thus, it represents an inclusive, predominantly knowledge-intensive intervention compared to conventional counterparts that focus on capital-intensive inputs such as fertilizer or hybrid seeds to increase yields.

Figure 7.6 All weeds or a healthy solution to sustainable development? A Push-Pull field in Western Kenya

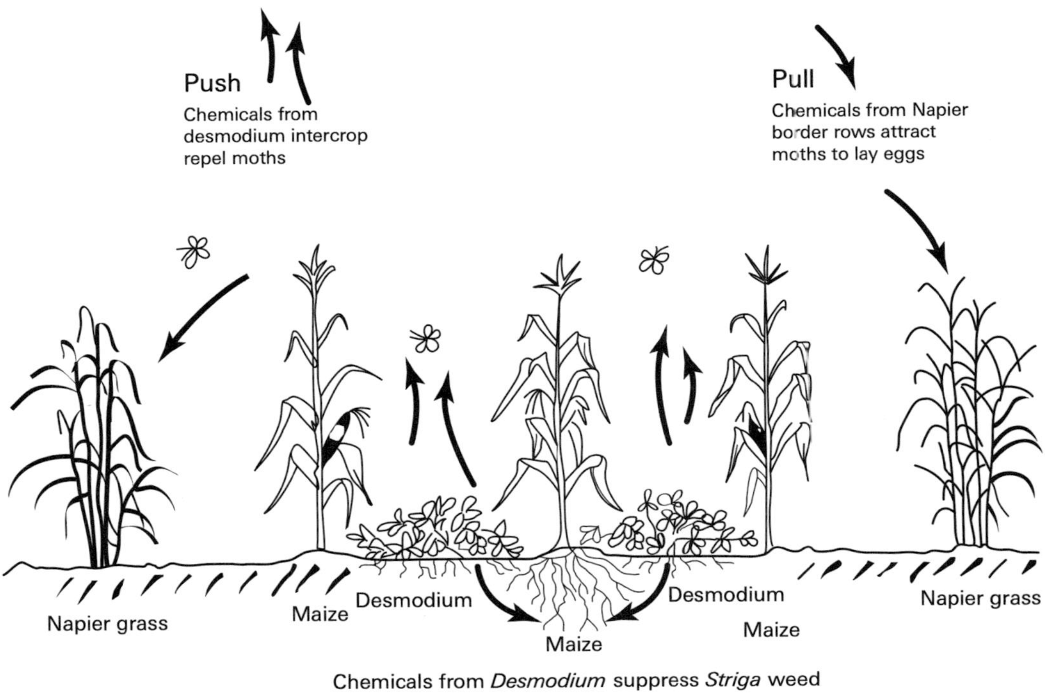

Figure 7.7
The Push and the Pull explained, The Gatsby Charitable Foundation, The Quiet Revolution: Push-Pull Technology and the African Farmer

CONTRIBUTING TO ACHIEVING THE SDGs

Together with Zeyaur Khan, we mapped Push-Pull's contributions to achieving the SDGs in Figure 7.8 (Ledermann and Khan 2017) (see Case Study 7.1). We identified six relevant SDGs: SDG 1, 2, 3, 5, 13, and 15. They are presented in order of relevance below.

SDG 2: Zero Hunger. Controlling the major pests of stem borer and Striga increases yields for smallholder farmers. In Kenya, maize yields increased from less than 1 ton per ha to at least 3.5 tons per ha. A further innovation of Push-Pull included integrating sorghum as a more drought-tolerant variety, with yields increasing from less than 1 ton per ha to 2.5 tons per ha.

SDG 1: No Poverty. From an income perspective, the labor-intensive investments by farmers into Push-Pull (with access to the initial companion plants of Napier grass and desmodium) pays off. Farmers more than double their investments (2.2:1), compared to negative returns from their traditional monocropping strategy. Revenue gains are almost three times larger at US$957 per ha per year for Push-Pull and US$351 for the control. As the companion plants are excellent livestock fodder, farmers who have access to dairy cows, for example, can double their milk production and further diversify their income sources.

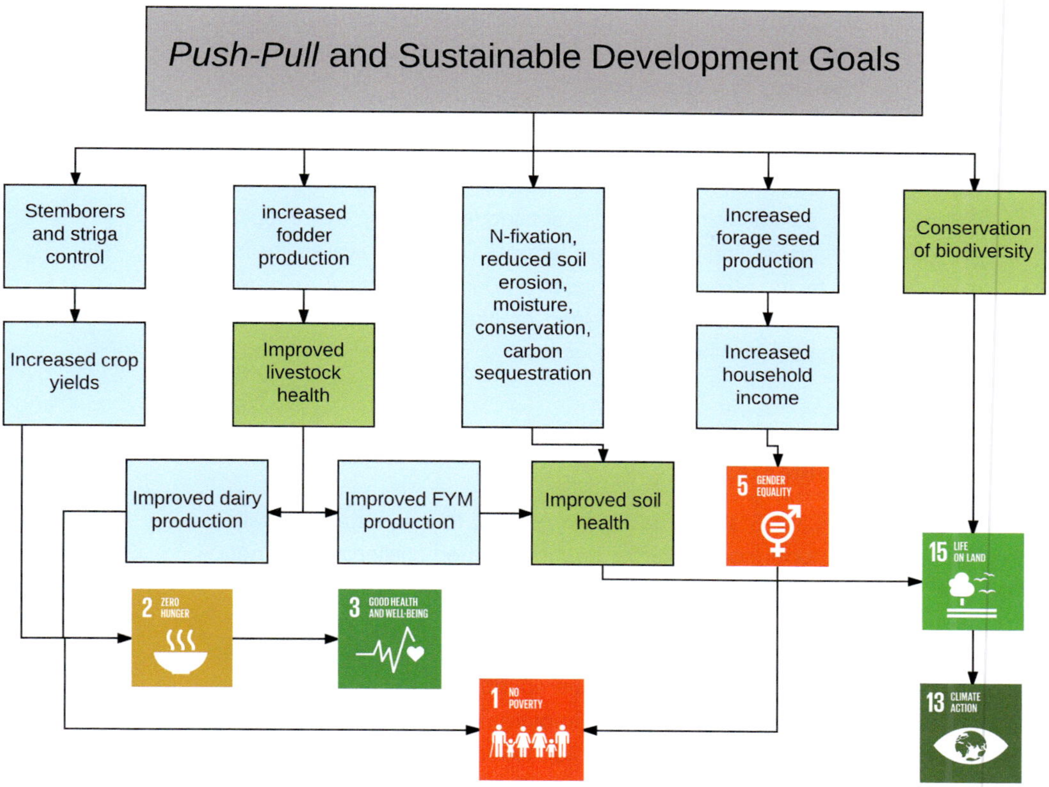

Figure 7.8 The logic of Push-Pull and its contribution to achieving the SDGs, Ledermann and Khan (2017)

SDG 3: Good Health and Well-Being: Farmers increase their domestic production of maize or sorghum and their income for spending on food items. If they have access to dairy cows or goats, the increase in milk production will translate into additional nutritional gains and a higher degree of self-sufficiency. Maize produced under Push-Pull also had lower levels of mycotoxins (Njeru et al. 2020), molds that are linked to acute poisoning and long-term cancer (WHO 2023).

SDG 13: Climate Action: Growing perennial desmodium as an intercrop negates the need to plow the field, which increases the overall carbon sequestered. Later adaptations of Push-Pull included climate-resilient versions, that use drought-tolerant green leaf desmodium and replaced Napier grass with Mulatto grass that performs better during drought stress (Ndayisaba et al. 2023).

SDG 15: Life on Land: The perennials planted have positive effects on the environment. Desmodium, as a leguminous plant, fixes nitrogen in the soil, which decreases the requirement for synthetic fertilizer on the field. Desmodium and Napier grass also help to control soil erosion. As a cover crop, desmodium improves soil moisture conservation (by shading the soil). Healthier livestock produce more high-quality farmyard manure (FYM), which further improves soil health. By not needing to use broad-spectrum synthetic pesticides, more beneficial insects can establish themselves, offering their services to manage pests.

SDG 5: Gender Equality: Scaling efforts of Push-Pull targeted women who in select regions were the majority of adopters as their husbands often worked as day laborers. While the initial investment in establishing a Push-Pull plot required additional labor, in subsequent years, the adoption proved to be labor-saving for women farmers. As seed availability is a major limiting factor (especially for income-poor households), *icipe* pursued female-led community-based seed multiplication efforts, allowing women to earn additional household income.

MISSING PIECES FOR SUSTAINABLE, LOCAL FOOD SYSTEMS TRANSFORMATION

The logic outlined in Case Study 7.1 takes a large step toward a transformation to more sustainable, local food systems. Two interrelated challenges are noteworthy, as they have revealed themselves as significant barriers to more widespread adoption:

First, as a knowledge-intensive technology that can be shared between neighbors, building capacity for adoption is paramount. Push-Pull outreach efforts ranged from the traditional farmer field day, to drama groups in local villages, to comic books for school children (see Figure 7.9), to radio programs. Besides the labor-intensive nature of adopting Push-Pull in the first year, farmers need access to capital to secure the relevant companion seeds for the intercropped legume and the border crop. If they are fortunate, these can be obtained as cuttings from other farmers nearby. But if they hear about it through a traveling troop of local drama actors or through a convincing radio program, a farmer's next step will be to secure access to these seeds. Best case, a farmer has the cash and can pay for the seeds from a local agrovet shop. But private sector actors have lower incentives to invest in perennial seed production compared to hybrid seeds that need to be purchased annually in order to generate higher yields. For this reason, the worst case is that farmers fail to find supply for the seeds and remain dependent on subsidized or free supplies offered through time-bound development projects. While seed multiplication represents a pertinent avenue, desmodium seeds only produced viable seeds in specific climatic conditions. The location of Push-Pull dissemination and its agro-ecological limits hence determine the viability of creating a self-sufficient food system transformation.

Second, intercropping a leguminous crop runs counter to the dominant strategies of female or poorer farmers for increasing nutrition security within their households. Their preferred approach is to plant edible beans or vegetables. Livestock fodder is of lesser use as marginalized households fail to own livestock. They can sell the desmodium and Napier grass at the market, but it involves an extra step. For these reasons, the missing holy grail appears be to intercrop an edible bean that holds the same properties as desmodium.

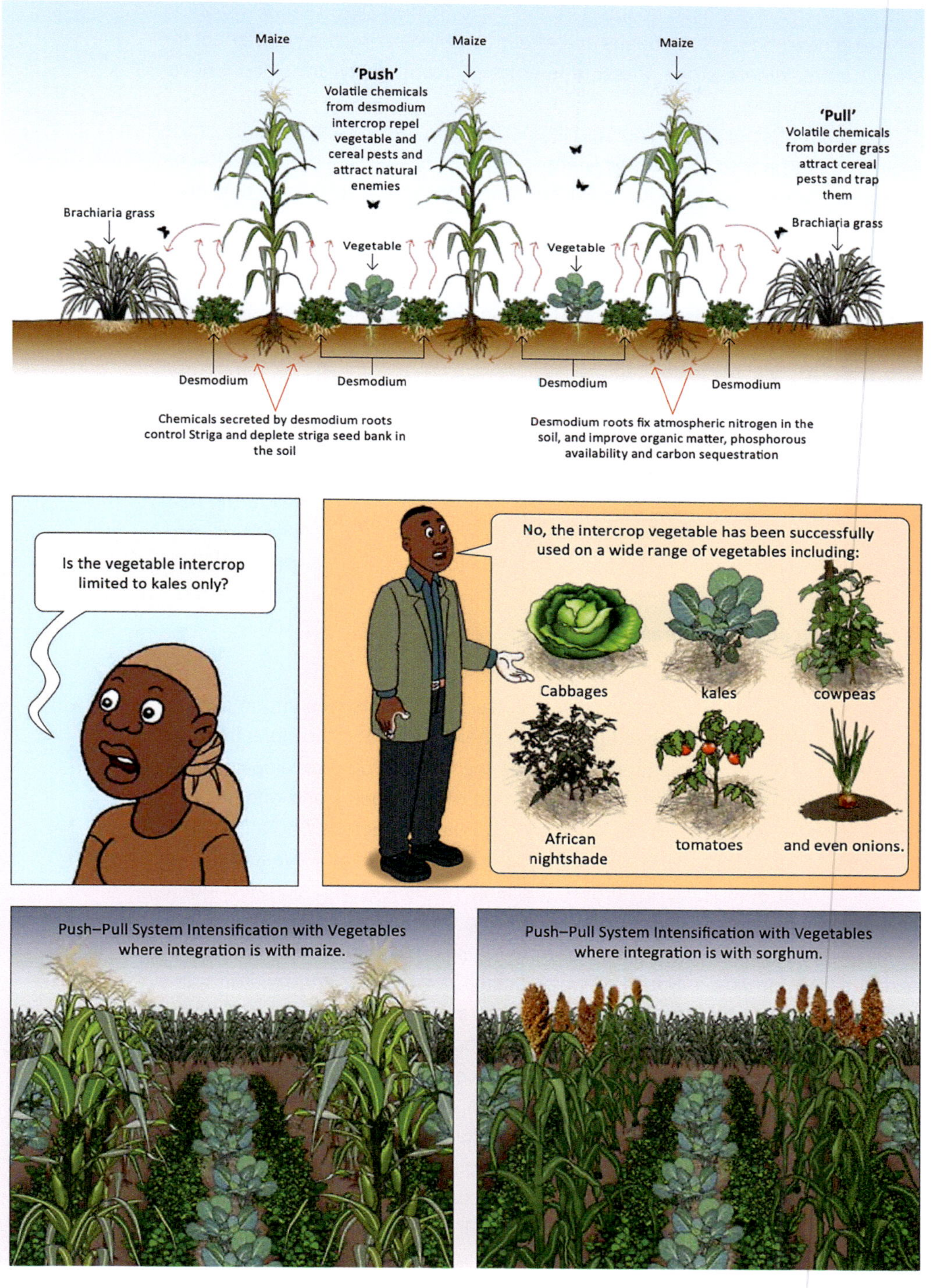

Figure 7.9
Comic book showing integration of vegetables into Push-Pull farming, *icipe*

Farmers themselves have adapted Push-Pull to fit their own needs, including replacing desmodium with beans and effectively eliminating the push effect for stem borers. Where striga pressure is not significant, such farmer-led innovations can be viable.

The team of scientists, dedicated to ensuring the viability and integrity of the Push-Pull mechanism, continuously are innovating. Nearly two decades after its original innovation, adaptations of Push-Pull are closer to reaching the holy grail. Although still searching for a leguminous crop that mirrors the biological pest control properties of desmodium, sustainable intensification has been advanced by relaxing the intensity of desmodium intercropping. As illustrated in the comic in Figure 7.9, vegetables intercropped without negating the volatile properties include African nightshade, kale, tomatoes, and onions.

IMPROVE YOUR PRACTICE: THE ROLE OF OUR FOUR CHANGEMAKERS

In this section, we translate the findings from the case of Push-Pull to the relevant changemakers. I highlight how each changemaker was essential in supporting the transformation and a future avenue to strengthen them.

Governance and policymaking

What made it work

At the origin of Push-Pull is world-class research that translates into impact on the ground. With over 150 scientific, mostly peer-reviewed publications from the Push-Pull team, the intervention rose above the ranks of other ecological intensification methods as the evidence generated of its impacts and effects were overwhelming. My former colleagues scored Push-Pull across *FAO's Elements of Agroecology* based on this research and found that it achieves very high scores around leveraging synergies, building resilience, and contributing to diversity (Agroecology Pool 2018). Leveraging donors, such as my previous employer, and other supporters who were seeking flagship programs that represent a transformation, Push-Pull was placed as an example for transformative action at the global levels, including at UN fora and reports (e.g., UN General Assembly 2010). As a faction within the UN Food and Agriculture Organization (FAO) pivoted toward sustainable and agro-ecological production in the aftermath of the global food crisis in the late 2000s, Push-Pull featured prominently in their publications and high-level fora. While continued research risks diminishing returns after the initial scientific breakthrough, it established the crucial foundation upon which any efforts at scaling (and raising additional donor funding) were built. The technology-agnostic nature of Push-Pull also enabled the *icipe* team to align itself with different waves of (agricultural) development priorities, from the MDGs to climate-smart farming.

What could strengthen the transformation?

The successful establishment of Push-Pull on the global stage translated into a continuous flow of donor funding. While efforts were made in Ethiopia and Kenya, Push-Pull struggled to gain a foothold in local and national government policies and priorities around sustainable transformation. Part of this challenge relates to a decline in government funding for agriculture itself. In other parts, it represents the ongoing challenge that most governments have yet to fully embrace a more transformative path to development. A prominent case is the government of Malawi, which heavily invested in input subsidies (including fertilizers) for farmers, as opposed to more knowledge-intensive technologies.

To take that next step in its transformative path, one would need to move beyond research and development, and implementation, toward advocacy. Innovative natural scientists can bridge skills gaps for implementation by hiring effective staff. Skills gaps are more pronounced in advocacy, as research teams are rarely well positioned to make the difference themselves.

A lesson from outside of my work in Push-Pull but exemplary of the limitations at hand: I supported and monitored a leading applied research program with a well-developed logical framework that captured the theory of change and objectives to be achieved. The multi-year plan translated into tangible results, including novel scientific discoveries that address knowledge gaps in the development literature. After publications are submitted and published, the final step (representing a final three months of the entire work plan), focused on sharing the findings with policymakers. Just as sustainability and handover to local stakeholders are usually an afterthought, engaging policymakers and translating the findings into a larger impact beyond the scientific community also demand planning from a project's start. Research findings compete in the marketplace of ideas. If our aim is to overcome the existing bias with the limited government funding available, it is crucial that (local) government stakeholders are embedded from the earliest stages to play an active role in project reviews and advocacy. Failure to do so is unlikely to rattle a national system that is locked in on business-as-usual.

Economy and finance

What made it work

When you engage in non-patentable research, such as Push-Pull, you lack the ability to extract rent from your innovation through patents or other forms of intellectual property rights. Beyond patents or property rights, most research institutes and NGOs are, by definition, non-profits, so they are limited in charging recipients or beneficiaries any co-contributions. This issue extends to their offering of services to farmers (e.g., training), where in-kind contributions (e.g., land) are the accepted standards for participating farmers. Broadly framed, in knowledge-intensive agricultural transformations, we deal with public goods deserving of donor funding. At the core of its success is the realization by donors that efforts like Push-Pull need such funding to pilot and validate the feasibility of their work.

Once we move toward scaling, the issue becomes more complex as other NGOs can hire organizations for capacity building and training of their staff (via a sub-award). They are still largely dependent on grant funding. Private actors in the seed industry have limited financial incentives to invest in further R&D efforts themselves for perennial crops. Business-as-usual also stifles research on other ecological avenues to improve maize productivity. *icipe* scientists, for example, identified that open pollinating varieties of maize send signals to attract beneficial insights in case they are infested with pests (Tamiru et al. 2011). Such an ingenious trait, however, was lost in decades of breeding focused on maximizing productivity. It is unlikely that the incentives exist to reintroduce such traits into future varieties, especially as some seed companies market their own pesticides. That is why public funding is crucial to support more advanced research that can illuminate these additional avenues for sustainable intensification.

What could strengthen the transformation?

While funding in agriculture is in retreat, funding gaps exist particularly in the missing middle (between US$10,000 to 1 million) in sub-Saharan Africa. New initiatives, such as FASA covered in Chapter 4, offer to invest up to US$200 million into agribusinesses across the continent. It is likely, however, that such important funding initiatives may fail to raise more agro-ecological, transformative options for two main reasons: lack of demand and low visibility of ecological alternatives.

On the first issue of demand, while Push-Pull has reached over 200,000 farmers in two decades, demand for seeds, for example, is too fragmented and struggles to generate significant aggregate demand. This issue showcases the limits of development-funded interventions, as they ultimately need to scale by receiving non-grant funding or being embedded into larger (governmental) initiatives. For-profit organizations that provide farming inputs and services to smallholder farmers at cost have a role to play. A more clustered approach that aims to stimulate higher demand in specific localities would be needed to generate enough momentum for a transformation that is attractive enough from an agribusiness perspective.

On the second issue of overall low visibility of ecological alternatives, an influx of capital is expected to chase reliable, proven investments in existing businesses. Ecological alternatives are more niche and lack a visible profile to reach an investor's desk. Dr. Julius Ecuru and his team have been at the forefront of providing startup funding to select competitive ecological businesses through their BioInnovate Africa Programme funded by the Swedish International Development Agency (SIDA) among others. Their supported projects include a diverse array across holistic development, including *Rhizobia-mycorrhizae-based bio-fertilizer* (BioInnovate Africa 2024b) or *Nanoencapsulated bromelain from pineapple waste for improved livestock production* (BioInnovate Africa 2024a). Interdisciplinary project teams include members from academia, the private sector, or government with the aim of bringing new innovations to market (Bioilnnovate Africa 2022). BioInnovate Africa provides a vehicle for world-class research to reach early market stages, and is embedding businesses as partners from the get-go. It represents one example to overcome the challenge of visibility and create more investable opportunities beyond the conventional agribusinesses.

Individual and collective action

What made it work

The *icipe* team invested a lot in understanding what the most effective and cost-effective avenues are for dissemination. As with most agricultural interventions, "seeing is believing." Donors and implementers are used to stretching funding as far as possible, reaching the maximum number of beneficiaries, and keeping the cost:beneficiary ratio low. A large push to reach farmers was through demonstration plots, supported through model or lead farmer concepts. These outreach methods were also the costliest. Therefore, leveraging synergies of allied project partners was important. A great example was the programming by Send-a-Cow (rebranded in 2022 to Ripple Effect). As captured in the NGO's original name, its projects focused on providing smallholder farmers with access to dairy cows (and other small livestock). Their own training emphasized best practices in livestock management, which included high-quality fodder supply. Push-Pull added to their holistic programming represented a win-win. Their beneficiaries acquired access to a proven sustainable intensification practice and needed quality fodder from the companion crops of desmodium and brachiaria.[4]

What could strengthen the transformation?

More than a decade ago, I received requests for financial support to scale Push-Pull in Zimbabwe. A similar request from the same farmer group reached us annually. As it rested outside our geographical focus areas, my foundation could not directly support them. We redirected them to our *icipe* colleagues who provided them with latest training manuals but not any funding.

With new funding received to scale out across our traditional focus in East Africa, we could reengage them. To my successor's surprise, who visited them in the field, it turned out that they had developed a sizable group of adopters of Push-Pull independent of any external financing. While the next chapter of collaborating with the group occurred after my departure, my lessons learnt include that spontaneous or self-led adoption is possible. Internally, it accentuated questions of negative externalities from development-funded dissemination. As voiced in the field by farmers and reinforced by a senior agricultural economist, covering the whole cost of seeds for farmers can be counterproductive. Neighbors, even as close as next door, might refuse to adopt a technology by themselves if they see others having received free in-person training and seeds. Or they might hear rumors of benefits that never existed. Even if an adopting neighbor has visible benefits – think doubling or tripling of yields – it might not be enough to overcome the bad aftertaste created by perceived inequities in beneficiaries.[5]

A greater focus on strengthening farmer adaptation and farmer-led innovation is warranted. Given the limited and terminal nature of funding, and the diversity in agro-ecologies demanding unique solutions to local challenges, farmers need to be encouraged to become innovators. As observed in Tanzania, farmers for too long (from colonial into post-colonial periods) have endured top-down advisory services that are prescriptive (Ledermann 2012). Even knowledge-intensive methods, such as applying FYM, can fail to create sustained support due to the top-down nature of its dissemination. From a farmer's perspective, a greater focus on *we know how* (and away from *they know how*) could go a long way in reactivating dormant, traditional knowledge and farmers' abilities to become innovators themselves. We will dive deeper into how change happens in Chapter 13.

Science and technology

What made it work

After funding R&D around Push-Pull for over a decade, a point comes where donors ask the question on the continued need or viability of such funding. "If the method doubles yields today, why do we need to keep researching?" was a refrain voiced initially in hallways and later in meetings. Even as a researcher myself, the questioning rubbed off on me. I learnt my lesson firsthand when a new pest – the Fall Armyworm – emerged across East Africa in 2016. On a trip to Malawi and Zambia, I was with Dr. Girma Hailu, an *icipe* colleague and he spotted the infestation wreaking havoc on maize fields grown in the government research stations (see Figure 7.10). A new pest – able to decimate entire maize harvests – reflected the need for continuous research to adapt proven technologies to local contexts and to emerging challenges. Fall Armyworm represented a highly potent challenger, worthy of our attention.

As documented by *icipe*, Alex and Ruphina Chingoli were among the 30 farmers who were chosen for initial trials to establish the viability of Push-Pull in Malawi. Alex established two plots – a new trial plot for Push-Pull and one where they continued monocropping maize. It did not take long for Fall Armyworm to appear on their farm, setting off a journey of farmer-led discovery and *icipe* research:

> Between December 2016 and January 2017, around the same time that the Push-Pull trials started, we noticed a new pest in our farms, and it was soon officially announced that the fall armyworm had arrived in Malawi. Of course, for us farmers, this new pest caused great concern, seeing that it was an added

Figure 7.10 Finding the latest culprit – a Fall armyworm in East Africa, *icipe*

problem to an already challenging situation in maize cultivation. As the season went along, I noticed something interesting: the fall armyworm appeared not to attack the Push-Pull plot. In sharp contrast, the control plot was heavily infested by the pest, as were neighbouring farms not using the technology. To me, this indicated that although Push-Pull was designed for other pests, the technology was having an impact on the fall armyworm. I shared my observations with the Push-Pull researchers, and I understand that they have now conducted verification studies and found that indeed, the technology is capable of tackling the fall armyworm. Naturally, the availability of a solution to reduce the damage of this invasive pest is good news for us.

(*icipe*, n.d.)

Over the following years, *icipe* successfully established the viability of Push-Pull to serve as an ecological method to control damage from Fall Armyworm, with the practice endorsed by leading donors and agricultural organizations (Prasanna et al. 2018). R&D was still needed. The research team's rigorous methodologies shortly after the Fall Armyworm's sightings helped it document Push-Pull's capacity to manage yet another pest (Midega et al. 2018).

What could strengthen the transformation?

Creating connections with aligned international research organizations operating across sub-Saharan Africa is important. The *Brachiaria Mulatto cv*, for example, integrated into a more climate-resilient version of Push-Pull was developed by CIAT – the International Center for Tropical Agriculture (CIAT n.d.). The forage grass was bred in 1989 and picked up by a commercial seed company. It can survive five to six months of dry spells (CIAT n.d.). Other organizations, such as IITA, shared similar missions in achieving sustainable intensification. Unfortunately, cross-pollinating research or collaborations are less common as the incentives appeared not to exist. With each institute promoting its own flagship technology, a lack of collaboration from the beginning meant that institutes continued to seek their own specific donors for follow-on funding for scaling.

Similar to academia, evaluations and promotions of these world-class scientists are based on the impact of their publications, and not necessarily the impact of their overarching project work. With the need to produce additional research outputs, a successful pilot can offer an opportunity to cross-subsidize research salaries alongside any implementation activities.

Geographical limitations emerge in the reach of research centers. Piloting and dissemination activities in development projects can concentrate around areas reachable within a day trip from a field station or a country office. Higher potential areas outside those zones are usually slower to be reached. Leveraging other institutes as a regional host or leader for further scaling can be beneficial, especially if they have a large representation in the field.

Too frequently, donor funding is not set up to address the above: project-based funding is more common than core contributions that permit greater liberty for engaging across institutes. Project-based funding fuels the need to achieve rapid impact, rather than focus on establishing synergies that pay off in the long run (see *Chapters* 13 and 16). Non-traditional donors, such as a family office that funded our scale-out, are much more likely to provide funding with fewer strings attached.

An overarching reform completed by 2022 across all major agricultural research institutes – the CGIAR – encourages greater collaboration, more pooled funding, and offers hope that we can overcome the challenge of who owns the research and innovation (CGIAR 2024). Linked to our silo-breaking from Chapter 1, increased collaborative efforts are needed to stimulate a greater transformation of agricultural food systems across a multitude of agro-ecological contexts. Continued fragmentation of efforts fails to meet the OECD Development Assistance Committee (DAC) (OECD 2021) guidelines of coherence (i.e., "How well does the intervention fit?") and the priorities of farmers seeking sustainable intensification. A larger reform of how merit is assessed within research teams would supplement such a shift.

Having established the role of subsidies and their rationale and drawbacks, let us focus next on what happens when we transition together.

CASE STUDY 7.2 MY SECOND JOB (IN AGRICULTURAL SUBSIDIES)

I got my start working as a server at an American diner – a novelty on a Swiss autobahn. My second job was more transformational: I worked consecutive summers as an assistant for the Department of Agriculture in the Canton of Solothurn, entering data forms from every farmer in the region to determine their annual subsidy payments. My father – who conveniently also worked in the same department – was the sole member of his family who did not end up continuing farming. He became an engineer working as a bureaucrat for farmers. In Switzerland, farmers have a love-hate relationship with subsidies and bureaucracy, and I found myself in the middle of it all. As a neighboring farmer remarked to me at the turn of the century, "we are not interested in becoming *Landschaftsgärtner* (trans: landscape gardeners). We are farmers."

This work experience shaped my understanding of the role and legality of subsidies, their incentives in creating new and maintaining green spaces, and the effects they have on perpetuating business-as-usual. In Switzerland, subsidies slowed down the push for small, unprofitable farmers to cease their farming operations (preventing the *Bauernsterben* (trans: deaths of farmers)). Combined with inspiration from my mentor, Dr. William Moseley, I pursued the topic further, investigating Brazil's successful fight against subsidies to US cotton farmers at the World Trade Organization (WTO) (Ledermann 2005, Ledermann and Moseley 2007).[6] My formative years engrained in me the role agricultural subsidies play in stimulating domestic agricultural production under the premise of food sovereignty and security.

The global food crisis from 2007 to 2008 and the global pandemic from 2020 to 2022 have put food systems back on policymakers' agendas. It also pushed them back on my desk. Subsidies can play an important role in stimulating a transformation, for example, through state support for underserved farming groups. Agricultural subsidies can also have the opposite effects as they promote more intensive, export-oriented production and offer a lock-in for business-as-usual.

With overall challenges of subsidies captured in Chapter 2, I focus here on how to move beyond business-as-usual and return to healthier agricultural development. As noted by Damania et al. (2023, 259):

> reforming environmentally harmful subsidies is critical – albeit not sufficient per se – for achieving the United Nations Sustainable Development Goals. Indeed, subsidy reform not only removes distorted incentives that undermine countries' ability to make progress toward these goals, but also can unlock significant domestic financing to facilitate and accelerate sustainable development efforts.

Subsidies offer an example of how the changemaker of *governance and policymaking*, can unlock the role of *economy and finance*, through redirecting existing public funding.

THE NEED FOR REFORM IN AGRICULTURAL SUBSIDIES

Five types of support are offered through agricultural subsidies: input subsidies, market price support, output payments, decoupled payments, and public goods provisions (Damania et al. 2023). Table 7.2 highlights the rationale behind them and the identified drawbacks. Ranked from most to least significant for our transformation, the clashing trade-offs identified include SDG 2, 10, and 15.

Table 7.2 Rationale, drawbacks, and trade-off from subsidies, adapted from Damania et al. (2023), 107

Type of agricultural support	**Rationale**	**Drawbacks**	**Main SDG trade-off**
Input subsidies	To reduce the price of capital-intensive inputs, such as fertilizer, seeds, or pesticides, to increase yields. Resultant intensification of production is expected to reduce pressure on expanding production (i.e., extensification).	Only effective where farmers lack access to credits and hence use below optimal-levels of inputs. Otherwise incentivizes overuse, creating inefficiencies and nutrient run-off. Even if targeted to the poor, risks perpetuating input dependency.	SDG 2 [-2; counteracting]
Market price supports	Bans or tariffs on imported agricultural products, increasing the domestic price. Aims to improve self-sufficiency by incentivizing farmers to produce more through higher prices.	Increased vulnerability to domestic shocks in food production. Incentivizes farmers to grow less-efficient crops and extend production onto more marginal lands. Result in habitat loss or deforestation.	SDG 15 [-2; counteracting]
Output payments	Direct subsidies provided to farmers to grow specific crops, aiming to increase production and resultant self-sufficiency	Same drawbacks as market price supports, yet more substantial in intensity. Wealthier farmers are also able to take largest advantage of support.	SDG 10 [-2; counteracting]
Decoupled payments	Delinked from production, they support farmers through income transfers. Can offer safety net for farmers, raising overall income.	Less negative effects on environment expected. Slow down potential transformation toward new food systems.	SDG 12 [-1; constraining]
Public good provision	Support in infrastructure (e.g., ameliorating road access for farmers) or research and development (e.g., including extension services)	By covering public goods, government support doesn't compete with private sector financing. Support in infrastructure, such as irrigation, can lead to overconsumption of water or lower levels of resilience.	SDG 6 [-1/0; constraining/ consistent]

RUSHED TRANSITIONS DON'T WORK

Rushed transitions do not work, especially if they are undertaken during a fiscal crisis. In July 2021, Sri Lankan farmers were out on the streets. They were protesting the decision by then-president Rajapaksa in April 2021 to ban the imports of synthetic fertilizers and transition its farmers to substitute the inputs with organic ones.[7] Since organic fertilizer or manure did not exist in sufficient quantities, the seven-month ban resulted in dramatic drops in crop yields. The protests contributed to the president stepping down, but the damage outlasted him.

Officially, the ban was motivated by the desire to move away from chemical fertilizers and their negative impacts on health and the environment. In reality, with a drop in tourism revenue the country was running out of foreign reserves, and essential goods and medicine became scarce. The compounding income shortfalls from agriculture left the country seeking international aid as the country defaulted on its debts (Park and Ratnaweera 2022). Rushed transitions do not work, especially if they are undertaken during a fiscal crisis.

TIMING IS EVERYTHING

A much slower approach, transitioning over a timeframe that spans a decade and more, Sikkim, India's least densely populated state successfully converted to organic farming from low-input agriculture (Krishna 2024). Timing the reform is important. We can also learn from successful fossil fuel subsidy reforms in India and Indonesia undertaken in 2014. Their reforms took place during a period of low oil prices and when the political will for reform was high (Benes et al. 2015; cited in Damania et al. 2023, 257). With higher oil prices, a reform would be more difficult to undertake. Importantly, both reforms had a lengthy planning and preparation period (Damania et al. 2023).

In agriculture, timing is especially critical. Farmers need to plan well in advance to adjust to subsidy reform, for example, as they purchase seeds and other inputs upfront (Damania et al. 2023, 257).

STEPS TO TRANSFORM AGRICULTURAL SUBSIDIES

Returning to my second job, we can see how Switzerland is passing its penultimate subsidy reforms. Adapting the timing and sequencing of fossil fuel subsidy reform from the World Bank's report (Damania et al. 2023) to an agricultural context in Figure 7.11, the main steps are: First, existing subsidies are assessed, and a reform package is designed. In Switzerland's case, it needed to be implemented through parliamentary approval and survive the future threat of a people's referendum. The subsidy was then subsequently reduced in phases until a return to market pricing was possible. Similar to efforts at decarbonization (Hallegatte et al. 2024), proactive and positive communication strategies were needed, as well as compensation and social protection. Complementary policies and reinvesting savings from the subsidy or revenue distributions followed. Highlighting the importance of geopolitical contexts, Switzerland's latest subsidy

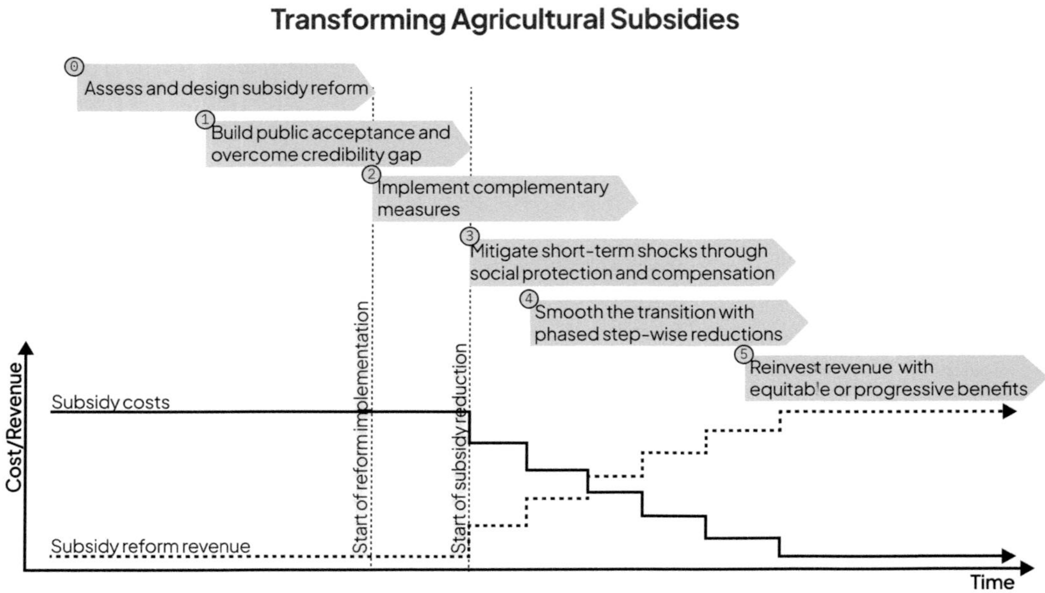

Figure 7.11 Transforming agricultural subsidies, adapted from Rentschler (2018) and Damania et al. (2023)

reform packages have been delayed because of the impact of the war in Ukraine on the agricultural market. In Switzerland's direct democracy, it is crucial to build public acceptance and overcome credibility gaps because of the oversized influence of the agrarian lobby both in parliament and in influencing public opinion. Failure to account for these power dynamics triggers a people's referendum, where progressive agrarian initiatives have recently failed to garner enough votes.

YOUR TURN

You are passionate in playing your part in transforming the local food system and reduce your own food waste (SDG Target 12.3). If scaled across your neighbors, what are the potential synergies and trade-offs across the SDGs and their targets? How can you enhance synergies and limit negative effects from trade-offs?

NOTES

1 Dr. Saul Newman's research has debunked some of the claims around longevity associated with these diets, as old age was attributed to poor record keeping (Carvan 2024).
2 I had the pleasure of first supporting the practice through my work at the Swiss Biovision Foundation, a long-time donor. I subsequently coordinated a scale-out of this African innovation across sub-Saharan Africa, with initial focal countries of Malawi and Zambia.
3 You can find a video of Push-Pull in action on our companion website, sustainabledevelopment.me.
4 The integration worked as livestock numbers were small and largely not free grazing. Agro-pastoralist communities where cattle or goats graze freely pose a problem to a

sustained establishment of Push-Pull, as the perennial plants make for a tasty treat during the offseason when other fields lay bare.

5 My observation relates to the importance of beneficiary selection. The less inclusive, the more likely you will face similar challenges. Especially in projects aiming to create evidence of its effectiveness – a must for pilot projects – the requirements for rigorous data collection need to be weighed against the ethical and sustained implications on development programming. For example, the use of randomized control trials (RCT) mandates that a treatment and control group be established. I witnessed first-hand the frustration with a control group of mango growers in Central Kenya and their underlying expectations that once research is completed, the same benefits will be transferred to non-beneficiaries. While RCTs represent the gold standard in many development agencies, more participatory approaches might be equally viable to generate insights without the costs to beneficiaries (and donors, as RCT designs cost hundreds of thousands of dollars). Examples include peer evaluations, as applied in Push-Pull by Martin Fischler, to establish the effectiveness of the intervention drawing on farmers' voices (Fischler 2010).

6 Receiving permission from the WTO to impose countermeasures, Brazil ultimately reached an agreement with the USA in 2014, ending the decade-long dispute (Office of the United States Trade Representative 2014).

7 Across the Global South, farmers have been effective in creating visibility from their mass protests in recent years. The same applies to the Global North, with one main difference: agribusinesses have much deeper pockets. In the USA, for example, more money is spent on lobbying on agriculture than on defense or oil and gas (Savage, Hancock, and Pooler 2024).

REFERENCES

Afshin, A., P. J. Sur, K. A. Fay, L. Cornaby, G. Ferrara, J. S. Salama, E. C. Mullany, K.H. Abate, C. Abbafati, Z. Abebe, M. Afarideh, A. Aggarwal, S. Agrawal, T. Akinyemiju, F. Alahdab, U. Bacha, V. F. Bachman, H. Badali, A. Badawi, ...C. J. L. Murray. 2019. "Health Effects of Dietary Risks in 195 Countries, 1990–2017: A Systematic Analysis for the Global Burden of Disease Study 2017." *The Lancet* 393 (10184): 1958–1972. https://doi.org/10.1016/S0140-6736(19)30041-8.

Agroecology Pool. 2018. "Push-Pull Project." Available at: https://www.agroecology-pool.org/wp-content/uploads/2018/10/fact-sheet-PP_final.pdf.

Avery, A. 2007. "'Organic Abundance' Report: Fatally Flawed." *Renewable Agriculture and Food Systems* 22: 321–323. https://doi.org/10.1017/S1742170507002116.

Badgley, C., J. Moghtader, E. Quintero, E. Zakem, M. J. Chappell, K. Avilés-Vázquez, A. Samulon, and I. Perfecto. 2007. "Organic Agriculture and tThe Global Food Supply." *Renewable Agriculture and Food Systems* 22: 86. https://doi.org/10.1017/S1742170507001640.

BioInnovate Africa. 2022. "About Us." Available at: https://bioinnovate-africa.org/about-us/ (accessed March 23, 2025).

BioInnovate Africa. 2024a. "Nanoencapsulated Bromelain from Pineapple Waste for Improved Livestock Production." Available at: https://bioinnovate-africa.org/nanoencapsulated-bromelain-from-pineapple-waste-for-improved-livestock-production/ (accessed March 23, 2025).

BioInnovate Africa. 2024b. "Rhizobia-Mycorrhizae-Based Bio-Fertilizer for Small Holder Farmers in Eastern Africa." Available at: https://bioinnovate-africa.org/rhizobia-mycorrhizae-based-bio-fertilizer-for-improved-livelihood-among-small-holder-farmers-in-eastern-africa/ (accessed March 23, 2025).

Biovision Foundation. 2024a. "CLEVER – Sustainable Consumption." Available at: https://www.biovision.ch/en/project/sustainable-consumption/.

Biovision Foundation. 2024b. "Switzerland." Available at: https://www.biovision.ch/en/region/switzerland/.

Brescia, S. 2017. *Fertile Ground: Scaling Agroecology from the Ground Up*. Oakland: Food First Books.

Carvan, T. 2024. "Dr Saul Newman Has Uncovered the 'Secret' to Living to 110." Australian National University. Available at: https://science.anu.edu.au/news-events/news/dr-saul-newman-has-uncovered-secret-living-110 (accessed March 31, 2025).

CGIAR. 2024. "One CGIAR." Available at: https://www.cgiar.org/food-security-impact/one-cgiar/ (accessed March 23, 2025).

Channell, J., H. Krause, Y. Qin, N. Baker, J. Coats, and A. Vaz. 2022. *Eliminating Poverty: The Importance of a Multidimensional Approach in Tackling SDG 1*. Oxford: citi and SOPHIA.

CIAT. n.d. "Brachiaria hybrid cultivar (cv.) Mulato II." Available at: http://ciat-library.ciat.cgiar.org/Forrajes_Tropicales/pdf/Posters/poster_31_epmr07.pdf.

CompostNow. 2025. "Easy, Clean Composting Services." Available at: https://www.compostnow.org (accessed March 23, 2025).

Damania, R., E. Balseca, C. de Fontaubert, J. Gill, K. Kim, J. Rentschler, J. Russ, and E. Zaveri. 2023. *Detox Development: Repurposing Environmentally Harmful Subsidies*. Washington, DC: World Bank. Available at: https://www.worldbank.org/en/topic/climatechange/publication/detox-development.

EPA. 2024. "About the 2021 Food Recovery Challenge National Award Winners." Available at: https://www.epa.gov/sustainable-management-food/about-2021-food-recovery-challenge-national-award-winners (accessed March 23, 2025).

Evans, J. 2021. "Nestlé Document Says Majority of Its Food Portfolio Is Unhealthy." *Financial Times*, May 31. https://www.ft.com/content/4c98d410-38b1-4be8-95b2-d029e054f492.

FAO. 2024a. *Compendium of Forgotten Foods in Africa*. Accra: FAO. Available at: https://openknowledge.fao.org/server/api/core/bitstreams/b815a827-71a6-4f89-8f55-10b1423b635b/content.

FAO. 2024b. "FAO Helps Foster New Food Frontiers with African Orphan Crops." Available at: https://www.fao.org/newsroom/detail/fao-helps-foster-new-food-frontiers-with-african-orphan-crops/en (accessed March 23, 2025).

Fischler, M. 2010. "Impact Assessment of Push-Pull Technology Developed and Promoted by Icipe and Partners in Eastern Africa." Available at: https://www.push-pull.net/Impact_Assessment.pdf.

Foroohar, R. 2023. "Big Companies Defy Expectations to Become Even Bigger." *Financial Times*, October 22. https://www.ft.com/content/0ddb8e87-feed-4506-a682-4af1386532bc (accessed May 28, 2024).

Hallegatte, S., C. Godinho, J. Rentschler, P. Avner, I. I. Dorband, C. Knudsen, J. Lemke, and P. Mealy. 2024. *Within Reach: Navigating the Political Economy of Decarbonization*. Washington, DC: World Bank.

icipe. n.d. "Push-Pull and Fall Armyworm: First Account." Available at: https://www.icipe.org/news/push-pull-and-fall-armyworm-first-account.

Independent Group of Scientists Appointed by the Secretary-General. 2019. *Global Sustainable Development Report 2019: The Future is Now – Science for Achieving Sustainable Development*. New York. United Nations. Available at: https://sdgs.un.org/gsdr/gsdr2019.

IUCN. n.d. "Scaling up Mountain EbA." Available at: https://iucn.org/our-work/topic/ecosystem-based-adaptation/scaling-mountain-eba.

Jacobs, A., and M. Ritchel. 2017. "How Big Business Got Brazil Hooked on Junk Food." *The New York Times*, September 16. https://www.nytimes.com/interactive/2017/09/16/health/brazil-obesity-nestle.html.

Kansanga, M. M., L. Shanmugasundaram, S. Ledermann, and D. Rain. 2025. "Nature-Inspired Solutions for Food Loss Prevention: Exploring Smallholder Farmers' Willingness to Adopt Solar-Powered Cold Storage." *Frontiers in Sustainable Food Systems* 9. https://doi.org/10.3389/fsufs.2025.1525148.

Kopainsky, B., T. Tribaldos, and S. T. Ledermann. 2017. "A Food Systems Perspective for Food and Nutrition Security beyond the Post-2015 Development Agenda." *Systems Research and Behavioral Science* 35: 178–190. https://doi.org/10.1002/sres.2458.

Krishna, G. 2024. "How One Indian State Went 100% Organic." Available at: https://reasonstobecheerful.world/how-one-indian-state-went-100-organic/.

Ledermann, S. 2005. *Agricultural Subsidies and the Doha Round: A Historic Breakthrough? Geography Honors Project*. Saint Paul, MN: Macalester College.

Ledermann, S. 2012. "Organic Revolution: Cotton and Its Impact on Poverty, Inequality and Sustainability." PhD thesis, Geography Department, Rutgers University.

Ledermann, S., and Z. Khan. 2017. "Push-Pull for Addressing the SDGs: Implementing Holistic Agricultural Development Technology." In IISD's SDG Knowledge Hub. https://sdg.iisd.org/commentary/guest-articles/push-pull-for-addressing-the-sdgs-implementing-holistic-agricultural-development-technology/.

Li, M., N. Jia, M. Lenzen, A. Malik, L. Wei, Y. Jin, and D. Raubenheimer. 2022. "Global Food-Miles Account for Nearly 20% of Total Food-Systems Emissions." *Nature Food* 3 (6): 445–453. https://doi.org/10.1038/s43016-022-00531-w.

Lynch, D. J. 2021. "Inside America's Broken Supply Chain." *The Washington Post*, October 2.

Masters, B. 2024. "Waste Less, Earn More: A New Market for Food Byproducts." *Financial Times*, November 26. https://www.ft.com/content/9da0429b-ab3d-42b0-81de-80659c393f03.

McIntyre, B. D., H. R. Herren, J. Wakhungu, and R. T. Watson (Eds.) 2009. *International Assessment of Agricultural Knowledge, Science and Technology for Development*. Washington, DC: Island Press.

McMullin, S., K. Njogu, B. Wekesa, A. Gachuiri, E. Ngethe, B. Stadlmayr, R. Jamnadass, and K. Kehlenbeck. 2019. "Developing Fruit Tree Portfolios That Link Agriculture More Effectively with Nutrition and Health: A New Approach for Providing Year-Round Micronutrients to Smallholder Farmers." *Food Security* 11 (6): 1355–1372. https://doi.org/10.1007/s12571-019-00970-7.

Midega, C. A. O., J. O. Pittchar, J. A. Pickett, G. W. Hailu, and Z. R. Khan. 2018. "A Climate-Adapted Push-Pull System Effectively Controls Fall Armyworm, Spodoptera Frugiperda (J E Smith), in Maize in East Africa." *Crop Protection* 105 (March): 10–15.

Ndayisaba, P. C., S. Kuyah, C. A. O. Midega, P. N. Mwangi, and Z. R. Khan. 2023. "Push-Pull Technology Enhances Resilience to Climate Change and Prevents Land Degradation: Perceptions of Adopters in Western Kenya." *Farming System* 1 (2). https://doi.org/10.1016/j.farsys.2023.100020.

Njeru, N. K., C. A. O. Midega, J. W. Muthomi, J. M. Wagacha, and Z. R. Khan. 2020. "Impact of Push–Pull Cropping System on Pest Management and Occurrence of Ear Rots and Mycotoxin Contamination of Maize in Western Kenya." *Plant Pathology* 69 (9): 1644–1654. https://doi.org/10.1111/ppa.13259.

OECD. 2021. *Applying Evaluation Criteria Thoughtfully*. Paris: OECD. Available at: https://doi.org/10.1787/543e84ed-en.

Office of the United States Trade Representative, 2014. "United States and Brazil Reach Agreement to End WTO Cotton Dispute." Available at: https://ustr.gov/about-us/policy-offices/press-office/press-releases/2014/October/United-States-and-Brazil-Reach-Agreement-to-End-WTO-Cotton-Dispute.

Park, B., and M. Ratnaweera. 2022. "Sri Lanka Farmers Count the Cost of Government Fertiliser Ban." *Financial Times*, December 27. https://www.ft.com/content/27a4fa03-74d7-409a-89e6-0f0391786c3e.

Pilling, D. 2021. "Making Chocolate Can Give Ghana a Taste of Prosperity." *Financial Times*, June 2. https://www.ft.com/content/dbd20f9f-b9f7-4bf4-86dd-1a8c84069f01.

Pilling, D. 2024. "Could T-Shirts Be the Way to Industrialise an African Nation?" *Financial Times*, August 28. https://www.ft.com/content/92b4f84d-229d-405b-85ec-7869529b7633.

Poole, N., J. Donovan, and O. Erenstein. 2021. "Viewpoint: Agri-Nutrition Research: Revisiting the Contribution of Maize and Wheat to Human Nutrition and Health." *Food Policy* 100: 101976. https://doi.org/10.1016/j.foodpol.2020.101976.

Prasanna, B. M., J. E. Huesing, R. Eddy, and V. M. Peschke (Eds.) 2018. *Fall Armyworm in Africa: A Guide to Integrated Pest Management*. Mexico: CIMMYT.

Pray, C., and S. Ledermann. 2016. "Genetically Engineered Crops and Certified Organic Agriculture for Improving Nutrition Security in Africa and South Asia." In H. K. Biesalski and R. E. Black (Eds.), *Hidden Hunger: Malnutrition and the First 1,000 Days of Life: Causes, Consequences and Solutions*. Berlin: Springer.

Rentschler, J. 2018. *Fossil Fuel Subsidy Reforms: A Guide to Economic and Political Complexity*. Abingdon: Routledge.

Savage, S. 2024. "Climate Change Is Pushing Up Food Prices — and Worrying Central Banks." *Financial Times*, July 2. https://www.ft.com/content/125e89c0-308a-492f-ae8e-6834847d1186.

Savage, S., A. Hancock, and M. Pooler. 2024. "The Global Power of Big Agriculture's Lobbying." *Financial Times*, August 21. https://www.ft.com/content/5f4e0538-10a4-4c8f-bc3c-28f255f20f0b.

Schader, C., A. Heidenreich, I. Kadzere, I. Egyir, A. Muriuki, J. Bandanaa, J. Clottey, J. Ndungu, C. Grovermann, G. Lazzarini, J. Blockeel, C. Borgemeister, A. Muller, F. Kabi, K. Fiaboe, N. Adamtey, B. Huber, U. Niggli, and M. Stolze. 2021. "How Is Organic Farming Performing Agronomically and Economically in sub-Saharan Africa?" *Global Environmental Change* 102325. https://doi.org/10.1016/j.gloenvcha.2021.102325.

Seufert, V., and N. Ramankutty. 2017. "Many Shades of Gray: The Context-Dependent Performance of Organic Agriculture." *Science Advances* 3 (3). https://doi.org/10.1126/sciadv.1602638.

Stein, A. J., and F. Santini. 2022. "The Sustainability of 'Local' Food: A Review for Policy-Makers." *Review of Agricultural, Food and Environmental Studies* 103 (1): 77–89. https://doi.org/10.1007/s41130-021-00148-w.

Stott, M. 2021. "Poverty Surge Sets Latin America Back Over a Decade, Says UN." *Financial Times*, March 4. https://www.ft.com/content/3a01895f-0456-4261-ac67-7ae0dc778b5f.

Tamiru, A., T. J. A. Bruce, C. M. Woodcock, J. C. Caulfield, C. A. O. Midega, C. K. P. O. Ogol, P. Mayon, M. A. Birkett, J. A. Pickett, and Z. R. Khan. 2011. "Maize Landraces Recruit Egg and Larval Parasitoids in Response to Egg Deposition by a Herbivore." *Ecology Letters* 14: 1075–1083. https://doi.org/10.1111/j.1461-0248.2011.01674.x.

Terazono, E. 2021. "Food Inflation Concerns Deepen as Prices Reach Highest Level Since 2014." *Financial Times*, February 4. https://www.ft.com/content/571a9d68-c8b5-4c56-a539-26fff81e9296.

Tilman, D., and M. Clark. 2014. "Global Diets Link Environmental Sustainability and Human Health." *Nature* 515: 518–522. https://doi.org/10.1038/nature13959.

UN General Assembly. 2010. *Report Submitted by the Special Rapporteur on the Right to Food, Olivier De Schutter*. New York: Human Rights Council.

United Nations. 2024. *The Sustainable Development Goals Report 2024*. New York: United Nations.

WHO. 2023. "Mycotoxins." Available at: https://www.who.int/news-room/fact-sheets/detail/mycotoxins (accessed March 23, 2025).

Willett, W., J. Rockström, B. Loken, M. Springmann, T. Lang, S. Vermeulen, T. Garnett, D. Tilman, F. DeClerck, A. Wood, M. Jonell, M. Clark, L. J. Gordon, J. Fanzo, C. Hawkes, R. Zurayk, J. A. Rivera, W. De Vries, L. M. Sibanda, … C. J. L. Murray. 2019. "Food in the Anthropocene: The EAT–Lancet Commission on Healthy Diets from Sustainable Food Systems." *The Lancet* 393: 447–492. https://doi.org/10.1016/S0140-6736(18)31788-4.

World Agroforestry, IFAD, EU, and CGIAR. n.d. "Nutritious Food Portfolios for Targeting Year-Round Food Harvest and Nutrient Gaps." Available at: https://apps.worldagroforestry.org/downloads/Publications/PDFS/LE19078.pdf.

Yaffe-Bellany, D., and M. Corkery. 2020. "Dumped Milk, Smashed Eggs, Plowed Vegetables: Food Waste of the Pandemic." *New York Times*, April 11.

8 Just economies

RECAP OF CHALLENGES AND WHY THIS HOTSPOT MATTERS

Using the latest data available (United Nations 2024) the following five challenges can be addressed by this transformation hotspot.

SDG 1: No Poverty

The numbers of the working poor are declining globally, but amount to nearly a quarter of a billion workers. While the global share of employed population living below US$2.15 per day decreased to 6.9 percent by 2023, in sub-Saharan Africa, for example, nearly one in three employees was working poor. Youth and women are more likely to be affected by low salaries (United Nations 2024, 8).

Creating more just economies aims to lift up those who are currently not receiving a fair wage to sustain the basic necessities for their livelihoods.

SDG 5: Gender Equality

Women perform most unpaid domestic and care work, spending 2.5 hours more each day working than their male counterparts. These gender gaps limit

DOI:10.4324/9781003400028-13

women's abilities to enter the labor market, contributing to their poverty and fueling inequality (United Nations 2024, 19).

Transforming towards more just economies entails addressing intra-household inequalities, as well as more systemic gender inequalities at a national scale.

SDG 8: Decent Work and Economic Growth

The labor rights to freedom of association and collective bargaining have declined across the globe from 2015 to 2022. In Asia, poor labor rights prevail because of suppressed union formation, leading to non-compliance rates that are up to 50 percent higher than the global average (United Nations 2024, 25).

Without improving these rights, progress towards more just economies is expected to be further restrained. Transforming economies can not only improve labor rights, but unlock potential for greater achievements in social justice.

SDG 10: Reduced Inequality

Returns to capital have increased, making up for losses incurred during the global pandemic. The labor income share of GDP, however, continues to be on a long-term decline, dropping from 54.1 percent in 2004 to 52.7 percent in 2021 – or a loss of US$568 per worker per year in their purchasing power. While labor productivity has increased, it has not translated into increased wages. These declines especially concern households that lack access to investments in capital markets where the larger growth has been observed (United Nations 2024, 28).

Improving the share of income gained from profits for workers through transforming more just economies can reduce a rise in inequalities.

SDG 16: Peace, Justice and Strong Institutions

Bribery is still too commonplace. One out of every five individuals worldwide is asked to pay or forced to pay a bribe to a public official at least once a year. Bribery is prominent in places where people already have less capital: bribery is three times more common in lower-income countries than in high-income countries (United Nations 2024, 41).

Reducing corruption through collective action and improved governance will go a long way towards sustaining a transformation towards just economies.

LIVING IN A MORE JUST WORLD ON A SAFER PLANET

Having focused on the example of a transformation in sustainable, local food systems, let us zoom out and look at how our transformation hotspots are positioned from a 30,000-feet perspective.

One of the most important concept to frame our understanding over the last 20 years – as we entered the Anthropocene – is the framework of planetary boundaries. Introduced in Chapter 2, they were originally developed by Rockström et al. (2009) and updated by Steffen et al. (2015) and Richardson et al. (2023). In 2024, an interdisciplinary consortium of researchers advanced them to build the complementary Earth-system boundaries (ESB) (Gupta et al. 2024). As we aim to achieve transformations toward just economies, a good starting point is the capacity of earth-systems to support a more equitable and just world.

The eight safe and just Earth-system boundaries

According to Gupta et al. (2024, 2), ESBs "go beyond planetary boundaries by combining elements from the local to global level with knowledge from biophysical and social science domains." They capture both safe and just boundaries that we should operate in, "[putting] a 'floor' on the minimum needed for people to live a life free from poverty"(Mooney 2024). Table 8.1 provides an overview of the boundaries, as well as their indicators. At the global level, seven of eight ESBs have already been crossed (Gupta et al. 2024, 8).

I focus next on synthesizing their Earth-system analysis of the just boundaries for the first four of the five domains crossed. Just boundaries are usually more stringent than the safe boundaries, as they incorporate intergenerational, intragenerational, and intraspecies justices.[1] Aligned with our map and the SDGs, their work calls for transformations in governance, consumption, economic systems, and technologies (Gupta et al. 2024, 6) and tackling poverty (SDG 1), inequality (SDG 10), gender inequality (SDG 5), economic growth (SDG 8), and peace, justice and strong institutions (SDG 16). At the core of their analysis is a push to address the drivers of ecological change, degradation, and vulnerability (Gupta et al. 2024, 6).

- *Biosphere: crossed.* Land degradation affects over three billion people. Restoring 15 percent more of our land can decrease harm to other species and to future generations. With 80 percent of biodiversity found in the territories of

Table 8.1 Safe and just Earth-system boundaries. Source: Gupta et al. (2024)

Domain	*Earth-system boundary*	*Indicator for safe and just boundary*
Climate	Global warming	Maximum of 1.0 degree C warming
biosphere	Natural ecosystem area	At least 50 percent should be largely intact, depending on spatial distribution
	Functional integrity	More than 20 percent of each km^2 should comprise natural or semi-natural vegetation
Freshwater	Surface water flow	Less than 20 percent monthly flow alteration
	Groundwater	Annual drawdown from natural and anthropogenic factors does not exceed recharge
Nutrients	Nitrogen	Surplus is less than 57 Tg per year
	Phosphorous	Surplus is less than 9 Tg per year
Air pollution	Aerosol	Annual mean interhemispheric aerosol optical depth difference is less than 0.15

Indigenous peoples, "those with the greatest responsibility for damaging biodiversity and the greatest capabilities" have the burden of action (Gupta et al. 2024, 11). These efforts hence need to include more marginal regions in wealthier countries, even if it comes with a lower cost-efficiency. We cover regeneration of nature in Chapter 12.

- *Nitrogen: crossed.* The most severe areas of excess nitrogen use are China, South and West Asia, North America, and Europe (see also Chapter 2). To further reduce greenhouse gas emissions, a sustainability transformation "[builds] on improved use and regenerative nutrient-conserving practices, [to ensure] equitable access, and [recycle] nutrients" (Gupta et al. 2024, 20). To promote equitable access, increasing fertilizer use in regions like sub-Saharan Africa is warranted from an intragenerational justice standpoint as the local nitrogen use boundaries are not exceeded.
- *Phosphorous: crossed.* From an intragenerational justice perspective, increasing access to phosphorous for tropical soils that are more prone to lack phosphorous from weathering, is warranted although current mining practices create harm to local communities (Gupta et al. 2024, 21). In contrast to nitrogen, phosphorous represents a non-renewable, finite resource. Despite improved access, agricultural soils will continue to deplete by 4 to 19 kg per ha annually (Alewell et al. 2020). Transformations should focus on improving efficiency and recycling of phosphorous (Gupta et al. 2024, 22).
- *Climate: crossed.* The 1.5° Celsius target referenced throughout, while safe from an intraspecies perspective, cannot address intergenerational and intragenerational justice (Gupta et al. 2024). Already at 1° Celsius warming sea-level rise threatens the livelihood of millions who have limited adaptation options. Over 500 million people will have at least one day each year with temperatures above 32 degrees Celsius, approaching the physiological limit for humans (Gupta et al. 2024). Urban residents without cooling, shade, or water at home or work face even more intense heat stress because of the heat island effect. In 2024, we breached the 1.5° Celsius threshold for the first time across the globe (Mooney, Tauschinski, and Bernard 2025). Declines in well-being due to high temperature days are expected to accelerate and not recover (Gallup and CITI 2022). Similar to fertilizer use, greater energy use in Africa can be justified, given their current low levels: a five-fold increase with natural gas would only push global emissions up by 1 percent (The Economist 2022).
- *Surface water flow and groundwater: crossed.* Some 70 percent of surface water withdrawals globally are used for irrigation in agriculture, increasing risks of land subsidence – the gradual or sudden sinking of land (Gupta et al. 2024, 24). Present generations are already impacted by excessive groundwater use. The transboundary nature of surface water use between upstream and downstream countries gives rise to issues of intragenerational justice. Governance and policymaking play a role in managing equitable water flows in both the

Global North (e.g., the lower Rhine flowing from Germany to the Netherlands) and the Global South (e.g., the Blue Nile flowing from Ethiopia through Sudan to Egypt) (Gupta et al. 2024). Regarding economy and finance, restrictions on water use, such as privatization or pay-to-use pricing, often reduce access for the poorest populations. Avenues to deal with trade-offs in water management include greater powers for states to recover water in times of emergency as well as reusing and recycling water (Gupta et al. 2024, 27).

Just planetary boundaries improve our understanding of poverty and inequality. Next, we will examine the vexing nature of poverty and how rural inequalities intersect with transformation.

THE VEXING NATURE OF POVERTY

Matthew Desmond, a prominent scholar of poverty and inequality in the United States, set off to address the vexing question of how the largest economy in the world has more poverty than its peers. His research shows that "America's poverty is not for a lack of resources. We lack something else" (Desmond 2023, 7). To tackle poverty, we need to move beyond economy and finance and leverage all changemakers. The lens of poverty (SDG 1) and inequality (SDG 10) reveals the engrained issues undermining our efforts to create just economies and improve human capabilities (see Chapter 9). As captured by Desmond (2023, 23):

> Poverty is often material scarcity piled on chronic pain piled on incarceration piled on depression piled on addiction – on and on it goes. Poverty isn't a line. It's a tight knot of social maladies. It is connected to every social problem we care about – crime, health, education, housing – and its persistence in American life means that millions of families are denied safety and security and dignity in one of the richest nations in the history of the world.

Translated into the context of our two transformation hotspots: to create more just economies and increase human capabilities, we need to focus not only on addressing poverty (SDG 1). We also need to improve good health and well-being (SDG 3), a quality education (SDG 4), address gender inequalities (SDG 5), ensure access to basic goods, such as clean water and sanitation (SDG 6) and access to affordable housing (SDG 11), focus on decent work and economic growth (SDG 8), and build industry, innovation, and infrastructure (SDG 9) that reduce inequalities (SDG 10), while ultimately working toward greater peace, justice and stronger institutions (SDG 16).[2]

Just economies need to be resilient to shocks. Perry, Aronson, and Pescosolido (2021, 3) argue, "with climate change leading to cumulating disasters, the Matthew Effect, where the rich get richer and the poor get poorer, will compound widening socioeconomic disparities."[3] As poverty manifests itself as vulnerabilities to shocks, investments in greater human capabilities increase resilience to future shocks.

Poverty reinforces the importance of silo breaking to advance sustainable development in both the Global South and the Global North. As reflected in Chapter 3, the framework of the SDGs and the subsequent agreed monitoring of progress explicitly account for national-level definitions of key indicators, such as poverty. In the USA, the absolute global poverty line used by the World Bank of US$1.90 per day would correspond to US$4.00 as a minimum to cover basic needs: twice as much as someone in the Global South. Based on this standard, Angus Deaton (cited in Desmond 2023, 18) found a staggering 5.3 million Americans living below US$4.00 a day back in 2018, leading him to conclude that "[t]here are millions of Americans, whose suffering, through material poverty and poor health, is as bad or worse than that of the people in Africa or Asia."

Ultimately, poverty is much more than not just having access to money. Desmond (2023, 78) argues "[poverty is] the condition of not having enough choice and being taken advantage of because of that." Exploitation traps people in poverty: exploitation on the labor market, exploitation on the housing market, or exploitation of the poor as consumers. Unbanked Americans, for example, spent US$1.6 billion to cash checks alone in 2020 (Desmond 2023). In the Global North, poverty also furthers loneliness and social exclusion (Huxhold, Suanet, and Wetzel 2022). Pathways out of the traps include collective action (e.g., at the level of unionization) and governance and policymaking (e.g., at curbing overdraft fees on checking accounts) (Desmond 2023). We need a holistic approach, leveraging all changemakers.

> When we ignore the role that exploitation plays in trapping people in poverty, we end up designing policy that is weak at best and ineffective at worst. When legislation lifts incomes at the bottom [...] without addressing the housing crisis, those gains are often realized instead by landlords, not wholly by the families the legislation was intended to help.
>
> (Desmond 2023, 78)

The New York Times reporter Jack Healy (2017) covered the challenges of US families after a backlash to a politician's comments that the uninsured should buy healthcare rather than the latest iPhone. He documented the essential nature of access to cellphones – as families struggle on minimum wage jobs without health insurance. Quoting one interviewee, "[a cellphone is] for work. It's for relationships, it's for kids, it's for my friends, everything. Life is hard to be by yourself" (Healy 2017). Their story serves as yet another reminder that digital divides and inequalities in access are not unique to the Global South. Addressing such inequalities form an integral part in fostering more just economies. Our next section captures the different inequalities present in rural spaces.

INEQUALITIES IN RURAL TRANSFORMATION

The following is adapted from my work (Ledermann, Pray, and Anderson 2024) to highlight three dimensions of rural inequality – socio-economic, gender, and spatial – and their roles in shaping rural transformations.

Inequality is most prominently understood in its socio-economic dimension as the income gap between the top and bottom percentiles within a society. Usually calculated using a Gini coefficient, income inequality ranges from 0 – perfect equality – to 1, representing maximum inequality. To reduce income inequality, you would want economic growth to translate into larger gains for the lower percentile in the income distribution. Economic growth from agriculture has been well established as being pro-poor: welfare gains are the largest for the poorest (Ligon and Sadoulet 2008). The introduction of new innovations widens the gap between early and late adopters. Early adopters tend to be wealthier, better-educated farmers with larger landholdings who can afford to take risks (Rogers 2003).

Inequalities can rise due to these increased profits by early adopters and since innovations diffuse within as opposed to between social strata. As Martin Ravallion (2005) showed, in unequal societies, economic growth fails to effectively reduce poverty. In sub-Saharan Africa, uneven land tenure and ownership risk exacerbating inequality (Dawson, Martin, and Sikor 2016). Higher productivity among wealthier farmers in closed agricultural markets can lead to lower local prices. While this helps poorer consumers, it can hurt smallholder farmers who lack the resources to adopt new technologies (Collier and Dercon 2014; Gollin, Lagakos, and Waugh 2014). Shocks are expected to further increase inequalities: adaptation to climate shocks, such as lower rainfall, is less likely if farmers are resource-poor (Amare et al. 2021). Wealthier farmers are more likely to be well-educated and have access to smartphones, allowing them to engage with digital extension services in response to COVID-19 (van Campenhout 2021; Wesolowski et al. 2012).

In rural development, gender inequality also looms large (see Chapter 14). The concept of "time-poverty" underscores the double burdens of productive and reproductive labor women face in their households. As farmers join cooperatives to improve their socio-economic livelihoods, women often face a third burden of organizational labor (Lyon, Mutersbaugh, and Worthen 2017). Gender inequalities are starkest at the intersections of poverty and marginalization: women and girls from the poorest caste, who have lower levels of education and suffer higher levels of violence (Kabeer 2015). Women face larger barriers in accessing and adopting new agricultural technologies (Meinzen-Dick et al. 2003; Peterman, Behrman, and Quisumbing 2010). Women are often left out of training, lack membership in relevant social networks, and are not reached by the male-dominated extension staff. Their lack of land ownership and small landholdings limits their ability to experiment and reduces their visibility to extension service providers (Ragasa 2014). Women are furthermore expected to be at greater risk of climate shocks (FAO 2018). The digital divide worsens gender inequality during crises. For example, unequal access to smartphones within households (GSMA 2024) will magnify gender disparities if future pandemics require social distancing.

Spatial inequalities also manifest in agricultural development, as the geographical location and distance of users of agricultural products and inputs matter. Poor transportation and communication infrastructure limits the reach of distant farmers to local markets. Country-level inequalities affect local inequalities, with infrastructure itself accounting for one-third of the differences observed (Kanbur and Venables 2005). Poor transportation and poor infrastructure increase input prices (Minten, Koro, and Stifel 2013), keeping market prices for crops and livestock low (Brenton, Portugal-Perez, and Regolo 2014). Spatial barriers limit access to new knowledge and increase the likelihood of incomplete information on relevant technologies and prices (Magruder 2018). Greater distances of learning through social networks reduce the availability of households to access new knowledge, including both absolute (e.g., physical proximity) and relative distance (e.g., membership in kinship groups) (Beaman and Dillon 2018; Takahashi, Muraoka, and Otsuka 2020).

VIGNETTE 1 COMBATTING (GENDERED) INEQUALITIES THROUGH CASH TRANSFERS IN BRAZIL

The World Bank classifies Brazil as a leader in expanding opportunities for women in Latin America, having passed policies on improving gender equality and protecting women (World Bank 2023). Women, however, still earn 20 percent less than men and their employment is often in lower-skilled jobs. The COVID-19 pandemic increased these gaps, as women's labor force participation dropped from 66 to 62 percent. Unemployment rates increased, while maternal mortality and gender-based violence increased (World Bank 2023).

A flagship program supported by the World Bank was Bolsa Familia. As of 2018, the program reached 47 million people, of which half are women. During the pandemic, the World Bank financed the expansion of Bolsa Familia with a US$1 billion project that added 14 million families (World Bank 2023). These emergency payments appear to have been effective in reducing inequalities to historic lows and declines in poverty. They did not affect labor force participation and temporarily reduced inequalities across race. The emergency funding was still limited in addressing prior inequalities. Black women, for example, were the most affected by the pandemic (Nazareno and de Castro Galvao 2023).

In summary, Bolsa Familia and its extension showcase the importance of cash transfers and emergency funds to offer a safety net during shocks. The case also reveals the importance of dealing with deeper intersectional challenges when addressing just economies.

ADDRESSING POLYCRISES DEMANDS ADAPTABILITY

Beyond these inequalities and the shocks from the pandemic, more regions in the world are dealing with a multitude of crises, so-called polycrises (United Nations Environment Programme 2024, 12–13). From droughts to mass unemployment to social unrest to war, insecurities translate into backtracking on sustainable development. Twenty percent of the world's extremely poor live in conflict-affected and fragile states (Sarabi 2023). By 2030, that figure is expected to exceed 60 percent (USAID 2022). Naheed Sarabi (2023) argues that peace, reconciliation, and political stability need to gain more political weight. Given the long duration over which transformations occur, stable political environments are needed. Achieving sustainable development demands greater emphasis on peacebuilding and peacemaking efforts.

I contend that the humanitarian-development-peace nexus provides significant analytical and practical value in effectively addressing such challenges. Its central tenet is the inclusion of "underlying causes of humanitarian and peacebuilding needs" in development programming (UNDP n.d.). Its efforts are aligned with the OECD Development Co-Operation Directorate (DAC) (2021) as donors integrate the coherence of the intervention (will it fit?), its relevance (is it doing the right things?), its sustainability (will the benefits last?), and the impact (will it make a difference?). Donors have shifted their humanitarian interventions toward an integration with the humanitarian-development-peace nexus. The goal is "to maximize impact and sustainability of programs across different kinds of assistance and to reduce the need for [humanitarian aid] over time" (USAID 2022, 19). Coalition building and understanding local dynamics to engage in joint planning and action are essential to achieve this goal. Humanitarian-development-peace coherence had been elevated, for example, into US government strategies, such as the US Government Global Food Security Strategy, the United States Strategy to Prevent Conflict and Promote Stability, and USAID Climate Strategy (USAID 2022) (see also Chapter 16).

Investing in fragile countries involves risks (Sarabi 2023). To reduce risks, multi-donor trust funds have been developed (Sarabi 2023; UN Sustainable Development Group 2016). To manage risk, improved capacity building, planning, and data collection are needed to support future funding flows. As I experienced myself, funding proposals on sustainable agriculture in fragile contexts have greater risks and reveal capacity gaps. Given the polycrises and absence of private capital, they also come with much greater rewards if they succeed.

To make funded projects in polycrises more effective, guidance was developed on how to make programming more flexible to account for shocks (USAID 2017). Established examples are a crisis modifier, whereby an agreement between the headquarters and a country mission is reached that in case of a severe drought, humanitarian funding will be released to address urgent needs in development programming (USAID 2017). Additional steps include integrating more flexible budgets and work plans, incorporating supplemental line items during the design phase, having more phased implementation, or drafting a crisis clause or provision (USAID 2017).

From a programming perspective, applying nexus thinking challenges us to work together and invest in our own capacities. Our efforts need to translate into strengthening resilience to reduce the vulnerabilities in fragile contexts.

Let me use the example of food and agriculture. According to the UN Food Systems Summit, "food systems in fragile settings are facing a poly crisis feeding off conflicts, economic crises, climate change, food insecurity, malnutrition" (WHO 2023). Climate change is expected to fuel further increases in inequalities (Emmerling et al. 2024). Agricultural production is especially sensitive to climate change. Projected impacts include a rise in chronic poverty, "highlighting the need for targeted measures to support agricultural resilience" (Dang, Hallegatte, and Trinh 2024, 24). To take action, a Humanitarian-Development-Peace Nexus Coalition was formed in 2021 to improve coordinated programming inclusive of peace and conflict (HDP Nexus Coalition n.d.). Reflecting all of our changemakers, the coalition's actions include advocacy, research, capacity building and knowledge sharing, with a central tenets of localization and community engagement (HDP Nexus Coalition n.d.).

MORE THAN JUST AN ACADEMIC DETOUR: DEGROWTH

Calls were increasing to implement degrowth, downscale, or decouple if we are to halt the economic juggernaut that consumes the environment in the process (see Case Study 8.1 on fast fashion). Soumaya Keynes (2024) took a closer look at the literature on degrowth. Besides a proclaimed lack of policy solutions or quantitative analysis, she cited Timothée Parrique of Lund University, who synthesized the common thread emerging out of a review of 115 definitions of degrowth: "a downscaling of production and consumption to reduce ecological footprints planned democratically in a way that is equitable while securing wellbeing."

I offer here two perspectives: Dr. Gilbert Rist, who argues for a more radical transformation, and Daniel Susskind, who chronicled our understanding of growth and strikes a middle ground.

Rist (2019, 257) argues that such new buzzwords unite stakeholders from different camps:

> theoretical critics of 'development', ecologists concerned about the state of the planet, citizens disgusted by the excess of consumption, libertarians eager to 'dump the system' – either through theoretical denunciation or through self-sufficient, eco-friendly production on the margins of society – supporters of tough government action in accordance with the polluter-pays principle, not to speak of individuals wishing to reduce their ecological footprint.

Given our embeddedness in the "cult of growth" (Rist 2019, 258),[4] Rist cites Latouche who calls for us to "decolonize our imaginary" and argues that we must "stop giving credence to economic 'science'," the subject of his latest book (Rist 2021).

■ How to Achieve Sustainable Development

According to Rist (2019, 260), avenues for alternative development are not going far enough:

> they scarcely diverge from the conventional discourse: participatory democracy, control of capital movements, debt cancellation, a reform of development aid, the removal of public goods (culture, water, nature, etc.) from the market, guaranteed prices for primary products and natural resources, stabilization of raw material prices, technology transfer to the countries of the South, and so on.

He argues for "abandoning the deep-rooted belief that economic growth can deliver social justice, the rational use of environment, or human well-being" (Rist 2007, 485). Economics should account for the lived diversity and open up different possibilities on how we interact with our ecology (Rist 2021).

It is ingrained in us to think of growth as improving our living standards. In his treatise, Growth, Susskind takes the causality of growth improving our living standards as his jumping-off point. He finds that hunter-gatherer communities might have enjoyed higher living standards than people at the start of the Industrial Revolution (Susskind 2024, 19). They even had around 125 more days off from work than the average worker today (Susskind 2024, 20).

Of our changemakers, science and technology – translated into innovation – proved to play a crucial role in explaining sustained growth (Susskind 2024). Advances in understanding growth have led to better management of its trade-offs:

> Now more than ever, we recognize that there is no economic law which says growth-promoting technologies must also ruin the environment, worsen inequality, hollow out local communities, and rely on technologies we cannot properly control. On the contrary, we understand that these harmful features are the consequence of technical choices taken by their inventors, knowingly or otherwise, in response to the particular incentives that have been put up in front of them.
>
> (Susskind 2024, 196)

Susskind positions himself as a pragmatist. He acknowledges that not all trade-offs from growth can be transformed to make "the world [a] fairer, greener, more respectful place" (Susskind 2024, 216), but dismisses the idea that we should reject growth in its entirety. Decisions to move away from fossil fuels – expected to reach their peak by 2030 (Sheppard 2023) – or wealth taxes to reduce inequality will weaken growth. But according to Susskind (2024, 221), it is a cost worth bearing. How much we are willing to bear these costs is getting tested over the next few years.

No matter which side of the debate you are on, we have much better tools available to model externalities to our economic development. Our companion website features tools, such as the Green Domestic Product, which is calculated using the standard measure of GDP minus social or environmental costs

(Thurm et al. 2024). As you build your own expertise in assessing co-benefits and trade-offs across the SDGs, I invite you next to explore how just economies matter in fast fashion.

CASE STUDY 8.1 UNWINDING THE THREAD OF FAST FASHION

For a decade, my research was consumed by cotton. I dedicated my early academic training to researching its role as a cash crop within global trade (Ledermann and Moseley 2005), and ended up writing my PhD dissertation on the local impact of organic cotton on poverty, inequality, and sustainability in Tanzania (Ledermann 2012). The 2004 documentary T-shirt Travels highlighted the effects of seemingly harmless charitable clothing donations from the Global North on markets in the Global South (Bloeman 2004). All the good and bad that makes up our clothes was captured in a popular public radio podcast – Planet Money (NPR 2013) – inspired by Petra Rivoli's book (Rivoli 2015).

While my consumption pattern is stuck on organic cotton, I have shifted my research away from cotton and its global value chains. As a consumer, I witnessed in the last decade the rise of fast fashion. Consumers in high-income countries, representing 7 percent of the global population, account for almost half of all global consumption (Kaplinsky 2022, 81). Although my ungainly body length means I rarely can buy anything off the rack, you still will see me wandering the floors of H&M, Uniqlo, or Zara.

Back on the streets, Temu and Shein have risen to the forefront. These fast fashion brands transformed factories from producing seasonal clothing lines – with up to a year of production lead – to small orders of new designs produced within seven days (Olcott and Hu 2024). Valuing speed over quality, Shein ran up a US$2 billion profit in 2023. Shein and Temu dominate entire factory towns in China: in Panyu, China, over 5,000 out of the 7,000 clothing factories were working for Shein (Olcott and Hu 2024). Working overtime, workers get paid a premium, averaging 7,000 to 12,000 Rmb per month (around US$1,000 to 1,500) (Olcott and Hu 2024).

Our consumption of fast fashion, fueled by social media trends and adverts, touches upon four dimensions of just economies. First, from a social justice perspective, while it has increased access to affordable clothing, the low prices and margins do not improve living conditions nor struggles for living wages for the workers. Second, not only do wealthier consumer spend more on clothing, consumers with greater incomes generate more clothing waste. Two-thirds of clothes end up in landfills near poorer neighborhoods, raising additional justice issues (Lundberg and Devoy 2022). Third, from an environmental perspective, cotton is input-intensive. The production of a single cotton shirt alone requires 5,400 bottles of water (Adarve Zuluaga 2024). Annual clothing consumption in the United Kingdom alone uses eight billion cubic meters of water. If everyone bought one pair of second-hand jeans in the UK, it could save one trillion bottles of water (Adarve Zuluaga 2024). Not only does garment production – outsourced to the Global South – pollute water, it exacerbates water stress in these regions (Adarve Zuluaga 2024). Fourth, poverty is omnipresent throughout the female workforce in the garment industry: every garment worker in Bangladesh and at least three out of four workers in Vietnam cannot earn living wages (Oxfam Australia 2019).

The clothing industry accounts for 8–10 percent of greenhouse gas emissions (Niinimäki et al. 2020). Similar to a transformation to sustainable, local food systems, its environmental deficit creates avenues to move the industry from a loser to a winner. A more circular economy, where we are emphasizing repairing, reusing, sharing, recycling or upcycling, can make a significant difference (Adarve Zuluaga 2024). So can higher quality clothing and resultant use of second-hand clothing. Extending the lifespan of clothes by nine months can reduce waste by 22 percent and water consumption by 33 percent (WRAP 2017; cited in Adarve Zuluaga 2024). Just reducing demand itself creates similar challenges present in just energy transitions. A reduction in demand for clothes, for example, will negatively affect the livelihoods of garment workers in the Global South in the short term. For a transformation within the fashion industry, we need similar investments from governments or retailers into reskilling or upskilling and offering social protection for workers (Adarve Zuluaga 2024). These efforts are essential as in a just transition – similar to the example of moving from coal mining to renewables in South Africa – some workers cannot benefit.

Advocacy groups are also demanding a fair living wage that covers basic needs for food, utilities, housing, healthcare, education, clothing, transportation, and opportunities for savings (Oxfam Australia 2019). As women continue to face an 18.5 percent gender pay gap (Environmental Audit Committee 2019; cited in Adarve Zuluaga 2024), a boycott would disproportionally affect them.

Achieving a transformation toward just economies needs investments for manufacturers to improve their standards and operations procedures to produce clothes more sustainably using ethical sources. Across the private sector, new standards have been developed to improve accountability. These include improved standards around climate reporting (from the International Sustainability Standards Board or the Task Force for Climate-related Financial Disclosures), guidance on SDG reporting (from the Global Reporting Initiative), or social and environmental risks (from the EU). Nelson and Ingram (2023) argue that a next step is to make such reporting mandatory, just like independent financial audits are required for businesses. Showcasing the importance of governance and policymaking, UK companies are being investigated for greenwashing in marketing to consumers (Thomas 2021), most recently the fashion industry (Competition and Markets Authority 2024).

In summary, the case of fashion illustrates the following three lessons for achieving just economies: First, our own actions in the consumption of fast fashion make a difference. Second, to leverage synergies and address trade-offs, we should engage the changemaker of governance and policymaking and alter how economy and finance works for the poor. Profits should translate into improved living wages. Third, we also need to encourage new innovations from science and technology to transform the manufacturing industry, including recycling and upcycling across the clothing industry.

YOUR TURN

Building off our first transformation, you also aim to contribute to achieving sustainable development by changing how you consume fast fashion aligned with SDG 12 in general and target 12.5 in particular ("By 2030, substantially reduce waste generation through prevention, reduction, recycling and reuse"). What individual and collective actions are available to you to reduce the amount of clothes ending up in landfills? What are the challenges that emerge in recycling or reuse of clothes in your area? How could these trade-offs be better managed?

NOTES

1 See our companion website for a breakdown of the different types of justice based on Gupta et al. (2024).

2 The resource wars in the Democratic Republic of Congo are indicative of the importance of SDG 16 if we are to address the other eight SDGs (Schipani 2022).

3 Not only did the rich get richer during COVID-19, so did the concentration of export revenue among the top companies increase. In developing countries, "the top 1 [percent] of exporting businesses within each country received between 40 and 90 [percent] of total export revenues for the nation as a whole" (Foroohar 2023).

4 Rist (2019, 258), polemically, asks, "How is it possible to give up exotic fruits, private cars, strawberries in winter, 'discovery' weekends in European cities, or a trip to the Maldives?"

REFERENCES

Adarve Zuluaga, M. 2024. "The Water Cost of Fast Fashion." Blog. Available at: www.oxfam.org.uk/oxfam-in-action/oxfam-blog/the.

Alewell, C., B. Ringeval, C. Ballabio, D.A. Robinson, P. Panagos, and P. Borrelli. 2020. "Global Phosphorus Shortage Will Be Aggravated by Soil Erosion." *Nature Communications* 11 (1): 4546. https://doi.org/10.1038/s41467-020-18326-7.

Amare, M., B. Shiferaw, H. Takeshima, and G. Mavrotas. 2021. "Variability in Agricultural Productivity and Rural Household Consumption Inequality: Evidence from Nigeria and Uganda." *Agricultural Economics* 52: 19–36. https://doi.org/10.1111/agec.12604.

Beaman, L., and A. Dillon. 2018. "Diffusion of Agricultural Information Within Social Networks: Evidence on Gender Inequalities from Mali." *Journal of Development Economics* 133: 147–161. https://doi.org/10.1016/j.jdeveco.2018.01.009.

Bloeman, S. 2004. *T-Shirt Travels*. Documentaray. PBS. Available at: www.pbs.org/independentlens/documentaries/tshirt.

Brenton, P., A. Portugal-Perez, and J. Regolo. 2014. *Food Prices, Road Infrastructure, and Market Integration in Central and Eastern Africa*. Washington, DC: World Bank.

Collier, P., and S. Dercon. 2014. "African Agriculture in 50 Years: Smallholders in a Rapidly Changing World?" *World Development* 63: 92–101. https://doi.org/10.1016/j.worlddev.2013.10.001.

Competition and Markets Authority. 2024. "Greenwashing: CMA issues Tailored Guide for Fashion Brands." Available at: https://www.gov.uk/government/news/greenwashing-cma-issues-tailored-guide-for-fashion-brands.

Dang, H.-A.H., S. Hallegatte, and T.-A. Trinh. 2024. "*Does Global Warming Worsen Poverty and Inequality?*" Washington, DC: World Bank.

Dawson, N., A. Martin, and T. Sikor. 2016. "Green Revolution in Sub-Saharan Africa: Implications of Imposed Innovation for the Wellbeing of Rural Smallholders." *World Development* 78: 204–218. https://doi.org/10.1016/j.worlddev.2015.10.008.

Desmond, M. 2023. *Poverty, by America*. New York: Crown.

Emmerling, J., P. Andreoni, I. Charalampidis, S. Dasgupta, F. Dennig, S. Feindt, D. Fragkiadakis, P. Fragkos, S. Fujimori, M. Gilli, C. Grottera, C. Guivarch, U. Kornek, E. Kriegler, D. Malerba, G. Marangoni, A. Méjean, F. Nijsse, F. Piontek, ... M. Tavoni. 2024. "A Multi-Model Assessment of Inequality and Climate Change." *Nature Climate Change* 14 (12): 1254–1260. https://doi.org/10.1038/s41558-024-02151-7.

Environmental Audit Committee. 2019. "Fixing Fashion: Clothing Consumption and Sustainability." House of Commons. Available at: https://publications.parliament.uk/pa/cm201719/cmselect/cmenvaud/1952/1952.pdf.

FAO. 2018. "Empowering Rural Women, Powering Agriculture." Available at: https://openknowledge.fao.org/server/api/core/bitstreams/e31ea1b4-3378-499e-81da-85552d69ceeb/content.

Foroohar, R. 2023. "Big Companies Defy Expectations to Become Even Bigger." *Financial Times*, October 22. https://www.ft.com/content/0ddb8e87-feed-4506-a682-4af1386532bc.

Gallup, and CITI. 2022. "Climate Change and Wellbeing Around the World." Available at: https://www.gallup.com/analytics/397940/climate-change-and-wellbeing.aspx.

Gollin, D., D. Lagakos, and M. E. Waugh. 2014. "Agricultural Productivity Differences Among Countries." *The American Economic Review (AER)* 104 (5).

GSMA. 2024. "The Mobile Gender Gap Report 2024." Available at: https://www.gsma.com/r/wp-content/uploads/2024/05/The-Mobile-Gender-Gap-Report-2024.pdf.

Gupta, J., X. Bai, D. M. Liverman, J. Rockström, D. Qin, B. Stewart-Koster, J. C. Rocha, L. Jacobson, J. F. Abrams, L. S. Andersen, D. I. Armstrong McKay, G. Bala, S. E. Bunn, D. Ciobanu, F. DeClerck, K. L. Ebi, L. Gifford, C. Gordon, S. Hasan, ... G. Gentile. 2024. "A Just World on a Safe Planet: A Lancet Planetary Health–Earth Commission Report on Earth-System Boundaries, Translations, and Transformations." *The Lancet Planetary Health* 8 (10): e813–e873. https://doi.org/10.1016/s2542-5196(24)00042-1.

HDP Nexus Coalition. n.d. "Principles of Action." Available at: https://hdpnexus.fightfoodcrises.net/principles-action (accessed March 23, 2025).

Healy, J. 2017. "No Health Insurance Is Hard. No Phone? Unthinkable." *The New York Times*, March 11.

Huxhold, O., B. Suanet, and M. Wetzel. 2022. "Perceived Social Exclusion and Loneliness: Two Distinct but Related Phenomena." *Sociological Science* 9: 430–453. https://doi.org/10.15195/v9.a17.

Kabeer, N. 2015. "Gender, Poverty, and Inequality: A Brief History of Feminist Contributions in the Field of International Development." *Gender and Development* 23: 189–205. https://doi.org/10.1080/13552074.2015.1062300.

Kanbur, R., and A. J. Venables. 2005. "Spatial Inequality and Development: Overview of UNU-WIDER Project." Department of Applied Economics and Management, Cornell University.

Kaplinsky, R. 2022. *Sustainable Futures: An Agenda for Action*. Cambridge: Polity Press.

Keynes, S. 2024. "What's Wrong with Research About 'Degrowth'?" *Financial Times*, September 19. https://www.ft.com/content/10fb32c7-67a0-48ff-9fe3-b6e7cd423004.

Ledermann, S. 2012. "Organic Revolution: Cotton and Its Impact on Poverty, Inequality and Sustainability." PhD thesis, Geography Department, Rutgers University.

Ledermann, S., and W. G. Moseley. 2005. "The World Trade Organization's Doha Round and Cotton: Continued Peripheral Status or a 'Historical Breakthrough' for African Farmers?" *African Geographical Review* 26: 37–58.

Ledermann, S., C. Pray, and J. R. Anderson. 2024. "Observations on Status and Trends of Agricultural Extension and Inequality in Uganda." *World Development – Sustainability* 4: 100147.

Ligon, E., and E. Sadoulet. 2008. "Estimating the Effects of Aggregate Agricultural Growth on the Distribution of Expenditures." Washington, DC: World Bank. Available at: https://openknowledge.worldbank.org/bitstream/handle/10986/9096/WDR2008_0021.pdf.

Lundberg, D., and J. Devoy. 2022. "The Aftermath of Fast Fashion." Available at: https://www.bu.edu/sph/news/articles/2022/the-aftermath-of-fast-fashion-how-discarded-clothes-impact-public-health-and-the-environment/1000/.

Lyon, S., T. Mutersbaugh, and H. Worthen. 2017. "The Triple Burden: The Impact of Time Poverty on Women's Participation in Coffee Producer Organizational Governance in Mexico." In *Agriculture and Human Values*: Dordrecht: Springer Netherlands.

Magruder, J. R. 2018. "An Assessment of Experimental Evidence on Agricultural Technology Adoption in Developing Countries." *Annual Review of Resource Economics* 10: 299–316. https://doi.org/10.1146/annurev-resource-100517-023202.

Meinzen-Dick, R., M. Adato, L. Haddad, and P. Hazell. 2003. "Impacts of Agricultural Research on Poverty: Findings of An Integrated Economic and Social Analysis." IFPRI.

Minten, B., B. Koro, and D. Stifel. 2013. "The Last Mile(s) in Modern Input Distribution: Evidence from Northwestern Ethiopia." IFPRI.

Mooney, A.. 2024. "Donald Trump's Re-Election Would Imperil Climate Action, Leading Scientist Says." *Financial Times*, September 11.

Mooney, A., J. Tauschinski, and S. Bernard. 2025. "World Breaches 1.5C Global Warming Target for First Time in 2024." *Financial Times*, January 10. https://www.ft.com/content/fd914266-71bf-4317-9fdc-44b55acb52f6.

Nazareno, L., and J. de Castro Galvao. 2023. "The Impact of Conditional Cash Transfers on Poverty, Inequality, and Employment During COVID-19: A Case Study from Brazil." *Population Research and Policy Review* 42 (2): 22. https://doi.org/10.1007/s11113-023-09749-3.

Nelson, J., and G. Ingram. 2023. "Scaling Private Sector Engagement in the SDGs." In Center for Sustainable Development at Brookings (Ed.), *The SDG Second Half: Ideas for Doing Things Differently*. Washington, DC: Center for Sustainable Development at Brookings.

Niinimäki, K., G. Peters, H. Dahlbo, P. Perry, T. Rissanen, and A. Gwilt. 2020. "The Environmental Price of Fast Fashion." *Nature Reviews Earth & Environment* 1 (4): 189–200. https://doi.org/10.1038/s43017-020-0039-9.

NPR. 2013. "Planet Money Makes a T-Shirt." Available at: https://apps.npr.org/tshirt/#/title (accessed March 23, 2025).

OECD. 2021. *Applying Evaluation Criteria Thoughtfully*. Paris: OECD. https://doi.org/10.1787/543e84ed-en.

Olcott, E., and T. Hu. 2024. "How China's 'Shein Village' Grabbed the Fast Fashion Supply Chain." *Financial Times*, September 17.

Oxfam Australia. 2019. "Made in Poverty: The True Price of Fashion." Available at: https://www.oxfam.org.au/wp-content/uploads/2021/11/Made-in-Poverty-the-True-Price-of-Fashion-Oxfam-Australia.pdf.

Perry, B. L., B. Aronson, and B.A. Pescosolido. 2021. "Pandemic Precarity: COVID-19 Is Exposing and Exacerbating Inequalities in the American Heartland." *Proceedings of the National Academy of Sciences* 118. https://doi.org/10.1073/pnas.2020685118.

Peterman, A., J. Behrman, and A. R. Quisumbing. 2010. "A Review of Empirical Evidence on Gender Differences in Nonland Agricultural Inputs, Technology, and Services in Developing Countries." In A. R. Quisumbing, R. Meinzen-Dick, T. Raney, A. Croppenstedt, J. A. Behrman, and A. Peterman (Eds.), *Food Policy*. Dordrecht: Springer.

Ragasa, C. 2014. "Improving Gender Responsiveness of Agricultural Extension." In A. R. Quisumbing, R. Meinzen-Dick, T. Raney, A. Croppenstedt, J. A. Behrman, and A. Peterman (Eds.), *Gender in Agriculture*. Dordrecht: Springer, pp. 411–430.

Ravallion, M. 2005. "Inequality Is Bad for the Poor." World Bank Policy Research Working Paper 3677. Available at: https://openknowledge.worldbank.org/server/api/core/bitstreams/5a910ead-d952-5fda-8665-794e3ca72602/content.

Richardson, K., W. Steffen, W. Lucht, J. Bendtsen, S. E. Cornell, J. F. Donges, M. Druke, I. Fetzer, G. Bala, W. von Bloh, G. Feulner, S. Fiedler, D. Gerten, T. Gleeson, M. Hofmann, W. Huiskamp, M. Kummu, C. Mohan, D. Nogues-Bravo, ... J. Rockström. 2023. "Earth Beyond Six of Nine Planetary Boundaries." *Science Advances* 9 (37). https://doi.org/10.1126/sciadv.adh2458. https://www.ncbi.nlm.nih.gov/pubmed/37703365.

Rist, G. 2007. "Development as a Buzzword." *Development in Practice* 17 (4–5): 485–491. https://doi.org/10.1080/09614520701469328.

Rist, G. 2019. *The History of Development: From Western Origins to Global Faith*. 5th edn. London: Zed Books.

Rist, G. 2021. *The Delusions of Economics: The Misguided Certainties of a Hazardous Science*. 1st edn. London: Zed Books.

Rivoli, P. 2015. *The Travels of a T-Shirt in the Global Economy: An Economist Examines the Markets, Power, and Politics of World Trade*. 2nd edn. Hoboken, NJ: Wiley.

Rockström, J., W. Steffen, K. Noone, Å. Persson, F.S. Chapin III, E. Lambin, T. M. Lenton, M. Scheffer, C. Folke, H. J. Schellnhuber, B. Nykvist, C. A. De Wit, T. Hughes, S. Van Der Leeuw, H. Rodhe, S. Sörlin, P. K. Snyder, R. Costanza, U. Svedin, ... J. Foley. 2009. "Planetary Boundaries: Exploring the Safe Operating Space for Humanity." *Ecology and Society* 14 (2).

Rogers, E. M. 2003. *Diffusion of Innovations*. 5th edn. New York: Free Press.

Sarabi, N. 2023. "The Need for More Risk-Taking." In Center for Sustainable Development at Brookings (Ed.), *The SDG Second Half: Ideas for Doing Things Differently*. Washington, DC: Brookings Institute.

Schipani, A. 2022. "How the DRC Became the Battleground of a Proxy War Over Precious Resources." *Financial Times*, October 26. https://www.ft.com/content/26125730-3113-4946-982a-989e2acbe472.

Sheppard, D. 2023. "World at 'Beginning of End' of Fossil Fuel Era, Says Global Energy Agency." *Financial Times*, September 11. https://www.ft.com/content/9df6003b-3760-4eee-b189-92c0247fa1a5.

Steffen, W., K. Richardson, J. Rockström, S. E. Cornell, I. Fetzer, E. M. Bennett, R. Biggs, S. R. Carpenter, W. de Vries, C. A. de Wit, C. Folke, D. Gerten, J. Heinke, G. M. Mace, L. M. Persson, V. Ramanathan, B. Reyers, and S. Sorlin. 2015. "Planetary Boundaries: Guiding Human Development on a Changing Planet." *Science* 347 (6223). https://doi.org/10.1126/science.1259855.

Susskind, D. 2024. *Growth: A History and a Reckoning*. Cambridge, MA: The Belknap Press of Harvard University Press.

Takahashi, K., R. Muraoka, and K. Otsuka. 2020. "Technology Adoption, Impact, and Extension in Developing Countries' Agriculture: A Review of the Recent Literature." *Agricultural Economics* 51 (1): 31–45. https://doi.org/10.1111/agec.12539.

The Economist. 2022. "The Green Continent: Powering Africa." November 5, pp. 50–52.

Thomas, H. 2021. "UK business is on notice for 'greenwashing' claims." *Financial Times*, August 18. https://www.ft.com/content/b907d46f-1ca3-4cc8-9a59-da921da656e7.

Thurm, B., V. Petrencu, S. Bürgin, A. Traiba, J. Widmer, J. -P. Danthine, and C. Gallopin. 2024. "Green Domestic Product." Available at: https://e4s.center/resources/reports/green-domestic-product/.

UN Sustainable Development Group. 2016. "Multi Donor Trust Funds Generic Terms of Reference." Available at: https://unsdg.un.org/resources/multi-donor-trust-funds-generic-terms-reference.

UNDP. n.d. "Humanitarian, Development and Peace Nexus." Available at: https://www.undp.org/crisis/humanitarian-development-and-peace-nexus.

United Nations. 2024. *The Sustainable Development Goals Report 2024*. New York: United Nations.

United Nations Environment Programme. 2024. *Navigating New Horizons: A Global Foresight Report on Planetary Health and Human Wellbeing*. Nairobi: United Nations Environment Programme.

USAID. 2017. *Shock Responsive Programming and Adaptive Mechanisms*. Washington, DC: USAID.

USAID. 2022. *2022 Resilience Policy Revision*. Washington, DC: USAID.

van Campenhout, B. 2021. "The Role of Information in Agricultural Technology Adoption: Experimental Evidence from Rice Farmers in Uganda." *Economic Development and Cultural Change* 69: 1239–1272. https://doi.org/10.1086/703868.

Wesolowski, A., N. Eagle, A. M. Noor, R. W. Snow, and C. O. Buckee. 2012. "Heterogeneous Mobile Phone Ownership and Usage Patterns in Kenya." *PloS One* 7. https://doi.org/10.1371/journal.pone.0035319.

WHO. 2023. "Food Systems Transformation Amid a Poly Crisis: Addressing Climate, Food Insecurity, Malnutrition, and Conflicts." Available at: https://www.who.int/news-room/events/detail/2023/07/25/default-calendar/food-systems-transformation-amid-a-addressing-climate-food-insecurity-malnutrition-and (accessed March 31, 2025).

World Bank. 2023. *The World Bank Annual Report 2023*. Washington, DC: World Bank.

WRAP. 2017. "Valuing Our Clothes: The Cost of UK Fashion." Available at: https://www.wrap.ngo/sites/default/files/2020-10/WRAP-valuing-our-clothes-the-cost-of-uk-fashion_WRAP.pdf (accessed March 23, 2025).

9 Human capabilities

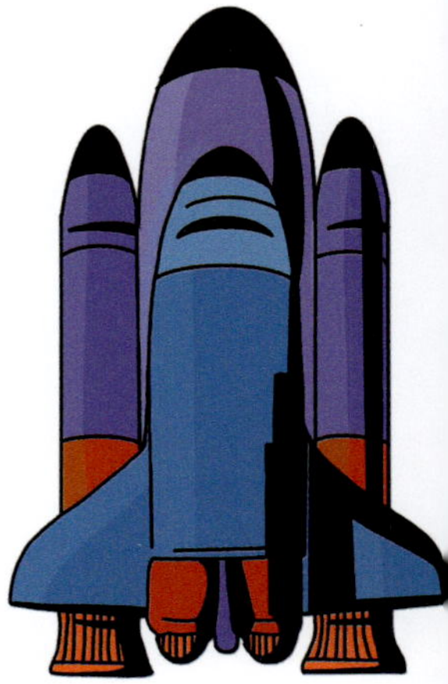

RECAP OF CHALLENGES AND WHY THIS HOTSPOT MATTERS

Using the latest data available (United Nations 2024), the following five challenges can be addressed by this transformation hotspot.

SDG 1: No Poverty

The global pandemic's long-term effects still are negatively impacting human capital (Schady et al. 2023). Government spending on essential services, such as education, health and social protection, are crucial to create a more level playing field for human capabilities to recover and flourish. A persistent 20 percentage-point gap exists in government spending on essential services between higher-income countries (approximately 60 percent) and lower-income countries (approximately 40 percent) (United Nations 2024, 9).

As Dr. Amartya Sen (1999) argued, addressing poverty moves beyond just economic development. We need to lift up people's capabilities to bring about change. Failure to address these spending gaps will translate into persistent inequalities between the Global North and the Global South regarding human capabilities.

SDG 3: Good Health and Well-being

A larger share of the world can access essential health services, yet 4.5 billion people remain without universal health coverage. Two billion people experience

DOI:10.4324/9781003400028-14

financial hardship from health care expenditures, causing 344 million people to experience extreme poverty (United Nations 2024, 12).

For human capabilities to flourish, we need a transformation of how health care is costed and provided.

SDG 4: Quality Education

As discussed earlier, educational outcomes have suffered post-pandemic, with reading and mathematics scores declining. Topics like climate change and sustainability are missing from high school curricula. At the center of addressing these challenges are educators. They are asked to do more with less (government funding). Limited funding also translates into poor qualification of teachers. An average of 15 percent of primary school teachers fall below their country's minimum qualifications (United Nations 2024, 16–17).

At the heart of a transformation in human capabilities is reinvesting in education to improve the quality offered to students.

SDG 6: Clean Water and Sanitation

Along with access to health and education, for human capabilities to fully come into play, we need to improve safe access to water. Even with significant progress made, over two billion people still lack access to drinking water, increasing the time-burden on women (see also Chapter 15). Some 3.5 billion people lack access to managed sanitation (United Nations 2024, 20).

Improving water access for all transforms human capabilities by reducing daily burdens, freeing up capacities and improving quality of life.

SDG 9: Industry, Innovation and Infrastructure

The opportunity for individuals to act upon their capabilities and translate them into a change in livelihoods often depends on their ability to access credit in order to start up their own small enterprise. Not only are these entrepreneurs more vulnerable to shocks, they continue to struggle because of a lack of credit and policies that favor medium-sized to large-scale enterprises. In Europe and North America, over 40 percent of small-scale industries accessed a loan or line of credit. In sub-Saharan Africa or Eastern and South-Eastern Asia, this percentage drops to 16.9 and 19.5 percent respectively (United Nations 2024, 27).

To translate human capabilities into action, entrepreneurs of all sizes need to have access to credit to translate their ideas into reality.

ON TRANSFORMATIVE EDUCATION POST-COVID-19

For the moonshot to happen, we need to re-focus our efforts on education and mental health. COVID-19 has widened equity gaps in educational attainment, reversing progress made. Rebecca Winthrop (2023) argues that we should not dial back our ambitions of lifelong learning embedded within SDG 4 in favor of focusing solely on closing the learning gaps.

She builds on her experience with the MDGs, where the focus was to increase access to primary schools. That goal was achieved, however, without significant improvements in the quality of education for the million of new students entering primary school. Beyond any digital tools deployed during the pandemic, Winthrop emphasizes the importance of engaging families to improve educational outcomes in transforming education. Empowering families helps deepen relationships between them and teachers, with her team's research showing that schools with deeper relational trust "are 10 times more likely to improve students' academic and well-being outcomes" (Winthrop 2023, 13).

Another focal point is mental health. At my university, I have noticed a significant increase in students who manage multiple conflicts. Integrating mental health and student well-being into our education can support them better in acquiring the educational skills needed (Winthrop et al. 2021). Mental health for parents is equally important (Schady et al. 2023).

To transform education, we all need to come together: teachers, students, parents, communities, and government leaders (Winthrop et al. 2021). Similar to other success stories, we need to agree on a shared vision that creates ownership across society. And in order not to repeat the failures from COVID-19, our pathway forward should build resilience and reduce inequalities (Winthrop et al. 2021).

BUILDING RESILIENCE THROUGH EDUCATION

Beyond closing education gaps, we need to move to leverage education to build resilience. As captured by the Independent Group of Scientists Appointed by the Secretary-General (2019, 49), education pays off in reducing vulnerabilities to climate change (Lutz, Butz, and Samir 2014): households are better prepared for disasters, and are more resilient when disaster strikes (Muttarak and Lutz 2014). Households with higher education levels can also cope better with shocks and recover faster (Muttarak and Lutz 2014).

Education historically offered an elevator ride up for lower-income families (Bernardi and Plavgo 2019). More educated households are also more likely to use clean energy sources (Pachauri and Jiang 2008). Lifelong learning, as proposed in the earlier section, will further increase the adaptability of workers who are reskilling based on shifting labor demands (Li 2022). Greater investment in STEM (science, technology, engineering, and mathematics) subjects, should be complemented by rebuilding social skills eroded during the pandemic. Ongoing efforts include the framework of Inner Development Goals, which focus on five skills to improve our ability to achieve sustainable development:

1. being – your relationship to self;
2. thinking – your cognitive skills;
3. relating – caring for others and the world;

4. collaborating – social skills;
5. acting – enabling change (Inner Development Goals n.d.).

These skills are needed to successfully achieve sustainable development by addressing complexity through systems thinking (Trilling and Fadel 2009). Revising curricula to place topics like climate change or planetary boundaries at the core are expected to pay dividends (Independent Group of Scientists Appointed by the Secretary-General 2019, 49). Companies, such a Rflect, co-founded by my friend Niels Rot, allow students to develop their inner development practice as part of their traditional university education (Stadler 2024).

Besides resilience as a concentration that will get you a job in sustainable development, working on your own resilience also pays off. Based on exit surveys, companies are hiring my graduate students because of how they strengthened their resilience, accentuated by how they adapted to the challenges they have faced in navigating a global pandemic, both inside and outside the classroom. Post-pandemic, resilience and adaptability continue to rank near the top of the list for employers.

Whenever discussing education, the gap in educational quality and access between the Global North and the Global South looms over us. Case Study 9.1 focuses on building resilience in a rural context in the Global South. Before we proceed, let us examine the critical need for water as a basic necessity for any investments in capabilities to grow.

WATER: THE LESS VISIBLE AGENT FOR IMPROVING HUMAN CAPABILITIES

Water flows through all the SDGs and their achievement (Center for Sustainable Development at Brookings 2023). For example, water plays a crucial role in achieving SDG 16, as the weaponization of water is on the rise (King 2024). Sustainable water management is crucial to reduce risks (Ovink et al. 2023). "[H]uman failure to understand, value, and manage water" is at the heart of current water crises (Center for Sustainable Development at Brookings 2023, 9). Inadequate knowledge, silo thinking, and top-down governance strategies have held back progress (Ovink et al. 2023).

The UN Water Conference in March 2023 marked a turning point, with the adoption of the Water Action Agenda for achieving the SDGs, featuring over 800 commitments, pledges or actions logged to date (UN DESA 2023).

Although my development work never focused on providing crucial sanitation infrastructure that improves in-home water access, pilot projects across East Africa integrated well-established interventions. I will focus on three to showcase the relevance of water as it relates to human capabilities.

The first low-cost interventions in the domain of water, sanitation, and hygiene (WASH) included the tippy tap, a handwashing station that is connected to a foot pedal. When activated, it tips the bucket of water, allowing the user to wash their hands with nearby soap. Anecdotally, I saw them more around project headquarters, and less in use in the countryside. Across the globe, these water stations effectively promote handwashing practices (Mbakaya, Kalembo, and Zgambo 2020). And innovation in these low-cost options is still ongoing. Oxfam, for example, has developed a new iteration of the tippy tap, that minimizes splashing water and muddy feet (Chadwick 2021).

The second low-cost intervention was an adapted version of micro-drip irrigation. Farmers could easily install them, allowing more targeted use of limited water resources, contributing to increased income, food security, and yields (Mupaso, Makombe, and Mugandani 2023). The drips were mainly powered by gravity, as opposed to a pump. The greater water use efficiency reduced the burden of procuring water for irrigation. I did worry about blockages. Similar to holiday lights with a broken bulb in the middle of the string, a blockage can cause the rest of the system to lose its value, if farmers could not purchase replacement lines. Local skills need to be built to rehabilitate installations, and a marketplace for repair materials should exist.

The third low-cost intervention was to install rainwater harvesting tanks. Depending on the project, these tanks were installed either at individual homes or at schools, and often included the use of rain gutters. They reduced costs and the gendered labor to secure water, and relieved the emotional stress from lacking reliable access to water. Beyond the burden placed on women in procuring water, distant access to water also increases their risk of becoming victims of gender-based violence (Pommells et al. 2018). Children are also vulnerable to dehydration, which can reduce their performance in school (Adan 2012). Access to safe drinking water at school and complementary school lunches would go a long way toward improving human capabilities.

LIFTING UP VOICES

If water is essential for capabilities to grow, more inclusive processes allow capabilities to reach new heights. Participatory and deliberative democratic practices are a key avenue to "elevate local agency by increasing people's direct participation in public debate on policy issues and community action" (Center for Sustainable Development at Brookings 2023, 5). For active participation that counteracts an observed shrinking of the civic space, we need to strengthen human capabilities and lift up the most marginalized. Working toward SDG 17 – Partnerships for Goals, we also need to invest more in collective leadership, meaning "investing in the development of people who have experienced inequity – and their allies who exercise agency and leadership across all levels of a system – and supporting them to collaborate and learn together in pursuit of justice and sustainability" (Center for Sustainable Development at Brookings 2023, 12).

Table 9.1 Building transformational capacities through donor funding. Source: 17 Rooms (2023)

From status quo...	... to investing more in...
Analyzing the evidence and deciding which intervention/project to spread	Prioritizing investments in agency, leadership development, and networks so that proximate actors can pioneer solutions and continuously improve over time
Deciding what strategies will work to accomplish objectives and searching for grantees who will pursue these strategies	Providing general operating support so proximate leaders can pursue strategies they believe make sense
Assuming that localization and attention to those formerly excluded means there's a lack of leadership capacity within communities	Recognizing that leadership already exists within people in marginalized contexts and that what's needed is to invest in fostering the development of that leadership
Investing in the leadership of a few high-profile people	Investing in people throughout ecosystems who can shift policy, practice, and culture
Prioritizing short-term interventions and scalability over more durable, long-term change.	Investing in agency and adaptive leadership development that strengthens and transforms systems sustainably.

Under the aegis of localization, investments in collective leadership development should support collaborations and co-learning (17 Rooms 2023, 5). To shift funding toward these capabilities, participants invited by the Brookings Institute and the Rockefeller Foundation (17 Rooms 2023, 5) identified five points for action for donors (see Table 9.1). Their proposed shifts – revisited also in Part IV – include moving beyond a focus on evidence and short-term impact, to investing in people and longer-term change. Ashoka, a leading non-profit, invested in over 4,000 Fellows and Young Changemakers in the belief that "all people have the right and ability to co-lead solutions that transform their societies for the better."

ADDRESSING OUR GAPS IN KNOWLEDGE

Knowledge gaps hold people back from achieving sustainable development. Knowledge gaps also limit our understanding of progress made or reduce accountability. Since the launch of the SDGs, policymakers and practitioners have been grappling with how to assess them. While discussions on indicators turn political, significant investments are rolling out to close our data gaps. For example, if we fail to have them on our radar, calls to invest more in local leadership and contexts can naturally be challenging.

Paradoxically, when we are collecting data, we might collect more than is needed. Dean Karlan and colleagues worked over the years to develop principles to guide us when creating data collection systems to fit their purpose, and not more (Gugerty and Karlan 2018). Silo thinking produces duplicate data collection efforts, as each agency or stakeholder is building their own knowledge base without sharing them across their domain. The International

Initiative for Impact Evaluation (3ie) (2024) provides a tremendous service to the sector with its development evidence portal where practitioners can check whether evidence is lacking in the first place before investing in (yet another) randomized-controlled trial (RCT) costing hundreds of thousands of dollars. Their mapping of resilience, for example, identified current research clustered around financial outcomes, with large knowledge gaps around which interventions are effective in strengthening our adaptive capacities to shocks (Berretta et al. 2023).

The budget lines for data collection efforts usually get squeezed in development project proposal writing. We can already foresee that limited top-down donor funding cannot fill the gaps alone. Empowering more local data collection efforts from the bottom-up will be useful. Not only more attuned to challenges that matter, local efforts can capture greater nuances around how human capabilities interact with achieving sustainable development. Moving beyond RCTs, which are biased to assess "micro-interventions at a local level that yield results that can be observed in the short term" (Alkire et al. 2018), we can invest in a more holistic, systematic development assessments.

One such initiative is the launch of the data cooperative by the Aapti Institute, Data 2X, and Open Data Manchester (2024): based on the cooperative model, it focuses on engaging communities to co-design their own data collection. These data efforts provide actionable insights for learning and improving within the cooperative and beyond. They have been piloted in places such as Gujarat, India, with a cooperative of Indigenous women farmers, or the urban residents of Helsinki, Finland, as they improved their city services.

Such initiatives can address our shared knowledge gaps and feed back into how we achieve sustainable development. They need to be combined with complementary research efforts to better understand how human capabilities are affected. An example is the four-decade-long work by Dr. Arline Geronimus, who in her book, Weathering: The Extraordinary Stress of Ordinary Life in an Unjust Society, discusses the effects of systemic injustices on the health of marginalized communities (Geronimus 2023). Her work on weathering has made significant contributions toward increasing our understanding of why young marginalized people are more vulnerable to chronic diseases and disability. Research like hers is historically underfunded. Her findings serve as an example of how closing knowledge gaps can improve our own policymaking to enable human capabilities to move forward.

Case Study 9.1 brings together our major themes: the role of education and learning, and how to translate it into resilience and action. It uses a leading methodology deployed by the Food and Agriculture Organization (FAO) to address our knowledge gaps around rural resilience, and a venture I started on how we leverage the knowledge gained to improve livelihoods. And at the heart of it all is building capacities through knowledge sharing and adaptation.

CASE STUDY 9.1 HOW TO FARM BETTER? BUILDING CAPABILITIES TO ADDRESS THE "SO WHAT?" QUESTION

Back in the early 2010s, two colleagues – Dr. John Choptiany and Benjamin Graeub – worked at the UN FAO. They developed and successfully scaled one of the first assessments of resilience of smallholder farmers and pastoralists. Its acronym was catchy: SHARP. The actual title represented more of a tongue twister: **S**elf-evaluation and **H**olistic **A**ssessment of climate **R**esilience of farmers and **P**astoralists. Based on a literature review, the team operationalized 13 indicators (adapted from Cabell and Oelofse 2012) for assessing household resilience. Table 9.2 presents their indicators and a sample question used in their SHARP questionnaire (Choptiany et al. 2015).

I provided early feedback across these indicators as part of a large-scale expert assessment. What the team ultimately created was comprehensive, accurate, and rigorous. It also took more than an hour per interviewee in the field, as the low literacy and complexity of the survey demanded a trained enumerator for each round. Figure 9.1 illustrates the deployment of the tool in Angola.

In my experience, any farmer (whether in the Global North or the Global South) that I engage with for research gets antsy after 45 minutes, as the farming tasks pile up (and the afternoon heat rises). Longer is a stretch. At the end of each session, the SHARP tool succeeded in its goal of providing a holistic assessment of a farmer's vulnerabilities. From a farmer's perspective, this begged the question of "so what?". It is one thing to understand where one performs well and which resilience dimensions one is most vulnerable in. It is another task to move beyond identifying weaknesses to finding suitable approaches to address those actual shortcomings. And the "so what?" can turn into frustration when it is followed by the realization that the development partner has no means or funding to provide any further support.

It is based on our observations from the development and scaling of the SHARP tool, that we ventured together in 2018 to form a startup – farmbetter – aiming to "empower farmers to build resilience to climate change." Our original product was a smartphone app that allows farmers to self-assess their resilience, drawing on the SHARP methodology, with the final screen asking farmers how they want to farm better. Combining an academic assessment of their resilience scores, along with their preferences (e.g., increasing yields or reducing labor), we matched their scoring with a database of over 1,000 peer-reviewed sustainable land management practices aggregated by WOCAT (2025), which serves as the official database for best practices by the UN Convention to Combat Desertification (UNCCD). Practices shared include, for example, how to use compost and inorganic fertilizer in saline soils.

The underlying assumption of the matching was that farmers (e.g., in Southern Turkey) who already needed to adapt to climate change have developed relevant practices for areas that are just seeing more severe impacts of climate change (e.g., in Western Kenya). To ensure a relevant match, the location of the best practices and the farmer's location is geo-referenced for agro-ecological characteristics (e.g., soil type, rainfall amounts).

Fast forward five years, and our startup was operational in Latin America (Colombia and Ecuador), sub-Saharan Africa (Burkina Faso, Kenya, Uganda, Tanzania), and South Asia (India, Nepal). Our journey to serving these farmers, however, was anything but straightforward. It reflects the challenges of leveraging digital ICT solutions for sharing knowledge-intensive solutions.

First, smartphone ownership in sub-Saharan Africa is rising (51 percent in 2022, projected 88 percent by 2030) (GSMA 2023), but mobile internet use lags (25 percent), and 5G access is minimal (less than 1 percent). A strong urban-rural bias also persists in smartphone ownership and mobile internet usage. To engage farmers, we subsequently pivoted away from a smartphone app (that included an offline function) to a WhatsApp chatbot.

Second, gender inequalities in smartphone use create challenges in diffusing knowledge-intensive agricultural practices in an inclusive matter. In low- and middle-income countries, women are 15 percent less likely than men to use mobile internet and 13 percent less likely to own a smartphone (GSMA 2024). Most women who lack mobile internet access live in sub-Saharan Africa and South Asia. Affordability and (digital) literacy are the top barriers preventing women from access to mobile internet (GSMA 2024). Our startup co-developed a large-scale five-year action research project called *AgriPath* to close knowledge gaps around gender barriers in digital agricultural advisory services.

Third, while significant advances have been made in using artificial intelligence or machine learning to improve digital services, they risk expanding language barriers between English-speakers and non-English speakers. To increase the inclusivity of our offerings, we had to invest significantly in translating the database of practices into local or sub-national languages. We also piloted options for farmers to share practices or insights themselves, further reducing barriers.

In summary, we started farmbetter as a venture to empower farmers to build their own resilience by learning from the experiences of others across the Global North and the Global South. We were sensitive to the risk of creating yet another tool, but we wanted to build upon what has worked before but is not accessible to most. As a case study of the role of the private sector in development, the willingness-to-pay for smallholder farmers who operate on limited income was lower than expected. Six years on, our social enterprise has pivoted to offer the services mostly to farmer organizations or cooperatives who provide their own extension services to farmers. While reducing the viability of reaching more marginalized farmers, our hope is that once barriers in access to smartphones and mobile internet are further reduced, our ability to support them will become more viable. Our venture will also benefit from complementary, transformative efforts across changemakers that empower farmers to move from being recipients of top-down information (or beneficiaries) to experimenting themselves. Given the context-based nature of agricultural development in particular, failure to build human capabilities will result in farmers asking a new follow-up question: "So what now?"

Table 9.2 Indicators for assessing the resilience of agro-ecosystems, adapted from Choptiany et al. (2015)

Indicator	*Sample SHARP question*
Socially self-organized	34. Are you a member of any groups, organizations or associations?
Ecologically self-regulated	5. Do you grow perennial crops (plants that can live several years)?
Appropriately connected	39. How often did you have access to information on market prices over the last season?
Functional and response diversity	52. How many different income sources did you have over the past year?
Optimally redundant	6. Livestock breeds/varieties: Do you have more than one breed of this specie?
Spatial and temporal heterogeneity	26. How many different types of soil can you observe on your field?
Exposed to disturbance	15. Over the last ten years, have you observed any changes relating to the weather? If yes, what changes have you noticed?
Coupled with local natural capital	23. In your farming system, do you use techniques and practices for water conservation?
Reflective and shared learning	15. Do you have access to information on cropping/ livestock practices? How do you get this information?
Globally autonomous and locally interdependent	48. Do you sell/trade some of those products directly with consumers?
Honors legacy	2. Describe the role that elders play within the community (e.g., caring for smaller children, assisting household or community decisions).
Builds human capital	2. Who has completed primary education? (gender)
Reasonably profitable	54. Do you have more saving than 5 years ago?

YOUR TURN

The UN captured in a promotional video for its Global Education First campaign in 2016 (see the link to the YouTube video on the companion website) how education "is a fundamental right and also the most powerful tool to achieve sustainable development." Improving education is said to contribute to achieving SDG 1 (171 million people lifted out of poverty), SDG 2 (12 million children saved

Figure 9.1
Field training on the SHARP tablet application for climate resilience assessment in Virei, Angola, John Choptiany/FAO

from stunting), SDG 3 (49 percent fewer child deaths and 66 percent decrease in maternal mortality), and SDG 5 (64 percent fewer early marriages and 59 percent fewer early pregnancies).

Looking at the SDG targets, identify co-benefits from SDG 4 to another SDG not listed. Can you also find any trade-offs?

If education is clearly a "win-win" in achieving sustainable development transformations, why are we still struggling with supporting it? What lessons can we learn from these challenges in financing education for other SDGs?

REFERENCES

Aapti Institute, Data 2X, and Open Data Manchester. 2024. "How to Build a Data Cooperative: A Practitioner's Handbook." Available at: https://aapti.in/blog/how-to-build-a-data-cooperative-a-practitioners-handbook/.

Adan, A. 2012. "Cognitive Performance and Dehydration." *Journal of the American College of Nutrition* 31 (2): 71–78. https://doi.org/10.1080/07315724.2012.10720011.

Agripath. "Project Introduction." Available at: https://www.agripath.net (accessed March 23. 2025).

Alkire, S., F. Bédécarrats, A. Deaton, G. Giraud, I. Guérin, B. Harriss-White, J. Heckman, J. Hickel, N. Kabeer, S. Morvant-Roux, J. Pearl, C. Renouard, F. Roubaud, J-M. Servet, and J. Stiglitz. 2018. "Buzzwords and Tortuous Impact Studies Won't Fix a Broken Aid System." *The Guardian*, July 16. https://www.theguardian.com/global-development/2018/jul/16/buzzwords-crazes-broken-aid-system-poverty.

Ashoka. "What Ashoka Does." Available at: https://www.ashoka.org/en-us (accessed March 23, 2025).

Bernardi, F., and I. Plavgo. 2019. *Education as an Equalizer for Human Development?* Nairobi: UNDP.

Berretta, M., S. Lee, M. Kupfer, C. Huang, W. Riddelhoover, D. Frey, F. Ahmed, B. Song, K. M. Edwards, J. Porciello, J. Eyers, and B. Snilstveit. 2023. "Strengthening Resilience Against Shocks, Stressors and Recurring Crises in Low- and Middle-Income Countries: An Evidence Gap Map." The International Initiative for Impact Evaluation. Available at: https://3ieimpact.org/sites/default/files/2023-05/Resilience-REAPER-EGM-brief.pdf.

Cabell, J. F., and M. Oelofse. 2012. "An Indicator Framework for Assessing Agroecosystem Resilience." *Ecology and Society* 17. https://doi.org/10.5751/ES-04666-170118.

Center for Sustainable Development at Brookings. 2023. *Forging Next Steps into the SDG Second Half*. Washington, DC: Brookings Institute.

Chadwick, V. 2021. "Oxfam Rolls Out Its Answer to the Tippy Tap." Devex. Available at: https://www.devex.com/news/oxfam-rolls-out-its-answer-to-the-tippy-tap-98836 (accessed March 23, 2025).

Choptiany, J., B. Graub, J. Dixon, and S. Phillips. 2015. "Self-Evaluation and Holistic Assessment of Climate Resilience of Farmers and Pastoralists (SHARP)." Rome: FAO. Available at: https://www.fao.org/3/i4495e/i4495e.pdf.

Geronimus, A. T. 2023. *Weathering: The Extraordinary Stress of Ordinary Life in an Unjust Society*. New York: Little, Brown Spark.

GSMA. 2023. *The Mobile Economy Sub-Saharan Africa*. Available at: https://www.gsma.com.

GSMA. 2024. *The Mobile Gender Gap Report 2024*. Available at: https://www.gsma.com/r/wp-content/uploads/2024/05/The-Mobile-Gender-Gap-Report-2024.pdf.

Gugerty, M. K., and D. S. Karlan. 2018. *The Goldilocks Challenge: Right-Fit Evidence for the Social Sector*. New York: Oxford University Press.

Independent Group of Scientists Appointed by the Secretary-General. 2019. *Global Sustainable Development Report 2019: The Future Is Now – Science for Achieving Sustainable Development*. New York: United Nations. Available at: https://sdgs.un.org/gsdr/gsdr2019.

Inner Development Goals. "Framework." Available at: https://innerdevelopmentgoals.org/framework/ (accessed March 31, 2025).

International Initiative for Impact Evaluation. 2024. "3ie Development Evidence Portal." Available at: https://developmentevidence.3ieimpact.org (accessed March 23, 2025).

King, M. 2024. "Climate Change and the Water Weapon: How Rising Temperatures Are Expanding the Footprint of Conflict." The Earth Commons. Available at: https://commonhome.georgetown.edu/issues/spring-2024/climate-change-and-the-water-weapon-how-rising-temperatures-are-expanding-the-footprint-of-conflict/ (ccessed March 23, 2025).

Li, L. 2022. "Reskilling and Upskilling the Future-ready Workforce for Industry 4.0 and Beyond." *Information Systems Frontiers* 1–16. https://doi.org/10.1007/s10796-022-10308-y.

Lutz, W., W. P. Butz, and K. C. Samir (Eds.) 2014. *World Population and Human Capital in the 21st Century*. Oxford: Oxford University Press.

Mbakaya, B. C., F. W. Kalembo, and M. Zgambo. 2020. "Use, Adoption, and Effectiveness of Tippy-Tap Handwashing Station in Promoting Hand Hygiene Practices in Resource-Limited Settings: A Systematic Review." *BMC Public Health* 20 (1): 1005. https://doi.org/10.1186/s12889-020-09101-w.

Mupaso, N., G. Makombe, and R. Mugandani. 2023. "Smallholder Irrigation and Poverty Reduction in Developing Countries: A Review." *Heliyon* 9 (2): e13341. https://doi.org/10.1016/j.heliyon.2023.e13341.

Muttarak, R., and W. Lutz. 2014. "Is Education a Key to Reducing Vulnerability to Natural Disasters and Hence Unavoidable Climate Change?" *Ecology and Society* 19 (1).

Ovink, H., S. Rahimzoda, J. Cullman, and A. J. Imperiale. 2023. "The UN 2023 Water Conference and Pathways Towards Sustainability Transformation for a Water-Secure World." *Nature Water* 1 (3): 212–215. https://doi.org/10.1038/s44221-023-00052-1.

Pachauri, S., and L. Jiang. 2008. "The Household Energy Transition in India and China." *Energy Policy* 36 (11): 4022–4035. https://doi.org/10.1016/j.enpol.2008.06.016.

Pommells, M., C. Schuster-Wallace, S. Watt, and Z. Mulawa. 2018. "Gender Violence as a Water, Sanitation, and Hygiene Risk: Uncovering Violence Against Women and Girls as It Pertains to Poor WaSH Access." *Violence Against Women* 24 (15): 1851–1862. https://doi.org/10.1177/1077801218754410.

Schady, N., A. Holla, S. Sabarwal, J. Silva, and A. Yi Chang. 2023. *Collapse and Recovery: How the COVID-19 Pandemic Eroded Human Capital and What to Do about It.* Washington DC: World Bank.

Sen, A. 1999. *Development as Freedom.* New York: Knopf.

Stadler, E. 2024. "Introducing the Inner Development Goals (IDGs)." Rflect.ch. Available at: https://www.rflect.ch/2024/12/05/introducing-the-inner-development-goals.html (accessed March 31, 2025).

Trilling, B., and C. Fadel. 2009. *21st Century Skills: Learning for Life in Our Times.* San Francisco: Jossey-Bass.

UN DESA. 2023. "Water Action Agenda." *United Nations.* Available at: https://sdgs.un.org/partnerships/action-networks/water (accessed March 23, 2025).

United Nations. 2024. *The Sustainable Development Goals Report 2024.* New York:United Nations.

Winthrop, R. 2023. "Participatory Policymaking to Transform Education Systems nd Meet SDG 4." In Center for Sustainable Development at Brookings (Ed.), *The SDG Second Half: Ideas for Doing Things Differently.* Washington, DC: Brookings Institute.

Winthrop, R., A. Barton, M. Ershadi, and L. Ziegler. 2021. *Collaborating to Transform and Improve Education Systems: A Playbook for Family-School Engagement.* Washington, DC: Brookings Institute.

WOCAT. 2025. "Sustainable Land Management Database." Available at: https://wocat.net/en/global-slm-database/.

10 Inclusive urban-rural development

RECAP OF CHALLENGES AND WHY THIS HOTSPOT MATTERS

Using the latest data available (United Nations 2024), the following five challenges can be addressed by this transformation hotspot:

SDG 7: Affordable and Clean Energy

As progress is made to increase the number of households that have access to electricity, poorer, more remote households are still most likely to have no power. With investments in off-grid or mini-grid solutions taking off (see also Chapter 11), it is important not to increase rural-to-urban disparities (United Nations 2024, 22).

A transformation to inclusive urban-rural development invests in innovative energy sources that include peri-urban and rural areas.

SDG 11: Sustainable Cities and Communities

With cities growing across the globe thanks to rural-to-urban migration, two challenges are emerging: first, while globally one out of four urban residents are living in slums, that number in sub-Saharan Africa rises to more than half of all urban residents. Second, to account for increased populations, sprawl is the

DOI:10.4324/9781003400028-15

norm as opposed to increasing higher density buildings. Such urban sprawl, as you can see, for example, when you fly into Los Angeles in the United States or into Addis Ababa in Ethiopia, converts rural land to the determinant of the environment. Cities grew over three times faster outwards as opposed to upwards (United Nations 2024, 30).

Inclusive urban-rural development focuses on how to improve the livelihoods of more than a billion urban population living in slums and considers smart urban planning that does not jeopardize rural areas.

SDG 12: Responsible Consumption and Production

Food waste is a major problem, especially in the Global North. In the USA, approximately 40 percent of food is wasted, equivalent to US$218 billion a year. Cities are the largest culprits: in New York City, 54 percent of the food waste is generated by households and another 20 percent by restaurants (Hoover 2017). While generating significant greenhouse gas emissions, the waste is further troubling given that 15 percent of New York City residents are food-insecure (New York City Council 2024).

Inclusive urban-rural development not only aims to reduce waste, but ensures improved access to food for those who are lacking it most.

SDG 13: Climate Action

In the past decade, we have recorded the ten hottest years on record. We also have experienced five times more natural disasters over the last half-century. The annual number of impacted individuals rose from an average of 29 million (2005–2014) to 133 million (2014–2022). Projections show that 1.5 medium- to large-scale disaster are expected every day between now and 2030 (United Nations 2024, 35).

An inclusive urban-rural development accounts for higher risks from natural disasters by improving disaster preparedness, such as leveraging early warning systems or green infrastructure made available to both urban and rural populations.

SDG 15: Life on Land

From 2000 to 2020, we lost almost 100 million hectares of forest across the globe. Agriculture is the largest driver of deforestation, accounting for nearly 90 percent of the loss through either expansion of arable areas or increased livestock grazing. Outside agriculture, urban and infrastructure development is the second largest driver, accounting for 6 percent of losses. Deforestation has multiple negative effects, including changes to microclimates, reduction in biodiversity through habitat loss, and increased erosion (United Nations 2024, 38).

Given the two largest drivers of agriculture and urban development, more inclusive urban-rural development fosters a sustainable food system, including more responsible consumption by reducing waste or transforming food waste into energy.

UNEVEN URBAN DEVELOPMENT

Let us start with a review of the state of urban development. Oswaldo de Rivero (2019) argues that the biggest crisis our civilization is facing across the Global North and the Global South is urban expansion. Unsustainable urbanization increases fossil fuel use, destroys agricultural land, and diverts water resources from farming (de Rivero 2019).

With 70–80 percent of the world's population projected to live in cities by 2050 (de Rivero 2019; Mukim and Roberts 2023), new solutions are needed for greater inclusive urban-rural development. De Rivero (2019) sees signs of strains in the global adoption of the "California model of expansion," with its emphasis on owning a second car, a second house, and a second shopping center within driving distance.

In China, the urban population has tripled since 1990, with the number of cities quadrupling: 700 new cities developed from scratch, with the others expanding. According to de Rivero (2019, 124), "[t]his colossal urban expansion has generated a deep physical-social imbalance." Some 400 Chinese cities experienced water shortages and car use multiplied 130 times, creating 100,000 additional kilometers of highway.

Seventy percent of global greenhouse gas emissions originate from cities (Mukim and Roberts 2023). Most of the emission occurs in cities in high- and upper-middle-income countries, with cities in lower-income countries only contributing 14 percent of global urban emissions (Mukim and Roberts 2023). The challenge is clear: we need to find pathways for fast-growing cities in lower-income countries to develop sustainably without following the historical trajectory of CO_2 emissions.

These lower-income cities are exposed the most to climate change hazards compared to their wealthier counterparts and experience lower levels of resilience. Hazards include floods, heat stress, tropical cyclones, sea-level rise, water stress, and wildfires (Mukim and Roberts 2023). As Figure 10.1 illustrates, large cities experience the highest climate hazard exposure. Cities in low- and lower-middle-income countries have high climate hazard exposure scores no matter their size. Medium and small cities in high-income countries have four times lower exposures to climate hazards (20 percent), reflecting their improved positioning.

Limited access to basic services, including healthcare and education, and inadequate infrastructure – water, electricity, waste management, and emergency services – contribute to lower levels of resilience (Mukim and Roberts 2023). Lack of vegetation and green spaces fuel urban temperatures, with the urban heat island effect causing a 10° C temperature increase compared to rural land surfaces (Mukim and Roberts 2023). To counter the extreme heat impacting billions of urban residents by 2050, experts propose "raising awareness and investment in nature-based and built cooling solutions [that] can save lives, improve health and safety, and enhance productivity" (Center for Sustainable

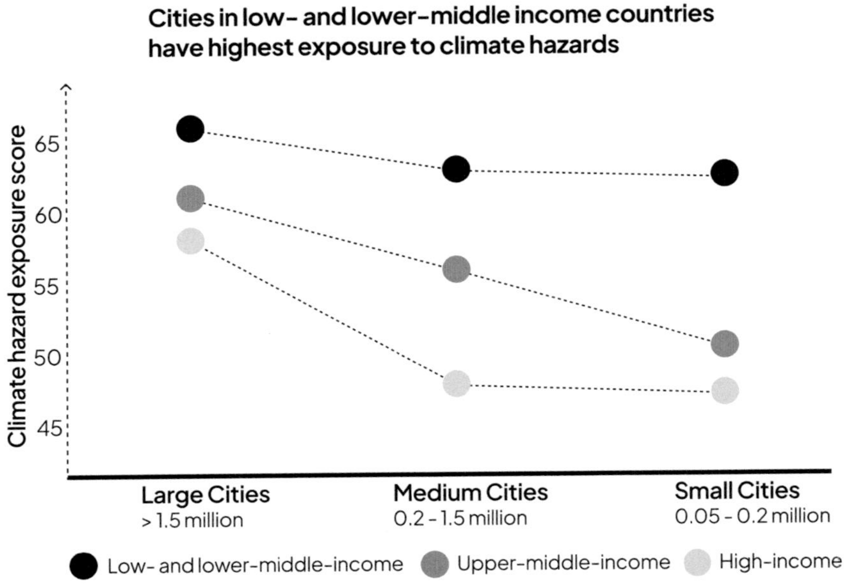

Figure 10.1 Cities in low- and lower-middle-income countries have the highest exposure to climate hazards, adapted from Mukim and Roberts (2023, 9), based on Moody's ESG Solution, Sub-Sovereign Physical Climate Risk Scores, October 2021

Development at Brookings 2023, 8). We will review such solutions in the section "*Greener, more resilient cities*."

The interconnection between urban-rural development is crucial. Extended droughts in rural hinterlands, for example, create a push effect for rural-to-urban migration, accelerating urban growth (Chlouba, Mukim, and Zaveri 2022). New migrants arriving in urban areas tend to cluster in vulnerable locations. Not only do they lack access to information on climate hazards, but once a disaster hits, they are the most vulnerable. Migrants lose more (in relative terms) in disasters and suffer from indirect effects through infrastructure disruptions (e.g., being cut off from roads) (Mukim and Roberts 2023). Without access to insurance and financial markets, they also are less likely to recover from shocks (Mukim and Roberts 2023). Migrants facing social exclusion and poverty are also less likely to engage in sustainable practices in urban areas (Abu et al. 2024).

These challenges are meaningful as "[c]ities have traditionally served as escalators out of poverty; however, more frequent climate change-related shocks may slow down these escalators" (Mukim and Roberts 2023, 29). A review by Abanokova et al. (2022) shows that poorer residents in larger cities across five countries in the Global South (Chile, Colombia, Ecuador, Indonesia, and Peru) historically reduced their poverty. With increased risks from flooding, poorer urban residents, however, are less likely to exit poverty (Mukim and Roberts 2023).

FROM AN URBAN BIAS TO MORE ADAPTIVE, INCLUSIVE CITIES

I remember well the first time I read Michael Lipton's work on urban bias (Lipton 1977). His scholarship argued that policies in developing countries were biased toward urban areas, resonating with my focus on a lack of investment in rural and agricultural development in favor of industrialization. As his work has been revisited and revised (Lipton 1984), his urban-to-rural dichotomy has stressed for me the need to overcome this juxtaposition to achieve sustainable development. I introduce here a reframing of the rural hinterland using a city regions perspective that reflects a rural-urban continuum (Forster and Escudero 2014). Later, I present three trends that offer to reshape how we live in cities.

CONNECTING THE HINTERLAND THROUGH CITY REGIONS

With agriculture as an entry point, Forster and Escudero (2014) use a landscape perspective to sketch out how urban, peri-urban, and rural areas can better integrate. A city regions lens can reshape how we tackle "food and nutrition security with climate action planning, disaster risk reduction, economic and community development, water, biodiversity, and other aspects of natural resource management" (Forster and Escudero 2014, 1).

While different concerns dominate urban or rural areas, their challenges and ecosystems are shared, interconnected across the rural-urban continuum. Rural-urban flows include people, production, commodities, capital, information, natural resources, waste and pollution As illustrated in Figure 10.2,

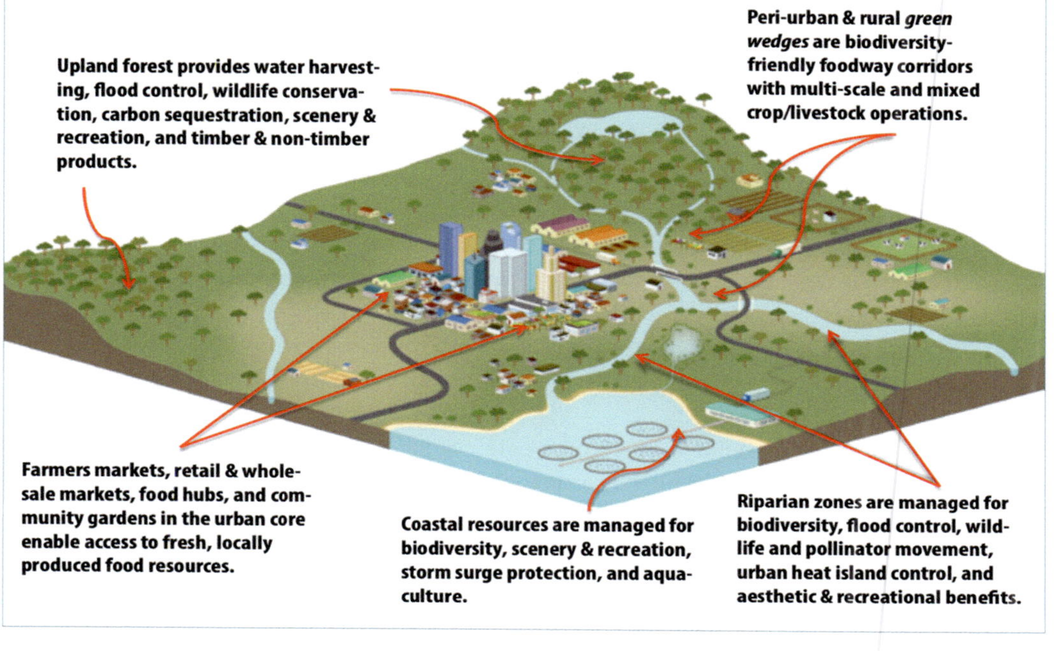

Figure 10.2 Flow of ecosystem services across rural and urban spaces, from Forster and Escudero (2014), 24

ecosystem services offer crucial benefit to urban populations beyond their urban area's boundary. Numerous challenges – such as air pollution – that are not limited to a specific postal code, demand improved coordination across city regions and even between cities (Mukim and Roberts 2023).

ECOSYSTEM SERVICE FLOWS AND FOOD

Over the last decade, the importance of the food system and its role in sustaining urban development while feeding inequities has become more prominent. We are more aware of food deserts in urban areas and food insecure laborers or farmers in rural areas. We also see all the waste piling up.

Waste and its disposal have a long history in urban development. Jacob Doherty's (2021) research into waste management in Kampala, Uganda, highlights how it is used to exert control over the informal sector, resulting in increased marginalization. He shows how changemakers – in his case, governance and policymaking – can also halt progress, as top-down efforts at "cleaning up" just push waste into less visible communities. Identifying avenues to process waste into energy and giving a voice to marginalized populations at the table of public waste management are crucial to achieving sustainable solutions. Informal actors, such as waste pickers (Dias 2016), can become a more visible part of the solution, as championed by *Re-Ciclo* in Fortaleza, Brazil (WRI Ross Center 2024).

From a food system perspective, efforts focus on recovering food waste into compost. In Quito, Ecuador, for example, more than half the waste generated is organic matter (C40 Cities 2016). The city has created a biogas plant to harness electricity from the waste, as well as processing 12 million tons of organic waste into compost to be used in local parks (C40 Cities 2016).

A skeptic's view of indoor farming

Urban agriculture has played a substantial role in addressing pockets of food insecurity across cities in the Global North and the Global South, including improving access to food and affordability (iPES Food 2017; Lawson 2016). If we take our productivity lens off, we can see that these spaces make significant contributions to the agri-*culture*: the urban plots offer opportunities for learning, education, and foster a sense of belonging. Combined with civil society efforts, these spaces can revitalize neighborhoods, address malnutrition brought on by food deserts, and offer productive outlets for reintegrating less visible residents (Santo, Palmer, and Kim 2016).

In contrast, I remain a skeptic regarding the potential for indoor farming to make a significant socio-cultural difference and reduce food insecurity. Products from modern indoor growing can be found in my neighborhood grocery store. In the need to cover their high running costs, they focus on high-value produce. In my store, this means lettuce and herbs. While limiting the use of pesticide or fertilizer, the profile of food produced and the premium price point limit its contributions to more inclusive, sustainable urban development.

My skepticism is further based on a fundamental flaw. Vertical, indoor farming operations shut out the most important free energy source available for plant growth: the sun. As captured by Dr. Bruce Bugbee (2015), the value of sunlight as energy for 1 acre of farmland is US$400,000. If we shut it out, we need to replace it with artificial lights, usually powered by fossil fuels, or the equivalent of 5 acres of solar panels for that 1 acre of production (Bugbee 2015). Since its first decade of boom, more indoor farming startups have gone bust, cracking under rising energy prices and an inability to generate profits (Walling and Lafleur 2023). For now, our own backyards, community or school gardens, and rooftops might serve us better – both for producing food, stimulating ecosystem services, and creating opportunities for agri-*culture* to work.

THREE TRENDS SHAPING RURAL-URBAN TRANSFORMATION

Beyond agriculture, let us review how technologies, urban planning, and green energy are offering to transform our city regions.

1 *The potential of smart cities*

Planners and policymakers have heralded the rise of *smart cities*. According to the Global Smart City Partnership Program of the World Bank (2024), these smart cities "leverage data and technology to integrate urban infrastructure and service delivery and provide solutions to achieve a citizen-centric approach." Smarter cities promise to close our knowledge gaps: more data can foster more inclusive solutions to address social injustices.

I first engaged with their potential advising one of my graduate student teams as they researched how smart cities in Thailand and Vietnam can create positive effects for marginalized populations. Their work on small and medium-sized enterprises (SMEs) piqued my interest, raising questions about who creates the smart city, who owns the data, and who decides what is being implemented based on the knowledge gained? These concerns are advanced by critics (Foth, Brynskov, and Ojala 2015; Greenfield 2013, 2017; Sassen 2011), who challenge smart cities' technocentric focus, business-driven origins, and surveillance mechanisms. They are concerned that we are creating cities that are closed systems as opposed to powered and shaped by the people who live in them.

AI has, of course, elevated the smart city discourse, with nine out of ten mayors stating that they want to leverage generative AI, yet only 2 percent of them are actually using it (North 2024). The introduction of new AI supercomputers, such as the launch of basketball court-sized one for use in Denmark (Novo Nordisk Foundation 2024), threaten to push the needle further from smart *cities* to *countries*.

2 The 15-minute city

City planners and policymakers have shifted their focus to making our urban neighborhoods more self-sufficient. Shocks like COVID-19, where social distancing limited our movement, further reinforced the push toward ensuring that necessary amenities are within reach. The 15-minute city principle is defined around the goal that all residents can meet most of their needs within a walk or bike ride of 15 minutes. Improved planning would ensure that essential services, grocery stores, shops, and parks are easily reachable for pedestrians or bikers, saving greenhouse gas emissions (Footprintnetwork n.d.).

The concept is credited to Carlos Moreno, a French professor and co-author of *The 15-Minute City: A Solution for Saving Our Time and Our Planet*. He was inspired by the challenges of the smart city and the need to reduce our car dependence in order to build a more human-centric city (Moreno, Gehl, and Thorne 2024). Paris's mayor – Anne Hidalgo – became one of the first to embrace it (Gongadze and Maassen 2023), closing main traffic arteries along the River Seine and re-utilizing them for pedestrians and bikes. The Dutch city of Utrecht has built it into their vision for 2040, Melbourne, Australia, for their 2017–2050 plan, and Ottawa, Canada, into its 25-year growth plan (see Figure 10.3) (Gongadze and Maassen 2023).

3 Expanding sustainable energy access

The city of Medellín in Colombia has emerged as a global model on how to transform a city: from the murder capital of the world into an eco-city marked by high degrees of resilience (Duran and Sofia 2017). Two hallmarks of the city's transformation were citizen engagement and investing in green infrastructure, such as building green corridors, which lowered temperatures in the city by 4 degrees Celsius (SE for All 2021).

In 2020, a local energy community was established as a pilot in a lower-middle-class neighborhood in Medellín, installing solar power on rooftops with support from the local utility company Empresas Públicas de Medellín (Cardena Alvarez et al. 2023). This energy supplier also actively worked on extending solar-powered access to rural areas that are not connected to the public grid (EPM 2024). While the pilot was successful, Cardena Alvarez et al. (2023) note that a just transition can be strengthened with a more bottom-up, community-led process.

In summary, smart cities and 15-minute cities offer prominent avenues to achieving sustainable development for urban residents. However, they also showcase the limits and challenges when applying our framework of city regions. For smart cities, the rural-urban digital divide threatens to make peripheral populations even less visible, reducing their urban access. For 15-minute cities, closing off traffic might be beneficial to wealthier neighborhoods, but can increase congestion and pollution in poorer neighborhoods on the periphery (Bou Sleiman 2024). Understanding these trade-offs is important when weighing up urban gains against creating a more inclusive city regions.

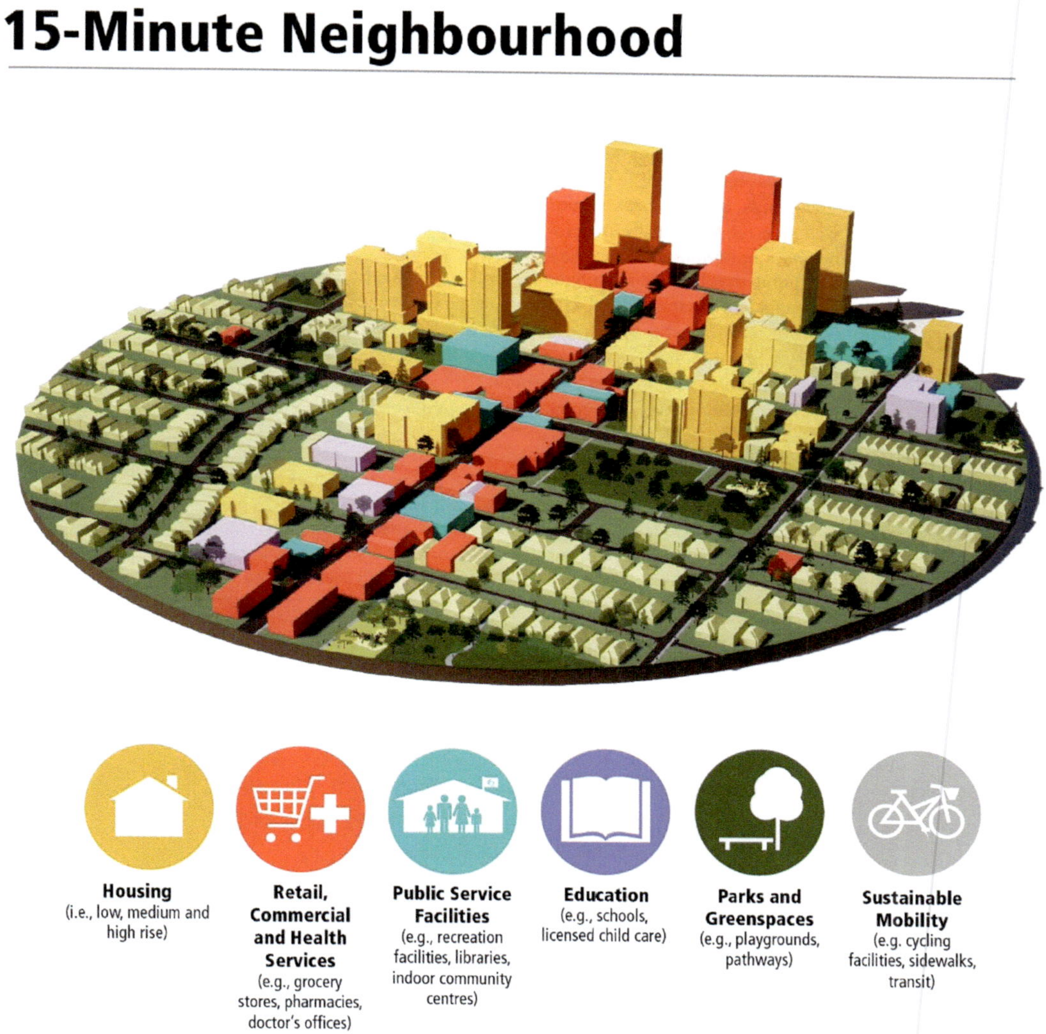

Figure 10.3
The plan for a 15-minute neighborhood in Ottawa, Canada, https://www.cpha.ca/public-health-and-planning-co-creating-15-minute-neighbourhoods-ottawa

GREENER, MORE RESILIENT CITIES

To improve the resilience of cities and make them more inclusive and greener, Mukim and Roberts (2023) propose the following five actions in Table 10.1: integration, information, incentives, insurance, and investments. An example is given for each action, drawing from the Global North and the Global South and a relevant synergy or trade-off.

Integration and a more resilient Bujumbura

Building on our discussion on integrating the rural-urban continuum and city regions, Mukim and Roberts (2023) offered the project of land certification around Bujumbura, Burundi, as an example for building resilience through

Table 10.1 Policy actions to create greener, more inclusive, and resilient cities. Source: Mukim and Roberts (2023)

Policy action	Description	Examples
Integration	Policy interventions and reforms that promote more compact cities and better integration of cities, both with other cities and with rural areas as a means of facilitating adaptation through migration and trade	Bujumbura, Burundi
Information	Policies and measures to improve the timely provision of credible information that helps people, businesses, and local governments better understand climate change-related risks, both across and within cities, and, in doing so, helps promote both mitigation and adaptation	Hong Kong, China
Incentives	Policy instruments that provide incentives for individuals and businesses to internalize negative environmental externalities, as well as institutional and other types of reform that provide government officials with incentives to work better together to address green, resilient, and inclusive development challenges. Incentives also include the removal of fossil fuel subsidies and other incentives that encourage activities with negative environmental externalities.	London, United Kingdom
Insurance	Policies and reforms that help people, businesses, and governments either to insure through the market or to self-insure against losses associated with climate change and other environmental shocks and stresses	New York City, USA
Investments	Investments by national and local governments in green, resilient, and inclusive urban infrastructure, including nature-based solutions, as well as measures to promote the crowding-in of private sector finance for such investments	Zhengzhou, China

greater integration. Informal settlements have grown 60 percent over the last four decades, mainly in high-risk flooding zones in the peri-urban areas (Mukim and Roberts 2023). The project, funded by the Swiss Agency for Development and Cooperation (SDC) and the EU, aimed to improve access to the land and housing markets, to ensure that settlements do not extend urban sprawl but increase the compactness (and subsequent integration) within the city.

The push for greater integration itself drew on lessons learnt from land tenure across the continent, including Madagascar (Elbow 2014). I was always struck when flying into Addis Ababa, Ethiopia, witnessing the persistent expansion of the city into the hinterland. Ethiopia's laudable efforts to build 300,000 low-income condominiums were fraught as its expansion infringed upon the tenure security of peripheral communities (Huang, Goodfellow, and Desta 2024). With land increasingly becoming a commodity, Leyronas, Coriat, and Nubukpo's (2023) work on the African commons reiterates the importance of managing trade-offs through more inclusive governance.

Other options to build resilience through integration for smaller cities include creating stronger connections to medium and larger cities, investing in public transit for historically underserved areas (Mukim and Roberts 2023, 311), and leveraging housing cooperatives (Simonneau et al. 2023).

Information and a more resilient Hong Kong

Hong Kong, China, has built an early warning system that allows people to reach safe spaces. Two crucial measures include investments in shelters at people's residences and building an adaptive transportation system. Investments in early warning systems have the largest return of investment across adaptation measures, with every dollar creating 10 dollars in return (Mukim and Roberts 2023). The intervention also leveraged synergies, integrating public education as a central pillar to increase its effectiveness (Kong, Kwan, and Pun 2020).

Other options to build resilience through information for smaller cities include building institutional capacities or upgrading building codes and urban design guidelines (Mukim and Roberts 2023, 310).

Incentives and a more resilient London

To reduce emissions from cars, congestion charging was introduced in London, UK. A trade-off is that the poorest households are least likely to afford such charges. Funds gained should hence be reinvested in public transportation (Mukim and Roberts 2023). Similar challenges exist for a fuel tax in Cape Town, South Africa, that is expected to increase urban inequality, hitting households with far commutes hardest (Liotta et al. 2022). New York City launched congestion pricing in January 2025, nearly doubling the speed of cars in notoriously jammed Manhattan (Roeder and Learner 2025).

Besides reducing time wasted in traffic, a systematic review showed that establishing congestion charging comes with health effects from lower road traffic injuries (Chamberlain et al. 2023). A similar concept of low emission zones was established in Tokyo, Japan, to lower the use of high-emission vehicles. It had the benefits of reducing occurrences of cardiovascular diseases from reduced air pollution (Chamberlain et al. 2023).

Other options to build resilience through incentives for smaller cities include offering subsidized housing or retrofitting buildings with tax credits (Mukim and Roberts 2023, 310).

Insurance and a more resilient New York City

New York's Metropolitan Transit Authority (MTA) was struck by Hurricane Sandy in October 2012. Causing US$4 billion in damages to MTA assets and infrastructure, the city opted to issue its first US$200 million catastrophe bond in 2013 "to transfer risk from storm surges, and more recently earthquakes" (Mukim and Roberts 2023). In case no disaster occurs over a set timeframe, the investor makes significant returns on their investment, whereas a disaster triggers a payout for the insured entity. Similar bonds have been used by corporations or even countries (e.g., Chile) to manage their risk of catastrophes (Braun and Kousky 2021). These financial measures need to build on synergies with climate actions, as recurring disasters will reduce investors' appetite, given the risk of consecutive losses.

Other options to build resilience through incentives for smaller cities include social protection or subsidizing insurance for residents (Mukim and Roberts 2023, 311).

Investments and a more resilient Zhengzhou, China

Improving how cities are absorbing and storing rainwater offers to further build resilience (Mukim and Roberts 2023). Ideally, the new infrastructure – such as permeable paving – mimics the hydrological cycle, as it can absorb, store, and release rainfall (Zaveri et al. 2021, 132–133). Popularized across China around a decade ago, the city of Zhengzhou is a leading pioneer in the country's effort to invest in natural infrastructure. The city invested over US$80 million into improved drainage and green spaces that can absorb excess rainfall (Barrett 2021). As the city was hit with a year's worth of rainfall in four days in 2021, the limits of the investments against extreme events emerged as flooding killed 71 people (Baragona 2021). The event highlighted the need to invest much more (US$14 million per square kilometer or over US$10 billion for Zhengzhou alone) (Barrett 2021). Given the sizable financing cost, it reinforces the need to make complementary low cost changes, such as improving green building standards that take the environment into account from the outset.

Other options to build resilience through investments for smaller cities include investing in affordable housing and access to renewable energy sources (Mukim and Roberts 2023, 311).

Following these success stories, Case Study 10.1 presents an extreme case of internal displacement and building better urban living conditions in two Somalian cities.

CASE STUDY 10.1 REDUCING VULNERABILITIES OF CLIMATE MIGRANTS IN SOMALIA

Climate change pushes people to move. To curb transnational migration flows, the Global North has increased their proactive messaging campaigns to reduce the pull effect of migration. With reduced appetite for humanitarian aid, efforts have also shifted toward building resilience to soften the push effects.

Missing from politicians' speeches and the geopolitical spotlight, climate change is also increasing the number of internal climate migrants. In sub-Saharan Africa an additional 86 million internal migrants are expected by 2050 (Rigaud et al. 2018). These internally displaced persons (IDPs), forced to abandon their land, frequently migrate to fragile cities that lack strong governance and are limited in capacity. It is in this development context that the UN and the Somali government launched the *Durable Solutions Initiative (DSI)* in 2016 to move toward longer-term solutions for IDPs in Somalia (Chawla 2020).

Annika Chawla reviewed the results from the *Midnimo I* project (translated as *unity* in Somali) implemented in collaboration with the International Organization for Migration (IOM) and UN-Habitat in the cities of Kismayo and Baidoa. The two cities are home to nearly half a million IDPs. Drawing on our changemakers, the project aimed to strengthen local, state, and federal policymakers to improve the integration of IDPs into society. A participatory project approach aimed to give IDPs a voice in decision-making (Chawla 2020).

At the heart of the project was co-creating solutions based on local community action plans (CAPs) as well as mobilizing funding. Capacity building was included for Somali government officials to use visioning exercises with communities to identify both short-term needs and long-term outcomes to increase self-reliance and build resilience. Outcomes included the construction or improvement of community infrastructure, such as schools, hospitals, or water, benefiting around 350,000 people (Chawla 2020). The project also improved how authorities and displaced communities engage with each other, empowering IDPs' voices through individual and collective action. A further milestone included a commission to deal with land rights.

As summarized by Chawla (2020), "[t]he DSI gives the people a voice in decision-making processes that shape their future – and offers a model for other cities that are, or soon will be, in similar circumstances." The DSI project showcases *in extremis* the effects of transforming our hotspot of rural-urban integration. An external evaluation identified the importance of training and capacity building to strengthen impacts (Axiom Monitoring and Evaluation Limited 2019). As nationalism is again on the rise, it also serves as a reminder that none of us – whether living in cities or villages – are immune from climate vulnerabilities. This chapter featured avenues to tackle them, by taking advantage of synergies and embracing a city regions perspective.

YOUR TURN

The example of a 15-minute city is gaining traction across the globe. What potential and challenges do you see in your closest urban area? Would they be addressed by urban planning that integrates the 15-minute city concept? What co-benefits and trade-offs can you identify? Who would be left out?

REFERENCES

Abanokova, K., H-A. Dang, S. Nakamura, S. Takamatsu, C. Pei, and D. Prospere. 2022. "Is Climate Change Slowing the Urban Escalator out of Poverty? Evidence from Indonesia and LAC." Background Paper. Washington, DC: World Bank.

Abu, M., S. N. A. Codjoe, W. N. Adger, S. Fransen, D. Jolivet, R. S. De Campos, M. F. Gavonel, C. Agyei-Asabere, A. H. Fábos, and C. Zickgraf. 2024. "Micro-scale Transformations in Sustainability Practices: Insights from New Migrant Populations in Growing Urban Settlements." *Global Environmental Change* 84. https://doi.org/10.1016/j.gloenvcha.2023.102790.

Axiom Monitoring and Evaluation Limited. 2019. "Midterm Evaluation of Midnimo Project to Maximize Peacebuilding Impact in Jubbaland, South West and Hirshabelle States, Somalia." Available at: https://www.un.org.peacebuilding/sites/www.un.org.peace building/files/documents/somalia_2018_project_evaluation_mid-term_0.pdf.

Baragona, S. 2021. "China's 'Sponge Cities' That Absorb Rainwater Pushed Past Limits." Available at: https://www.voanews.com/a/east-asia-pacific_chinas-sponge-cities-absorb-rainwater-pushed-past-limits/6208875.html.

Barrett, E. 2021. "China's Latest Floods Put Its Climate-Friendly 'Sponge Cities' to the Test." *Fortune*. https://fortune.com/2021/08/07/sponge-city-concept-zhengzhou-flooding-china-climate-change/.

Bou Sleiman, L. 2024. "Displacing Congestion: Evidence from Paris." Available at: https://www.dropbox.com/scl/fi/h9zmhaqut3w93w2sm8ie0/JMP_LBS_2022.pdf?rlkey=vn9l8t0f6yngei7j3ds67vt5g&e=1&dl=0.

Braun, A., and C. Kousky. 2021. *Catastrophe Bonds*. Wharton Risk Center Primer. Available at: https://esg.wharton.upenn.edu/wp-content/uploads/2023/08/Cat-Bond-Primer-July-2021.pdf.

Bugbee, B. 2015. "Vertical Farming: Turning Fossil Fuels into Food." Available at: https://bpb-us-e1.wpmucdn.com/blogs.cornell.edu/dist/8/8824/files/2020/03/PSC-departmental-seminar-Vertical-farming-Turning-fossil-fuels-into-food.pdf.

C40 Cities. 2016. "Cities100: Quito – Creating Electricity and Compost from Organic Waste." Available at: https://www.c40.org/case-studies/cities100-quito-creating-electricity-and-compost-from-organic-waste/.

Cardena, A., J. Pablo, J. George, J. G. Quiroz, J. A. E. Walker, J. M. E. Forero, S.O. Arango, and C. D. Restrepo. 2023. *Rethinking Energy Communities for a Just Transition*. Transactive Energy Colombia.

Center for Sustainable Development at Brookings (Ed.) 2023. *Forging Next Steps into the SDG Second Half*. Washington, DC: Brookings Institute.

Chamberlain, R. C., D. Fecht, B. Davies, and A. A. Laverty. 2023. "Health Effects of Low Emission and Congestion Charging Zones: A Systematic Review." *Lancet Public Health* 8 (7). https://doi.org/10.1016/s2468-2667(23)00120-2.

Chawla, A. 2020. "Fragile Cities Are Being Inundated with People Fleeing the Impacts of Climate Change: How Can They Cope?" Available at: https://ensia.com/articles/climate-migrant-urban-refugee-idp-displacement-fragile-cities/ (accessed July 16, 2024).

Chlouba, V., M. Mukim, and E. Zaveri. 2022. "How Do Climate Change-Related Stressors Affect Urban Form?" Background Paper, Washington, DC.

de Rivero, O. 2019. "The Myth of Development: Non-Viable Economies and the Crisis of Civilization." In *Development Essentials*. 2nd edn. London: Zed Books Ltd.

Dias, S. M. 2016. "Waste Pickers and Cities." *Environment and Urbanization* 28 (2): 375–390. https://doi.org/10.1177/0956247816657302.

Doherty, J. 2021. *Waste Worlds: Inhabiting Kampala's Infrastructures of Disposability*. Berkeley, CA: University of California Press.

Duran V., and P. Sofia. 2017. "Medellín: A Model for Cities Worldwide." Blog. Available at: https://blogs.worldbank.org/en/latinamerica/medellin-model-cities-worldwide.

Elbow, K. 2014. "Lessons on Decentralized Land Tenure Management and Property Rights Surveys." LandLinks. Available at: https://www.land-links.org/2014/06/lessons-on-decentralized-land-tenure-management-and-property-rights-surveys/ (accessed March 23, 2025).

EPM. 2024. "Energía Solar Rural." Available at: https://www.epm.com.co/clientesyusuarios/energia/tipos-de-energia/energia-solar-rural.html#.

Footprintnetwork. n.d. "15-Minute City." Available at: https://overshoot.footprintnetwork.org/portfolio/15-minute-city/.

Forster, T., and A. G. Escudero. 2014. *City Regions as Landscapes for People, Food and Nature*. Washington, DC: EcoAgriculture Partners, on behalf of the Landscapes for People, Food and Nature Initiative.

Foth, M., M. Brynskov, and T. Ojala (Eds.) 2015. *Citizen's Right to the Digital City*.

Gongadze, S., and A. Maassen. 2023. "Paris' Vision for a '15-Minute City' Sparks a Global Movement." Available at: https://www.wri.org/insights/paris-15-minute-city.

Greenfield, A. 2013. "Against the Smart City." Pamphlet.

Greenfield, A. 2017. *Radical Technologies: The Design of Everyday Life*. London: Verso.

Hoover, D. 2017. "Estimating Quantities and Types of Food Waste at the City Level." NRDC. Available at: https://www.nrdc.org/sites/default/files/food-waste-city-level-report.pdf.

Huang, Z, T. Goodfellow, and M. K. Desta. 2024. "Homeownership and Tenure (In)Security in Fractured Urban Peripheries: Ethiopia's Mass Housing Programme." *Housing Studies*: 1-23. https://doi.org/10.1080/02673037.2024.2427668.

iPES Food. 2017. "What Makes Urban Food Policy Happen? Insights from Five Case Studies." International Panel of Experts on Sustainable Food Systems. Available at: https://www.ipes-food.org/_img/upload/files/Cities_full.pdf.

Kong, V. W. W., J. S. H. Kwan, and W. K. Pun. 2020. "Hong Kong's Landslip Warning System—40 Years of Progress." *Landslides* 17 (6): 1453–1463. https://doi.org/10.1007/s10346-020-01379-6.

Lawson, L. 2016. "Agriculture: Sowing the City." *Nature* 540: 522–524. https://doi.org/10.1038/540522a.

Leyronas, S., B. Coriat, and K. Nubukpo. 2023. *The Commons: Drivers of Change and Opportunities for Africa*. Washington, DC: AfD and World Bank.

Liotta, C., P. Avner, V. Viguié, H. Selod, and S. Hallegatte. 2022. "*Climate Policy and Inequality in Urban Areas: Beyond Incomes*." Washington, DC: World Bank.

Lipton, M. 1977. *Why Poor People Stay Poor: A Study of Urban Bias in World Development*. Cambridge, MA: Harvard University Press.

Lipton, M. 1984. "Urban Bias Revisited." *Journal of Development Studies* 20 (3): 139–166. https://doi.org/10.1080/00220388408421910.

Moreno, C., J. Gehl, and M. Thorne. 2024. *The 15-Minute City: A Solution to Saving Our Time & Our Planet*. Hoboken, NJ: Wiley.

Mukim, M., and M. Roberts. 2023. *THRIVING: Making Cities Green, Resilient, and Inclusive in a Changing Climate*. Washington, DC: World Bank.

New York City Council. 2024. "Emergency Food in NYC." Available at: https://council.nyc.gov/data/emergency-food-in-nyc/.

North, M. 2024. "How 5 Cities Are Making Innovative Use of Generative AI." Available at: https://www.weforum.org/stories/2024/07/generative-ai-smart-cities/.

Novo Nordisk Foundation. 2024. "Danish Centre for AI Innovation." Available at: https://novonordiskfonden.dk/en/projects/danish-centre-for-ai-innovation/.

Rigaud, K. K., A. de Sherbinin, B. Jones, J. Bergmann, V. Clement, K. Ober, J. Schewe, S. Adamo, B. McCusker, S. Heuser, and A. Midgley. 2018. *Groundswell: Preparing for Internal Climate Migration*. Washington, DC: World Bank.

Roeder, O., and S. Learner. 2025. "First US Congestion Pricing Scheme Brings Dramatic Drop in NY Traffic." *Financial Times*, January 15. https://www.ft.com/content/c229b603-3c6e-4a1c-bede-67df2d10d59f.

Santo, R., A. Palmer, and B. Kim. 2016. "Vacant Lots to Vibrant Plots: A Review of the Benefits and Limitations of Urban Agriculture." Working Paper. Johns Hopkins Center for a Livable Future. Available at: https://clf.jhsph.edu/sites/default/files/2019-01/vacant-lots-to-vibrant-plots.pdf.

Sassen, S. 2011. "The Future of Smart Cities." Available at: http://opentranscripts.org/transcript/future-of-smart-cities/.

SE for All. 2021. "Creating a Greener, Cooler and Healthier Medellín." Available at: https://www.seforall.org/stories-of-success/creating-a-greener-cooler-and-healthier-medellin.

Simonneau, C., B. Bon, E. Denis, S. Leyronas, and I. Sory. 2023. "Housing and the Future of Rural Land-Based Commons." In *The Commons: Drivers of Change and Opportunities for Africa*, edited by S. Leyronas, B. Coriat, and K. Nubukpo. Washington, DC: AfD and World Bank.

United Nations. 2024. *The Sustainable Development Goals Report 2024*. New York: United Nations.

Walling, M., and K. Lafleur. 2023. "Indoor Farming Startups with Hundreds of Millions File Bankruptcy." *Fortune*.

World Bank. 2024. "Global Smart City Partnership Program." World Bank Group. Available at: https://www.worldbank.org/en/programs/global-smart-city-partnership-program (accessed February 14, 2025).

WRI Ross Center. 2024. "Re-Ciclo." Available at: https://prizeforcities.org/project/re-ciclo (accessed March 31, 2025).

Zaveri, E., J. Russ, A. Khan, R. Damania, E. Borgomeo, and A. Jägerskog. 2021. *Ebb and Flow: Vol. 1. Water, Migration, and Development*. Washington, DC: World Bank.

11 Green energy access

RECAP OF CHALLENGES AND WHY THIS HOTSPOT MATTERS

Using the latest data available (United Nations 2024), the following five challenges can be addressed by this transformation hotspot.

SDG 7: Affordable and Clean Energy

Renewable energy – especially wind and solar – has not only become more affordable, it represents a larger share of the energy mix consumed. In six years alone, its share increased from 22.9 percent in 2015 to 28.2 percent in 2021. With the goal in sight to triple capacity by 2030, two challenges emerge: (1) renewable energy consumption in heating and transportation remains limited, growing at a much slower pace. In transport, for example, the share of total energy consumed by renewables increased from a low level of 3.5 percent in 2021 to only 4.4 percent in 2021. (2) Progress in improving renewable energy access is geographically uneven. While the wealthiest countries have increased the amount of watts per capita from renewables by 382 watts (a 55 percent increase from 691 to 1073 watts from 2015 to 2022), in the least developed countries, the renewable watts per capita increased merely by 13 watts (from 27 watts in 2015 to 40 watts in 2022) (United Nations 2024, 23).

DOI:10.4324/9781003400028-16

A transformation improving green energy access focuses on closing gaps between the Global North – where renewables have taken off – and the Global South – where uneven access to electricity persists and renewable options remain limited.

SDG 9: Industry, Innovation and Infrastructure

We are reducing CO_2 emissions intensity from manufacturing, with emissions growing at a slower pace than global GDP growth. From 2015 to 2021, each dollar gained from manufacturing tracked with almost 20 percent lower emissions (dropping from 0.49 kg of CO_2 to 0.40 kg of CO_2). These trends mask that the overall emissions still increased to a record 37.4 gigatons. Driven by coal, our overall emissions highlight the need to increase the availability of renewable energy for the private sector (United Nations 2024, 27).

Improving green energy access also entails enabling manufacturing companies to access green energy sources and an upgraded energy grid.

SDG 10: Reduce Inequality

Over two billion people lack access to clean energy sources for cooking. Some 21 percent of the global population (primarily in sub-Saharan Africa) are projected to continue using wood, charcoal, or biomass for their cooking, creating health burdens for women and children (United Nations 2024, 22). Besides cooking, we need to consider the multi-dimensional use of household energy, as polluting stoves are utilized for room heating, hygiene, insect repulsion, illumination, and income generation (Grabher et al. 2023).

Both low- and high-tech options are needed to tackle the threat of climate change, reduce air pollution that causes unnecessary deaths each year, and address household needs.

SDG 13: Climate Action

The shock from the war in Ukraine on the global energy markets triggered a rise in energy prices. To reduce the impact on consumers, governments increased their fossil fuel subsidies. They rose from a pandemic low of US$372 billion in 2020 to a record high of US$1.5 trillion in 2022. These subsidies not only kept fossil fuel prices artificially low, they harmed the competitiveness of green energy alternatives (United Nations 2024, 35).

A green energy transformation includes revising governance and policymaking to ensure that the limited public funding is directed towards renewable outlets.

SDG 14: Life below Water

As global temperatures rise, so do ocean temperatures. Fragile coral reefs play a crucial role in ensuring a healthy marine life, including supporting fishing communities across the world. Between 2023 and 2024, we had 13 consecutive months of the highest temperatures ever recorded. These record high ocean temperatures contribute to coral bleaching events, threatening the survival of both the reefs and the surrounding communities' livelihoods (United Nations 2024, 37).

Transformations on supporting green energy access will contribute to the reduction of global warming, contributing to address transnational challenges and unlocking positive synergies across the board.

THE INTERSECTIONAL CHALLENGE OF CLIMATE CHANGE

Climate change affects all aspects of our lives. It is an intersectional challenge, presenting economic, technical, societal, and political issues. As summarized by the Center for Sustainable Development at Brookings (2023, 3), climate change represents:

"*an economic issue* challenging institutions' ability to distribute necessary capital and equitably distribute opportunities"
"*a technical issue* requiring rapid development of context-calibrated technologies at global scale"
"*a societal issue* contributing to instability in many geographies and disproportionately affecting vulnerable populations"
"*a political issue* requiring diverse stakeholders to agree on specific problems and common solutions"

THE ROLE OF OUR CHANGEMAKERS

Improving green energy access is one transformation where the role of the changemaker – governance and policymaking – looms large, especially as we balance regulatory frameworks with social and environmental interests (Jones 2021). Science and technology have continuously improved, and calls for action are not dying down. What started visibly with Skolstrejk from Greta Thunberg has grown into a global movement, including activists from the Global South (UNEP 2019). Green energy access has strong linkages and synergies across the SDGs (UN 2021; Sustainable Energy for All 2024). Given the challenges in mobilizing finance to improve green access, the private sector will need to step up the most to bring about a transformation.

NAVIGATING (SHIFTING) PRIORITIES FROM THE GLOBAL NORTH

Daniel Cohan, in his book, *Confronting Climate Gridlock: How Diplomacy, Technology and Policy Can Unlock a Clean Energy Future*, covers how countries have failed to follow through on their commitments from the Paris Agreement (Cohan 2022)[1]. Seven parties to the agreement play an oversized role, as they make up two-thirds of global greenhouse emissions: China, the USA, the EU, Russia, Japan, and Brazil. Cohan argues that the USA – given its economic growth derived from the use of fossil fuels – should position itself as a leader to get us out of this gridlock. Mirroring our changemakers, Cohan (2022, 3) proposes three concrete steps:

1. *Diplomacy* (to "create ... an international playing field that motivates U.S. action and leverages it to accelerate action abroad").
2. *Technology* ("that make[s] emission reductions achievable and affordable across all sectors of the energy economy").
3. *Policy* (to "accelerate the development and adoption of those technologies domestically and abroad").

To overcome climate gridlock, collective action is needed that creates incentives, establishes policies that minimize free riders, and rewards countries that take more decisive action (Cohan 2022).

With the burning of fossil fuels accounting for over three-quarters of global emissions, decarbonization remains at the core of climate efforts (Cohan 2022). Cohan's strategy for decarbonization includes: (1) increased energy efficiency to shrink emissions across all sectors; (2) the use of clean electricity to replace fossil fuels; and (3) using carbon-sinks to offset any remaining emissions to achieve a net-zero balance. Both demand (i.e., consumption) and supply (i.e., technology) sides need to be enabled. We risk overestimating supply-side factors and technology's impact, as discussed in Chapter 2. Stimulating demand or meeting existing demand are equally important (Cohan 2022).

In a global endeavor, domestic policy shifts matter – and not just in the USA which dominates climate headlines. A vacuum left behind by the USA provides opportunities for China to take over global climate leadership (Murtaugh 2024). Australia, witnessing climate change, from the dying off of the Great Barrier Reefs to record-breaking droughts, has experienced continued political turmoil in efforts to reduce carbon emissions (Cave 2018). Lobbying efforts – estimated at over US$3 million for the coal industry versus around US$135,000 for environmental groups (Cave 2018) – were lopsided in favor of more conservative candidates. The 2022 election saw historic levels of support for candidates in favor of tackling the climate crisis more aggressively. The winner, Prime Minister Anthony Albanese, has since made significant commitments to increase Australia's carbon reductions, hoping to make his country's spot among the G20 as the top per capita emitter of greenhouse gases a sin of the past (Turnbull 2023).

USING DEMAND TO FINANCE ENERGY TRANSFORMATION...

Conservatives contend that renewable energy relies too heavily on government subsidies to succeed. Instead of stimulating new demand, the lower hanging fruit is to identify where unmet demand for renewable energy exists. One such initiative is the *Clean Energy Demand Initiative* – co-led by the U.S. State Department in partnership with the Clean Energy Buyers Association, which aimed to increase the visibility of private-sector demand for clean energy. A sustainable clothing manufacturer in Southeast Asia, for example, wants to draw on solar power for its manufacturing plant. The initiative engaged the respective governments and relevant stakeholders in the energy sector to jump-start a transformation.

Instead of leveraging more donor funding, the main drive was to match existing demand with potential clean energy suppliers. Renamed the *Corporate Energy Demand Initiative* under the second Trump administration, the initiative initially avoided more dogmatic efforts as it mirrored the interests of US businesses (and ultimately consumers) to drive demand for a transformation. In its first three years, "over 100 companies have indicated demand for clean energy in 14 markets, translating into hundreds of billions of dollars in potential clean energy investment" (U.S. Department of State 2024). Such positioning could provide political capital to sustain similar public-private platform across changes in administrations. It also reflects how diplomacy extends beyond providing funding, including capacity building or standard setting (e.g., in the case of mining).

... AS THE WORLD HOLDS ITS BREATH FOR CLIMATE FINANCING

A minor breakthrough was reached at COP 29 in Baku, Azerbaijan, to raise climate financing commitments to US$300 billion annually by 2035. For Martin Wolf (2024), the financing deal "is too little, too late." COP 29 also opened up pathways for carbon credits to be traded across countries, expecting to support additional financing from polluters to countries in the Global South (UNFCCC 2024), at the risk of large-scale land grabs in Africa (Bryan 2023) or the Amazon region (Pooler 2024). The carbon credits market has also been plagued by recent scandals of increased emissions or having sold offsets that don't exist (Follow The Money 2023; Greenfield 2023), creating a crisis of confidence across the sector (Dvorak 2023; Bryan 2024).

Beyond financing, the looming question is how many countries are stepping up their next Nationally Determined Contributions (NDCs) to stay within the 1.5 degree warming. Cutting greenhouse gases in half by 2030 and achieving net-zero carbon dioxide emissions by 2050 gives us a 50 percent chance to reach that goal (Plumer 2023). We officially crossed the 1.5° Celsius threshold globally for the first time in 2024. To get us on the right track, we need to leverage diplomacy and finance, but also engage our other two changemakers.

A *LOCAL* TRANSFORMATION

With all the global turbulence, we can't lose sight that ultimately, the energy transformation will be local. According to experts from the Center for Sustainable Development at Brookings (2023), "to deliver clean and reliable energy access globally will only be achieved if energy solutions move beyond technical realms and become human-centered and locally calibrated." As captured by my former mentor, Dr. Robin Leichenko, and her colleague, Dr. Karen O'Brien, a successful transformation to address climate change occurs across practical, political, and personal spheres, including a shift in socio-cultural norms and strengthening our individual and collective agency (Leichenko and O'Brien 2019, 180–183).

For a human-centered and locally calibrated effort, let us return to Medellín and its progress and potential from a justice perspective. Leveraging energy communities to drive change from the bottom-up requires greater access to external financing for households and lowering the barriers through new legislation for communities to connect to the grid (Cardena Alvarez et al. 2023). A more bottom-up approach in creating energy communities has the power to support distributional justice beyond just providing improved energy access. Additional avenues include reductions in energy costs, creating new income sources from feeding renewable power to the grid, and improving infrastructures. Procedural justice can be strengthened through greater citizen participation, building their social capital and community resilience. Recognition justice in energy communities entails promoting diversity and inclusion of marginalized communities (Cardena Alvarez et al. 2023).

Such just transformations are expected to have a significant effect on inequalities. Creating green jobs offers opportunities to take advantage of the synergies from climate change (SDG 13), reducing inequality (SDG 10), and creating decent work and economic growth (SDG 8). With youth unemployment on the rise across the Global South, leapfrogging through an energy transformation can create new job opportunities and improved livelihood prospects for the next generation to come (Center for Sustainable Development at Brookings 2023).

With the potential of improving green energy access and its relationship to justice established, let us move to our most in-depth look at attempts to achieve a just energy transformation in South Africa in Case Study 11.1.

CASE STUDY 11.1 JUST ENERGY TRANSITION IN SOUTH AFRICA

The history of just transitions

The term *just transitions* emerged out of US trade unions' efforts in the 1980s to secure workers' rights and livelihoods as the economy shifted away from pollutive industries (e.g., nuclear and atomic industries) (Otlhogile and Shirley 2023). The term was later linked with the environmental justice movement (see Chapter 5), creating alliances between the movements (Otlhogile and Shirley 2023). Researchers also apply just transitions as a theoretical framework and a governance strategy (Wang and Lo 2021).

By 2015, the preamble of the United Nations Framework Convention on Climate Change (UNFCCC) (2015, 2) Paris Agreement explicitly referenced the "imperative of a just transition of the workforce." Translated from its more radical origins into the pavilions of international climate conferences, Otlhogile and Shirley (2023, 3) define just transition as "the set of principles, strategies, and practices required to transition workers and their surrounding communities into thriving economies that provide dignified, productive and ecologically sustainable livelihoods, participatory and democratic governance, and ecological resilience."

Increasing inequalities post-COVID-19 threaten an increase in vulnerabilities for marginalized communities. Given their increased vulnerabilities to climate change, emerging movements argue that a transition to a low carbon environment needs to occur in a fairer economic system (Otlhogile and Shirley 2023). Echoing the spirit of the SDGs, these movements "aim to build economic and political power from the bottom up to embed the shift from extractive and exploitative economies to sustainable economies with principles of fair participation, fair distribution, and fair compensation and leave no one behind" (Otlhogile and Shirley 2023, 2).

What follows is an exploration of one of the four main fields where just transition has been applied: the Just Energy Transition. The other fields are Just Rural Transition, Just Urban Transition, and the Just Circular Economy Transition (Otlhogile and Shirley 2023, 3). As defined by Otlhogile and Shirley:

> [A]cknowledging that a transformation of the global energy sector from fossil-based to zero-carbon by the second half of this century sits at the core of limiting severe climate change, the Just Energy Transition specifically focuses on energy sectors, both upstream and downstream, to secure workers' rights, to provide social protection for affected workers and communities, to remove barriers to integrating fossil fuel workers into clean energy sectors, and to build the necessary workforce capacity to deliver clean energy goals. It is often expanded to include delivering universal, reliable, affordable energy access.
>
> (2023, 3)

Enter JETP

The launch of the *Just Energy Transition Partnership* (JETP, pronounced Jet-P) between South Africa and France, Germany, the United Kingdom, the United States, and the European Union was met with great fanfare at COP 26 in November 2021. As the first of several JETPs,[2] it received considerable media attention and was hailed as a success story by leveraging governance and economy and finance to bring about a sustainable energy transformation by 2027. To address the estimated US$2.4 trillion annual funding shortfall needed to combat global warming by 2030, the partnership sought to leverage public funds to lower investment risk and draw in private capital (*The Economist* 2022).

In 2022, the South African government formed the 2022 *Presidential Climate Commission* and its framework for a Just Transition (Presidential Climate Commission Towards a Just Transition 2023). For South Africa, the donor governments committed to provide US$8.5 billion to contribute to the overall US$85 billion needed (The Presidency Republic of South Africa 2022).

It seemed to be an ideal case for public-private partnerships, as South Africa was already looking to transition toward renewables away from coal, but lacked funding. South Africa also has ideal climatic conditions for solar and wind energy to thrive. Its large dependency on coal (80 percent of its energy production) and single operator with Eskom, the electricity utility, should ensure a manageable and effective setup for change (Sguazzin and Burkhardt 2023).

Mismanagement at Eskom, however, has successively worsened the energy supply across South Africa. In February 2023, the South African president declared a national state of disaster to "address an electricity crisis so dire that day-old chicks are freezing to death, supermarket owners are rushing to sell meat before it spoils and many businesses have been forced to shut down" (Chutel 2023).

And it was getting worse: by April 2023, the number of planned blackouts because of lack of supply were already higher than all of 2022 combined. Since the announcement of JETP in 2021, 157 days in 2022 and 215 days in 2023 have been affected by blackouts, compared with an average of 28 days of blackouts from 2018 to 2021 (esp.info 2023). Beyond negative impacts on individual livelihoods (imagine storing your food in a freezer that constantly loses power), power cuts are estimated to have reduced South Africa's GDP by 7 to 8 percent in 2022 alone (*The Economist* 2023). Khaya Sithole, a South African analyst, lamented the failure to maintain the energy grid: "Maintenance does not amount to a new project, so if it's not a new project, you're not going to have a P.R. exercise where you're cutting a ribbon" (Chutel 2023). These setbacks underscore both the urgent need for action and the challenges of transformation.

It did not help that the former head of Eskom, Andre de Ruyter, was poisoned in a suspected attack on his life because he revealed corruption and mismanagement within Eskom and beyond (Pilling 2023). According to his estimates, the failing energy company lost around US$55 million a month because of theft (Sguazzin and Burkhardt 2023). "You can put solar on your roof, live in a gated estate, have private security, own a holiday home. But if 90% of the population are suffering, how is that sustainable?" said Andre de Ruyter, former head of Eskom, South Africa's electricity utility (cited in *The Economist* 2023).

A model for energy transformation that stumbled...

For the International Partners Group that are the leading financiers of the Just Energy Transition Partnerships, the deal with South Africa was a template for subsequent deals in Indonesia, Senegal, and Vietnam (Magid 2023). These investments are made as a partial counter to China's increasing influence in the Global South (Putsch 2022). But in South Africa, with the constant blackouts, domestic pressure resulted in the government only closing one coal-fired power plant and pledging to keep others operating as long as energy shortages continue (Sguazzin and Burkhardt 2023).

Thankfully, by 2024, load shedding appears to be a ghost in the past. Better governance and state investments in improved infrastructure have helped overcome the gaps left behind by the private sector. Overall energy demand was lower because of independent investments in renewable energy that reduced the pressure on the grid (Habershon 2024). Showing the utility of renewables, the cost of solar panels and battery storage in South Africa has declined over recent years. Private investments in solar over a span of two years almost tripled the amount of energy (*The Economist* 2024b). For commercial customers, private solar power costs are 20 to 40 percent lower than the cheapest electricity obtainable from Eskom (*The Economist* 2024b).

... and raises questions around its justness that it was meant to address

From a justice perspective, however, Andre de Ruyter's words on who has access to renewables still echo. And keeping the coal-powered plants on line past 2030 is estimated to come not only with increasing economic costs, but also to cause over 15,000 additional deaths from air pollution (van Diemen 2024).

Otlhogile and Shirley (2023, 3) argue that any African transition needs to address the global power imbalances, as well as the uniqueness of the African labor force. In stark contrast to the successful transition away from coal in the German Ruhr region, where workers were offered early retirement, the African work population, on average, is significantly younger and rarely organized in labor unions (Otlhogile and Shirley 2023, 6). They furthermore lack social protection and labor rights, as they operate outside the formal economy.

As a whole, Black South Africans continue to suffer as "[a]partheid has essentially persisted in economic form" (Goodman 2017). Women experience higher vulnerabilities from climate change. We need to account for such persistent gender inequalities by integrating gender-sensitive approaches, including developing their social capital (Lewis and Browne 2023).

A closer look of the JETP deal shows that it might fall short of accounting for the above challenges. Loans make up 90 percent of the total JETP funding to support the transition away from coal-fired power plants to private-sector renewables. Within that share, JETP funding is heavily biased (also 90 percent) toward the improvement of infrastructure as part of the energy transformation (Magid 2023). The *just* focus of the deal is minimal: only 0.14 percent are allocated to skills development and 0.19 percent to social investments and inclusion (Magid 2023).

Given the current low-income jobs offered in coal production, the workers in the coal belt in Mpumalanga Province are increasingly worried about their role post-transition (Magid 2023). Eskom CEO, Dan Marokane, argued that the shutdown of a coal plant in Komati, Mpumalaga, in 2022 was equivalent to an "'atom bomb' for jobs" (*The Economist* 2024a), even with nearly half a billion of World Bank funding dedicated to convert the plant for renewables and reskilling. With unemployment in South Africa soaring above 30 percent by 2023 (Magome 2023), unions are protective, given the uncertainty that displaced mining workers would find jobs (Cock 2019). As captured by Pravin Gordhan, South Africa's minister overseeing Eskom: "You've got to put the 'just' part first ... You've got to take care of societies, businesses, individuals, people working in mines, people working in power stations. Don't just promise them training. Get them training on the ground" (Smyth and Pilling 2023).

Coal has also played a large role in the apartheid struggle of South Africa (Sguazzin and Burkhardt 2023). A fear also exists that the dismantling of state control over energy will increase inequalities in access to energy, as well as increased debt burdens (Magid 2023). The threat to remove funding because of slow progress (or reversal of action on coal-powered plants) falls hollow. Europe relies on non-renewable fuels during Russia's war in Ukraine (Putsch 2022) and the US administration declared a national energy emergency to bolster fossil fuel production (Guarna and Burger 2025).

The following account details the tense relationship between Germany, a donor country and member of the G7, and South Africa, a founding member of the BRICS block and 2025 host of the G20:

> Gwede Mantashe, the coal-supporting minister of mineral resources and energy, told the [German] Green Party politician [Habeck] that South Africans didn't want to be the West's guinea pig for the global energy transition. Habeck offered to withdraw the JETP funding, according to people familiar with the conversation.
>
> (Sguazzin and Burkhardt 2023)

From an environmental perspective, both mining and burning of coal are at the heart of environmental justice issues. As noted above, keeping the coal-powered plants on longer will increase deaths from pollution. Cock (2019, 860) argues that these inequalities

> [also encompass people's] lack of universal access to critical resources such as clean water and air, fertile land, clean, affordable energy, and in their vulnerability to the extreme weather events such as droughts, heat waves, crop failures and floods associated with climate change.

Poor, black communities are living in the closest proximity to both mines and coal power stations, resulting in negative direct effects due to "loss of their health due to air pollution, and dealing with forced removals, social dislocation and dispossession" (Cock 2019, 864). Negative effects extend to their backyard food production and livestock, with soils depleted, access to water limited. About 2,000 deaths per year, as well as nearly 10,000 chronic bronchitis cases in children aged 6–12 are observed in the heartland of Mpumalanga, where the coal mines and coal power plants are located (Cock 2019, 864).

CHANGEMAKERS FOR A BRIGHTER FUTURE?

As stated at the outset of this chapter, science and technology are not our main limiting factors for achieving improved access to green energy. We need to activate or reactivate government, private sector, and civil society to get us there. Improved governance and investments in South Africa have resulted in at least bringing the load shedding to a halt. After losing its majority for the first time post-Apartheid, the African National Congress (ANC) formed a coalition government with the Democratic Alliance (DA) during the 2024 Presidential Election (Imray and Magome 2024), bolstering confidence in a turnaround. Funding has also increased, with 2.4 out of the US$8.5 billion raised by the end of 2023 (Mahlaka 2024). European countries are stepping up to fund gaps left behind from a US retreat (Ainger, Yokoyama, and Harsono 2025). Regarding individual and collective action, Cock (2019, 863) argues that "the most powerful driver of a transformative just transition would be co-operation between the labour and environmental movements, a 'red-green' alliance." These alliances were at the heart of driving successful transitions elsewhere.

To get us there, however, awareness of the JETP needs to increase nationally. A survey of South African households by Roberts, Struwig, and Zondi (2024) found that 65 percent have never heard the term *just transition* and similar numbers have heard nothing or only little about the Just Energy Transition. Yet 62 percent approve or strongly approve moving from coal to solar or wind energy. While benefits are clear, respondents are worried about paying even more for electricity or losing their jobs. Reskilling and (financial) support for affected populations received over 70 percent support across the board.

In conclusion, JETP kicked off with a lot of fanfare with a focus on a technological solution to an energy crisis. But now, financing needs to step up. With the new US administration withdrawing from financing deals, opportunities exist for other parties to take the lead. The Ford Foundation, for example, is funding a five-year initiative to foster just energy transitions across the continent (Ford Foundation 2025). African leaders are also stepping up their commitment to electrification by committing US$35 billion to increasing energy access, including solar mini-grids (Bearak 2025). The success of JETP ultimately depends on the changemakers' abilities to put the lives affected and the local at the heart of its transformation.

YOUR TURN

Using the example of JETP in South Africa, what are the synergies and trade-offs you can identify at the SDG Goals and Targets level? What actions have been undertaken to address trade-offs? What other actions could be taken, including by leveraging changemakers?

Based on your assessment, what do you think is the key to unlock climate gridlock moving forward?

On a personal level, what form of energy do you consume the most on a daily basis? And what steps can you take?

NOTES

1 With thanks to my former graduate students and their podcast covering the book.
2 South Africa was followed by deals in Indonesia and Vietnam. While the JETP deal in Asia struggles with similar financing challenges, new financing opportunities emerge through carbon credits generated from emissions avoided by shutting down coal plants earlier than planned (*The Economist* 2024a).

REFERENCES

Ainger, J., E. Yokoyama, and N. Harsono. 2025. "US Retreat Sparks Push to Save $45 Billion Climate Deals." *Bloomberg Law*. Available at: https://news.bloomberglaw.com/health-law-and-business/us-retreat-sparks-new-push-to-save-45-billion-climate-deals.

Bearak, M. 2025. "Inside a New Plan to Bring Electricity to 300 Million in Africa." *The New York Times*, January 27.

Bryan, K. 2023. "The Looming Land Grab in Africa for Carbon Credits." *Financial Times*, December 5. https://www.ft.com/content/f9bead69-7401-44fe-8db9-1c4063ae958c.

Bryan, K. 2024. "Staff Criticise Green Targets Verifier Over Carbon Credits." *Financial Times*, April 13.

Cardena A., J. Pablo, J. George, J. G. Quiroz, J. A. E. Walker, J. M. E. Forero, S. O. Arango, and C. D. Restrepo. 2023. *Rethinking Energy Communities for a Just Transition*. Medellín: Transactive Energy Colombia.

Cave, D. 2018. "Australia Wilts from Climate Change. Why Can't Its Politicians Act?" *The New York Times*, August 21. https://www.nytimes.com/2018/08/21/world/australia/australia-climate-change-malcolm-turnbull.html.

Center for Sustainable Development at Brookings. (Eds.) 2023. *Forging Next Steps into the SDG Second Half*. Washington, DC: Brookings Institute.

Chutel, L. 2023. "Dead Chicks and Rotting Meat: South Africa Declares 'Disaster' Over Power Cuts." *The New York Times*, February 9. https://www.nytimes.com/2023/02/09/world/africa/south-africa-power-energy-crisis.html.

Cock, J. 2019. "Resistance to Coal Inequalities and the Possibilities of a Just Transition in South Africa." *Development Southern Africa* 36(6): 860–873. https://doi.org/10.1080/0376835x.2019.1660859.

Cohan, D. S. 2022. *Confronting Climate Gridlock: How Diplomacy, Technology, and Policy Can Unlock a Clean Energy Future*. New Haven, CT: Yale University Press.

Dvorak, P. 2023. "He Pioneered Carbon Offsets to Save Tropical Forests. Now the Market Is Collapsing." *The Wall Street Journal*, October 30. https://www.wsj.com/science/environment/he-pioneered-carbon-offsets-to-save-tropical-forests-now-the-market-is-collapsing-18a5bc54.

esp.info. 2023. "EskomSePush (ESP)." Available at: https://esp.info (accessed August 31, 2025).

Follow The Money. 2023. "Showcase Project by the World's Biggest Carbon Trader Actually Resulted in More Carbon Emissions." Available at: https://www.ftm.eu/articles/south-pole-kariba-carbon-emission.

Ford Foundation. 2025. "Ford Foundation Supports Efforts to Center Justice in Africa's Energy Transition." Available at: https://www.fordfoundation.org/news-and-stories/news-and-press/news/ford-foundation-supports-efforts-to-center-justice-in-africas-energy-transition/ (accessed March 31, 2025).

Goodman, P. S. 2017. "End of Apartheid in South Africa? Not in Economic Terms." *The New York Times*, October 24. https://www.nytimes.com/2017/10/24/business/south-africa-economy-apartheid.html.

Grabher, H. F., H. Rau, S. T. Ledermann, and H. Haberl. 2023. "Beyond Cooking: An Energy Services Perspective on Household Energy Use in Low and Middle Income Countries." *Energy Research & Social Science* 97. https://doi.org/10.1016/j.erss.2023.102946.

Greenfield, P. 2023. "Revealed: More Than 90% of Rainforest Carbon Offsets by Biggest Certifier Are Worthless, Analysis Shows." *The Guardian*, January 18.. https://www.theguardian.com/environment/2023/jan/18/revealed-forest-carbon-offsets-biggest-provider-worthless-verra-aoe.

Guarna, O., and M. Burger. 2025. "Demystifying President Trump's 'National Energy Emergency' and the Scope of Emergency Authority." Sabin Center for Climate Change Law. Available at: https://blogs.law.columbia.edu/climatechange/2025/02/14/demystifying-president-trumps-national-energy-emergency-and-the-scope-of-emergency-authority/ (accessed March 31, 2025).

Habershon, E. 2024. "Joy and Relief as South Africa Manages to Keep Its Lights On." BBC. Available at: https://www.bbc.com/news/articles/c4ng1ygxzddo.

Imray, G., and M. Magome. 2024. "South Africa's President Ramaphosa Is Reelected for Second Term After a Dramatic Late Coalition Deal." Available at: https://apnews.com/article/south-africa-president-election-parliament-b71d8f3de7b3bb4be76f94e8d75f0daf.

Jones, S. 2021. "Alpine Nimbyism Freezes Swiss Green Energy Dreams." *Financial Times*, July 5. https://www.ft.com/content/062ae877-66c8-4782-8838-3a57b3873a1b.

Leichenko, R. M., and K. L. O'Brien. 2019. *Climate and Society: Transforming the Future.* Cambridge: Polity.

Lewis, F., and M. Browne. 2023. "Shaping Ecosystem-Based and Community-Based Adaptation Policies and Programmes to Effectively Incorporate Empowerment of Women." In Presidential Climate Commission Towards a Just Transition (Ed.), *Supporting a Just and Climate-Resilient Transition in South Africa.* Available at: https://www.climatecommission.org.za/just-transition-framework.

Magid, M. 2023. "The Failure of the Just Energy Transition Fund in South Africa." *EUObserver*, September 26. Available at: euobserver.com/africa/ar3f9e910d (accessed October 3).

Magome, M. 2023. "South Africa's Unemployment Is a 'Ticking Time Bomb.' Anger Rises with Millions Jobless." *AP*, August 14.

Mahlaka, R. 2024. "Slow Start to Funding for SA's Just Energy Transition, but Acceleration Expected". *Daily Maverick*, July 17. https://www.dailymaverick.co.za/article/2024-07-17-slow-start-to-funding-for-sas-just-energy-transition-but-acceleration-expected/.

Murtaugh, D. 2024. "Trump's Return Gives China a Second Chance at Climate Leadership." *Bloomberg*, November 7. https://www.bloomberg.com/news/newsletters/2024-11-07/trump-s-return-gives-china-a-second-chance-at-climate-leadership.

Otlhogile, M., and R. Shirley. 2023. "The Evolving Just Transition: Definitions, Context, and Practical Insights for Africa." *Environmental Research: Infrastructure and Sustainability* 3.

Pilling, D. 2023. "Scandal at South Africa's Eskom: the CEO and the Cyanide-Laced Coffee." *Financial Times*, March 1. https://www.ft.com/content/5fe8291d-9895-4272-9e0a-eefa279116f4.

Plumer, B. 2023. "Climate Change Is Speeding Toward Catastrophe. The Next Decade Is Crucial, U.N. Panel Says." *The New York Times*, March 20. https://www.nytimes.com/2023/03/20/climate/global-warming-ipcc-earth.html.

Pooler, M. 2024. "Brazil Minister Warns Carbon Credit Buyers to Beware Fraud." *Financial Times*, August 3. https://www.ft.com/content/ea246444-7bd4-4d1f-8bf0-8adcd3a7eba4.

Presidential Climate Commission Towards a Just Transition. 2023. "Just Transition Framework." https://www.climatecommission.org.za/just-transition-framework.

Putsch, C. 2022. "Habecks Plan für die Energiewende in Afrika." *Die Welt*, December 9.

Roberts, B., J. Struwig, and T. Zondi. 2024. *Public Perceptions and Attitudes Relating to Climate Change and the Just Transition in South Africa: High-Level Findings from a 2023 Nationally Representative Survey.* Presidential Climate Commission (PCC). Available at: https://hsrc.ac.za/wp-content/uploads/2024/09/HSRC-PCC-Just-transition-headline-findings-report.pdf.

Sguazzin, A., and P. Burkhardt. 2023. "How 60 Million South Africans Are Being Failed by the West's Green Ambitions." *Bloomberg*, April 27.

Smyth, J., and D. Pilling. 2023. "Two Tales from South Africa's Energy Transition." *Financial Times*, November 28.

Sustainable Energy for All. 2024. *Powering the Sustainable Development Goals.* Available at: https://www.seforall.org/data-stories/powering-the-sustainable-development-goals.

The Economist. 2022. "There Is a Better Way to Help Poor Countries Fight Climate Change." November 17.

The Economist. 2023. "Fear of the Failed State." May 27.

The Economist. 2024a. "Dethroning Fossil Fuels." November 11.

The Economist. 2024b. "The Light Continent." July 18.

The Presidency Republic of South Africa. 2022. *South Africa's Just Energy Transition Investment Plan (JET IP).* Available at: https://assets.bbhub.io/company/sites/63/2022/11/south-africa_s-just-energy-transition-investment-plan-2023-2027.pdf.

Turnbull, T. 2023. "Has Australia Cleaned Up Its Act on Climate?," BBC News, September 7. Available at: https://www.bbc.com/news/world-australia-65606208.
U.S. Department of State, 2024. "The Bureau of Energy Resources Highlights Efforts to Advance Energy Security and Decarbonization at COP 29." Available at: https://www.state.gov/the-bureau-of-energy-resources-highlights-efforts-to-advance-energy-security-and-decarbonization-at-cop29/.
UN. 2021. *Leveraging Energy Action for Advancing the Sustainable Development Goals*. Available at: https://sdgs.un.org/sites/default/files/2021-06/2021-UN_POLICY%20BRIEFS-063021.pdf.
UNEP. 2019. "Thundering Youth Call on Power to Act Now." Available at: https://www.unep.org/pt-br/node/24644?_ga=2.185523689.1734540025.1731012729-13844932.1.17292.1329.
UNFCCC. 2024. "COP29 UN Climate Conference Agrees to Triple Finance to Developing Countries, Protecting Lives and Livelihoods." Available at: https://unfccc.int/news/cop29-un-climate-conference-agrees-to-triple-finance-to-developing-countries-protecting-lives-and.
United Nations. 2024. *The Sustainable Development Goals Report 2024*. New York: United Nations.
United Nations Framework Convention on Climate Change (UNFCCC). 2015. Paris Agreement. Available at: https://unfccc.int/process-and-meetings/the-paris-agreement.
van Diemen, E. 2024. "Delaying Coal Power Plant Shutdowns Could Push Costs to R90bn, Warns Eskom CEO." *Daily Maverick*, June 7. https://www.dailymaverick.co.za/article/2024-06-07-delaying-coal-power-plant-shutdowns-could-push-costs-to-r90bn-warns-eskom-ceo/.
Wang, X., and K. Lo. 2021. "Just Transition: A Conceptual Review." *Energy Research & Social Science* 82. https://doi.org/10.1016/j.erss.2021.102291.
Wolf, M. 2024. "Climate Change Is a Global Problem — It Requires a Global Solution." *Financial Times*, November 26. https://www.ft.com/content/a0cfcf55-6b4d-4aeb-a3ec-b3db33575047.

12 Regeneration of nature

RECAP OF CHALLENGES AND WHY THIS HOTSPOT MATTERS

Using the latest data available (United Nations 2024), the following five challenges can be addressed by this transformation hotspot.

SDG 6: Clean Water and Sanitation

To effectively manage water crises, policymakers, scientists, and practitioners apply the concept of integrated water resources management (IWRM). IWRM is defined as "a process which promotes the co-ordinated development and management of water, land and related resources, in order to maximize the resultant economic and social welfare in an equitable manner without compromising the sustainability of vital ecosystem" (Global Water Partnership Technical Advisory Committee 2000, 22). IWRM better positions countries to deal with challenges ranging from a loss of biodiversity, climate change, or water pollution. As an integral part of achieving the SDGs, IWRM adoption, however, has been slow. While widespread in the Global North, countries in Central Asia, Africa, and Latin America are lagging behind in utilizing IWRM (United Nations 2024, 21).

Regeneration of nature requires improved cross-sectoral coordination and capacity-building among policymakers, the private sector, and the scientific

DOI:10.4324/9781003400028-17

community. It leverages financing to support integrated initiatives that yield significant improvements.

SDG 13: Climate Action

Financing also plays an integral part in achieving the goals regarding climate mitigation and adaptation. Developed countries had at long last reached by 2022 their annual pledge of US$100 billion to the cause. The next step is to increase it to US$300 billion by 2035 (UNFCCC 2024). National climate action plans in the Global South, however, remain under-funded, with an estimated US$6 trillion needed by 2030 for developing countries alone (United Nations 2024, 34).

To achieve a transformation in the regeneration of nature, we need to leverage solutions that bring in the private sector. We also need to translate the biggest causes of global warming into solutions to it (such as agriculture).

SDG 14: Life below Water

Overfishing and increased pollution are major drivers in the decline of global fishing stocks. Two out of three fish caught in the Southeast Pacific are overfished. In the Eastern and Central Atlantic, it is one out of two. In these marine regions, the proportion of fish stock within sustainable levels has continuously declined, with overfishing contributing to reductions in biodiversity and deteriorations in ecosystems. Reversing these trends is possible. In the Pacific Northeast or Atlantic Northeast, sustainable fishing stocks have increased, with only two out of ten fishes caught from overfishing. Improved fisheries management – such as quotas – can contribute to rebuilding our fish stocks, with a regeneration translating into improved food security and livelihoods for coastal populations (United Nations 2024, 36).)

The health of global fish populations shows the multi-dimensionality of sustainability challenges and the crucial role played by policymaking and governance of the commons, such as the ocean. 2023 marked the adoption of a landmark high seas treaty – the UN Marine Biodiversity of Areas Beyond National Jurisdiction Agreement. The milestone was reached after nearly two decades of negotiations to protect 30 percent of the oceans by 2030 (High Seas Alliance 2023).

SDG 15: Life on Land

Our unsustainable consumption habits are increasing the threats to species across the globe. More than one out of four currently known species are threatened (United Nations 2024, 39). Deforestation for agricultural gains drives carbon emission and increases droughts (Ionova 2024). Deforestation also increased extinction risks. These challenges are present beyond just the Amazon region. Eastern and South-Eastern Asia have seen the largest increase in species at risk of extinction from 1993 to 2024 (United Nations 2024, 39).

Identifying solutions that both increase biodiversity but also can sustainably intensify production without drawing on more land are needed. As captured by the European Central Bank executive, Frank Elderson, "[d]estroy nature and you destroy the economy" (Arnold 2023). New delayed EU law supports efforts to halt deforestation, requiring documentation for imports that they were not produced by destroying forests (e.g., cocoa) (Hancock and Bounds 2024).

SDG 17: Partnerships for the Goals

Beyond financial support for climate change or oceans, UN Secretary Guterres has proposed US$500 billion additional financing to get the SDGs back on track. As debt loads in the Global South are reaching unsustainable levels, increased amounts of limited national budgetary resources are spent each year on servicing debt instead of investing in local sustainable development efforts (United Nations 2024, 43). By 2022, debt interest payments by the poorest countries reached a combined US$23.6 billion, double what it was in 2015 (United Nations 2024, 43). The world's largest bilateral creditor, China, has offered rescue loans for some of its largest borrowers under the Belt and Road Initiative (Kynge 2023). Emergency loans from international lenders are also used to pay overdue loans from bilateral (e.g., to China) or private lenders (Cohen 2024). As countries are struggling in the aftermath of COVID-19 and high interest environments, the number of sovereign debt defaults since 2020 has exceeded the total defaults in the past two decades (McDougall 2023). "Given the fiscal environment, countries are tempted to open up additional land for generating economic gains, as opposed to investing in nature. Actors should combine a focus on regenerating nature and economic sustainability" (United Nations 2024).

OUR EVOLVING RELATIONSHIP WITH NATURE

Since starting our journey on the 2030 Agenda, Jungcurt, Kosolapova, and Bonaccorsi (2024) find that our understanding of nature and our relationship to our environment have shifted. They highlight two specific instances: the Global Biodiversity Framework and ecosystem services.

The Global Biodiversity Framework "recognizes that humans are part of nature and that ecosystems are also livelihoods [representing] a departure from older concepts that pitched conservation objectives against the rights of those living the areas to be conserved" (Jungcurt, Kosolapova, and Bonaccorsi 2024). This shift sidesteps the pitfalls of green colonialism, whereby environmentalism is used to recolonize spaces abroad as opposed to tackling challenges in our own backyards. Or as noted by Krenak (2020, 25), our own backyards give way to concrete: "what starts as a park ends up as a parking *lot.* They've got to find somewhere to park the endless stream of cars they produce out there."

The African continent instead served as a place for Western fantasies: perceived as an unoccupied Eden that needs protection from Africans, the resultant interventions were parks and nature reserves (Blanc 2022).

Regarding ecosystem services, discourse has shifted beyond their monetary value (e.g., *payment-for-ecosystem services*) toward a "broader concept of nature's contributions to people, which recognizes non-monetary benefits [and] negative consequences of human-environment interactions" (Jungcurt, Kosolapova, and Bonaccorsi 2024). Interacting with nature can offer healing (Sacks 2019). As became painfully obvious from the global pandemic, zoonotic diseases are also more likely to emerge as we increase our encroachment on nature.

Krenak takes us a step further. Krenak (2020, 6) argues that we have to move away from our anthropocentric view of "ourselves as one thing, and Earth another ... There's a lot more to Earth than us, and biodiversity doesn't seem to be missing us at all."

> My communion with what we call nature is an experience long scoffed at by city folk. Rather than see any value in it, they poke fun at it: "He talks to trees, he's a tree-hugger; he talks to rivers, he contemplates the mountains." But that's my experience of life. I don't see anything out there that is not nature. Everything is nature. The cosmos is nature. Everything I can think of is nature.
>
> Mother Earth puts us to sleep and wakes us up again with the rising sun; she lets the birds sing, the currents and winds flow. She creates this wonderful world for us to share, and what have we done in return? What we are going through now might well be the act of a loving mother who has decided [her] child needs to be taught a lesson. "Hush, now." That is what the Earth is saying to humanity.
>
> (Krenak 2020, 6–7)

What do we find if we look below the surface? Western positivism is remarkably silent. Back in 2015, I tangentially was engaged in efforts to increase awareness of the UN International Year of Soils campaign (FAO 2015a). As business-as-usual focuses on how to maximize growth above ground, we neglected research to understand life below ground (think microbes, such as bacteria, or mega-fauna, such as earthworms). Prominent efforts in the USA, such as the Marin Carbon Project, were initially self-funded, opening themselves up for critiques about its scientific rigor (Velasquez-Manoff 2018). Not only are data limited or missing, there are few dedicated programs on soil biodiversity (FAO et al. 2020). This lack of research limits our understanding of soil as a non-renewable resource and its living components that form the basis for our food, feed, medicines, ecosystem services, and fuel (FAO 2015c).

The UN campaign was successful in creating greater awareness of the extractive nature of agriculture (e.g., see the animated film from FAO 2015b). Yet we know little about what happens below our feet. One square meter of soil can hide over 10,000 miles of fungal networks underneath (Velasquez-Manoff 2018). Professor Simard's work on the topic is featured later in the chapter. Two inches below the surface, we might find 200 different species of mites delivering crucial services to healthy soils (Jarvis 2018). To dive below the surface is a humbling experience. As Klein (2017), covering fungi-farming ants captured it: "How Ants Figured Out Farming Millions of Years Before Humans."

THE POTENTIAL OF SUSTAINABLE LANDSCAPE MANAGEMENT

In Chapter 9 I introduced the World Overview of Conservation Approaches and Technologies (WOCAT) database that captures local, peer-reviewed sustainable land management (SLM) practices and approaches. SLM addresses SDG Target 15.3, which aims to "*combat desertification, restore degraded land and soil, including and effected by desertification, droughts and floods, and strive to achieve a land degradation-neutral world.*" Our goal should be to move beyond reversing degradation toward regeneration of land. In that context, SLM contributes to mitigation and adaptation to climate change, supporting countries' achievements of their nationally determined contribution targets. It furthermore advances the conservation of biological diversity through ecosystem-based adaptation, captured under the UNFCCC (Article 2) and the UN Convention on Biological Diversity (UNCBD, Article 1) (UNCCD, n.d.).

Building off our work on the role of local knowledge in Chapter 6, indigenous knowledge systems play a large role in achieving the above. Seguada and Banerjee (2024, 1) define indigenous knowledge as "a body of knowledge, practice and belief about the relationship of humans with each other and with other non-human living beings, passed down from human ancestors over generations and yet thriving in many parts of the world." Having gained recognition within the UNFCC and UNCBD, the challenge remains of how to mainstream more diverse forms of knowledge given their historical neglect and continued higher credence given to Western science. Major steps include working with "respect, curiosity, openness, and dialogue to break cultural stereotypes that create hierarchies where Western science is placed at the top and all the 6,900 documented [indigenous knowledge systems] somewhere at the bottom" (Seguada and Banerjee 2024, 2). Given the context-dependence of working with land, SLM practices need to be inclusive of specific (agri-)cultural contexts. Otherwise, they are set up for failure.

I introduce here two Case Studies: a large-scale intervention in Indonesia funded by the World Bank, and a hyper-local intervention close to my childhood home's backyard that was primarily financed through public funds.

CASE STUDY 12.1 FROM A WORLD BANK-FUNDED SUSTAINABLE LAND MANAGEMENT PROGRAM IN INDONESIA...

Indonesia's 17,504 islands are highly vulnerable to climate change (World Bank 2023a). Since 2016, the World Bank has supported Indonesia through the Sustainable Landscapes Management Program (SLMP) to support the country's transition to low-carbon, diverse, and resilient landscapes (World Bank 2023b). Deforestation and fires drive Indonesia's carbon emissions. Creating more resilient and yet prosperous landscapes is at the heart of the intervention. Beyond the third-largest tropical forests that offer a substantial carbon sink (94.1 million hectares), Indonesia is also home to the world's largest tropical peat land areas (14.9 million hectares), and one-fifth of the world's mangroves (3.3 million hectares) (World Bank 2025a). Indonesia's goal is for the forests to become a net carbon sink by 2030, and reduce overall greenhouse gas emissions by over 30 percent by 2030. Beyond forests, new funding was approved in 2022 for a US$400 million loan to improve coastal resilience through mangroves (World Bank 2025b).

The SLMP offers an integrated approach that includes stakeholders from civil society, local communities, and the private sector. Its approach includes five domains:

1. Sustainable Development of Indonesia's Priority Landscapes
2. Improved Land Administration and Governance
3. Sustainable Financing for Landscape Management
4. Equitable and Inclusive Growth within Rural Landscapes
5. Improved Management of Forest Resources (World Bank 2025a).

With local impacts, climate change demands local action. The World Bank applies a "locally led climate action" framework, that addresses *information* ("Are people aware of the climate benefits and risks?"), provides *incentives* ("Are people willing to make changes?"), and creates *instruments* ("Are people able to make changes?") (World Bank 2023a).

Based on the implementation of the SLMP, the World Bank identified four main lessons aligned with our changemakers:

> First, strong governmental leadership is essential, particularly at local level. Second, decentralized forest management is key to balancing protection with local economic activity. Third, land-tenure security for local communities and Indigenous peoples is critical for forest protection and livelihoods. Fourth, solutions should be tailored to the different challenges across land uses.
>
> (World Bank 2023b, 27)

These lessons, especially the fourth one, reiterate the need for context-specific interventions and the viability of building upon local knowledge to achieve global change – a topic we return to in Chapter 13.

CASE STUDY 12.2 ... TO A LOCALLY-FUNDED PROGRAM IN MY BACKYARD FOR REGENERATING NATURE

Since 1765, German agricultural expansion has resulted in the draining of over 90 percent of its bogland (Heinrich Böll Stiftung 2023). Switzerland, like its Northern neighbor, raided its bogland of valuable peat soil during industrialization. I do not have to travel far from my family's home in Switzerland to reach the Lake Burgäschi, where pile dwellers inhabited the shores of the moor lake 6,000 years ago. The lake is still there, but the adjacent bogland was in a state of decline for most of my childhood.

In 2017, the environmental NGO *Pro Natura* launched a collaborative project to regenerate the bogland area *Chlepfibeerimoos*. The NGO previously purchased land rights to the bogland. Located across two cantonal boundaries, the project was a collaboration with the relevant environmental agencies of the Cantons Berne and Solothurn. The costs of re-establishing the bogland were more than US$700,000. Financing was mobilized from public funds from the Cantons, foundation money from regional energy producers, and donations from the Pro Natura members (Pro Natura Solothurn n.d.-b).

ENVIRONMENTAL VALUE OF THE PEAT BOGLAND

Bogs capture and bind around a third of the world's carbon dioxide. Occupying approximately 5 percent of the world's major ecosystems, they capture nearly twice as much carbon as all the forests combined (which covers 31 percent of the globe; see Table 12.1) (Heinrich Böll Stiftung 2023). Resultant bogland destruction had detrimental impacts as it released carbon dioxide back into the atmosphere.

SOCIETAL VALUE OF THE PEAT BOGLAND

The 8 hectares hold a large cultural value. Twenty plant species exist solely within this landscape; another 13 are rare. Another 28 species are at risk of extinction (Pro Natura Solothurn n.d.-a). They also offer important educational and sensitization opportunities for current and future generations.

WHY IT WORKED AND HOW IT RELATES TO THE SDGS

This local case of a raised bog highlights how actions by civil society, combined with finance, policymakers and science and technology, come together to support a transformation to regeneration of nature. Such a transformation contributes to crucial SDGs, including SDG 6 (with most freshwater coming directly or indirectly from wetland areas), SDG 13 (with regenerated bogs capturing more CO_2 than all other major ecosystems), SDG 14 (by supporting well-functioning coastal wetlands), and SDG 15 (with 40 percent of all species living in wetlands) (Heinrich Böll Stiftung 2023).

Table 12.1 How bogland trumps them all. Source: Heinrich Böll Stiftung (2023)

Ecosystem	Size in million hectares	Share (%)	Billion tons of carbon stored	Share (%)
Bogland	500	5	600	34
Grassland	3750	36	588	34
Forest	3300	31	372	21
Deserts	3000	28	191	11

THE RAPID RISE OF NATURE-BASED SOLUTIONS

If sustainable land management sounds dusty, the latest buzzword in town is nature-based solutions, representing win-win situations by creating economic growth through improving the environment. Coined by the International Union for Conservation of Nature (IUCN) around two decades ago, these solutions offer to "address societal challenges through actions to protect, sustainably manage, and restore natural and modified ecosystems, benefiting people and nature at the same time" (IUCN, 2025). Peatland restoration, captured earlier, represented one of the most impactful solutions assessed (UNEP and IUCN 2021).

REGENERATIVE AGRICULTURE

Regenerative agriculture has also risen to the forefront as a nature-based solution (Miralles-Wilhelm and Iseman 2021). Adopting regenerative practices offers to increase food production and capture carbon in the soil. As enhanced soil carbon storage itself has been monetized, the practice of regenerative agriculture offers a viable investment opportunity for impact investors who are aiming to achieve the triple bottom line of financial, social, and environmental returns. Regenerative agriculture even competes on grocery shelves with old staples, such as organic farming. The practice is argued to hold under-explored health benefits (LeZaks and Ellerton 2021) and can be marketed to eco-conscious consumers.

To illustrate how private sector funding is being leveraged, I provide you with two investments: one from the Global North and one from the Global South.

Investing in regenerative agriculture in the Global North

One of the largest players is Indigo Ag. Its *Carbon by Indigo* program allows farmers to get paid for adopting regenerative agricultural practices that are subsequently sold as certified soil carbon credits to companies looking to go carbon-neutral (e.g., to Microsoft; Indigo Ag 2024c). It has enrolled three thousand US farmers and their seven million acres as of 2024 (Indigo Ag 2024b). The agricultural practices paid for include conversion from tillage to no-till and planting cover crops, which offer payments of around US$25 per acre per year. Payments increase over time as "companies purchasing carbon credits want proof that carbon will remain stored" (Indigo Ag 2024a).

By offering farmers the chance to earn an additional income on top of their incomes from selling produce, Indigo Ag facilitates for farmers the win-win promise that is at the core of nature-based solutions. These carbon credit offerings by Indigo Ag are on top of its much longer business in biological inputs and more recent inclusion of logistical services, with an aim of offering integrated solutions to farmers to transform their production (Marston 2023).

Across the USA, by adopting conservation practices by farmers and ranchers, the quantifiable ecosystem benefits are estimated to reach nearly US$14 billion (Miralles-Wilhelm and Iseman 2021). The total potential for carbon sequestration is estimated at 190 million metric tons, with nitrogen runoff reduced by 1.6 billion pounds, and phosphorous runoff by 0.8 billion pounds (Miralles-Wilhelm and Iseman 2021, 43). Investors as diverse as the WWF (2025) Impact branch and the Swiss bank UBS (Ramachandran 2023) are touting the benefits of regenerative agriculture as an investment portfolio. Benefits beyond carbon credit sales include greater financial resilience, cost savings on inputs, increased land value, and access to government support or tax incentives (Bashian et al. 2024). These are in addition to payments for carbon credits from third parties or premiums paid by food corporations or retailers.

These attributes make US regenerative farmland an attractive outlet for investments by pension funds. An expected increase in scarcity of land has further fueled the doubling of US land investments from 2020 to 2023 (Savage 2024b). All the rush for regenerative agriculture has revealed one of the biggest challenges: a shortage of regenerative farmers to keep the boom going (McMahon 2024).

Investing in regenerative agriculture in the Global South

Investments are also ramping up into regenerative agriculture across the Global South. For example, a US$200 million fund is planned by Vietnam-based Mekong Capital for "supporting growth-stage funding for regenerative ag[riculture] and sustainable forestry in the lower Mekong region of Southeast Asia" (Terzo 2024). While US investments can flow directly into agricultural land, this investment fund aims to support business to scale their regenerative product offerings and disseminate relevant practices to reach 1 million beneficiaries, regenerate over 200,000 hectares of land, and reduce over 10 million tons of CO_2 emissions (Vo 2024).

Investors looking to invest in large-scale farming operations are more frequently found in Latin America. The Land Group, for example, owns 40,000 hectares of land and has plans to expand its regenerative farming across the region (Hailstone 2024). Throughout the Global South, capacity building for farmers is important for regenerative agriculture to take-off.

Don't blink (or plow) or...

...poof. As hinted at earlier, Indigo Ag's payment system for carbon credits pays out more over time as evidence grows that carbon is actively being stored in the soil. Farmers are expected to reach their maximum storage after 20 years of adopting agricultural practices (Lal 2004).

This time horizon raises two issues. First, there is a physical limit to how much carbon can be stored in soils, muting those who see endless potential (Savage 2024a). "Carbon capture in soils has been promoted by the livestock industry as 'a get out of jail free card,'" according to Pete Smith (Savage 2024a). His research showed limitations in carbon sequestration in grasslands (Wang et al. 2023). Second, and more severely, if a farmer opts to return to plowing (e.g., to control an invasive weed that fails to respond to ecological measures), most gains in carbon storage might be negated. How much carbon is released depends on the length that regenerative agriculture has been practiced, the type of rotation and crops grown, the soil type and amount of organic matter, and the depth of the plowing method (with minimal tillage methods, such as using rippers, having less effect). For the entire soil, 20 to 60 percent soil carbon losses are expected (Hu et al. 2024; West and Post 2002). Thus, storing carbon in the soil through regenerative agriculture depends on continued good agricultural practices. A reversal would negate much of the progress made, highlighting the need to use complementary approaches to mitigate climate change.

Revisiting the win-win proposition

It is important to capture that just as farmers are motivated by incentives, so are private companies and investors motivated by profits. Under increased scrutiny, corporations are also interested in demonstrating their commitments to sustainability. As cautioned by Wilson, Hendrickson, and Myers (2024, 11), the *win-win* discourse in renewable agriculture allows companies to "sidestep more fundamental questions about the role of industrial development in environmental degradation and social inequities." Given the need to certify carbon credits, similar barriers exist to organic production: smaller farmers are less likely to produce at a scale large enough to afford certification costs.

BREAKING OUT OF YOUR COMFORT ZONE

Similar to the criticism raised earlier by proponents of agroecology of organic agriculture, the above measures stop short of a more radical transformation of our production system and our relationships with nature.

If we look closely enough and move beyond the obvious, we can find examples of how local communities have innovated and adapted to regenerate nature.

Three examples illustrate our relationship with water, forests, and their combination in agriculture. Part of their fabric is greater complexity and a move toward working *with* as opposed to *controlling* nature.

Regeneration through water management

Erica Gies (2022) captured how local groups have improved their water systems by *working with* water as opposed to *controlling* it. She has traveled the globe to see how local communities successfully manage increased floods from climate change. She found that those communities expanded the needs beyond immediate human ones. At the core of a transformation is changing our relationship with water, from a *what* (a commodity) to a *who* (a friend or relative).

Giving greater agency to water, Gies argues, means addressing water's needs for slow phases. The resulting slow water movement emphasizes the importance of wetlands, floodplains, or forests, demands systems thinking, and a local bottom-up approach that includes aspects of social justice. It draws on lesser-known scholarship of historical ecology and socio-hydrology. She cites examples of success stories in Seattle, where allowing Thornton Creek to move beyond its concrete channels reduced flooding and improved ecosystems, in Kenya, where tree planting helped attract rainfall and bring back groundwater, or the United Kingdom, where the return of beavers prevented droughts and flooding (Bioneers 2024b).

Regenerating nature through forests

Reforestation is omnipresent, representing one of the most common pathway for carbon compensation. It was well understood that the composition of trees matters, with monocultures grown to maximize profits resulting in lower levels of biodiversity and ecosystem benefits. Dr. Suzanne Simard has researched extensively the forests of British Columbia, Canada. Similar to Erin Gies' work, Dr. Simard found forests regulate water flow with snow melts, as water gets absorbed by the healthy soil. As more land was cleared for logging, snow melts much faster, resulting in flooding. A lack of ability to store water also increased drought risks and resultant wildfires. Soil organic matter plays a crucial role in absorbing snowmelt. Approximately 60 percent of it is destroyed due to compacting from the use of heavy machinery during logging (Bioneers 2024a).

Dr. Simard launched the *Mother Tree Network* to work within silviculture using their *regenerative forest stewardship* concept that promotes "ecosystem-based forest management, guided by Indigenous communities, using indigenous and western science, that prioritizes ecosystem health, including carbon sequestration and storage, in addition to other eco-cultural and socio-economic goals" (The Mother Tree Network 2024). Her practice is based on her groundbreaking research that established a wood-wide web – underground exchange between trees through networks of mycorrhizal fungi (Simard et al. 1997). Over the years, she has found exchanges of carbon, water, nutrients, and alarm signals among trees (Jabr 2020). Her work argues that older trees play a significant role in nurturing or *raising* younger ones, leading to her most recent research of keeping a percentage of mother trees to improve forest regeneration efforts (Jabr 2020).

Regeneration of nature through agriculture

Combining the above, permaculture or successional agroforestry have been framed as popular options that can effectively regenerate nature by linking agricultural production with improving soil health and reduced input-dependency. Permaculture tries to create self-sustaining farm production systems "mimicking the patterns and relationships found in natural ecosystems, which can be achieved by producers if they observe and interact with the agroecosystem, apply self-regulation, accept feedback, integrate biodiversity, [and] use small and slow solutions ..." (FAO 2022). Attempts to adopt permaculture that I've witnessed included capturing and reusing water during farm operations, relying heavily on adoption of sustainable land management practices.

Permaculture nearly always incorporates trees combined with crops or livestock. Successional agroforestry uses trees as the entry point to create a multi-strata growing system that utilizes intercropping and multi-purpose species (Young 2017). Research colleagues from the Swiss Organic Research Institute (FiBL) together with the Bolivian organization Ecotop undertook a long-term research trial to better understand the biophysical and livelihood impacts of a transition from monoculture cocoa cultivation to a more diverse approach. They found that farm-level resilience was highest under such successional systems, with organic cocoa systems producing the largest impact on livelihoods (Bhullar et al. 2021). Related approaches include syntropic agriculture, attributed to Ernst Götsch, who aimed to use natural succession processes combined with heavy pruning to regenerate soils and nature (de Oliveira n.d.). Muthuri et al. (2024) successfully developed a "farmer first" agroforestry systems in Ethiopia, Rwanda, and Uganda that leveraged all four changemakers, including governance and policymaking, as well as integration into markets.

In summary, breaking out of your comfort zone pays off for you to move beyond conservation toward regeneration of nature. With a strong value proposition across all sustainability dimensions, the hotspot offers a vital avenue for achieving sustainable development.

YOUR TURN

What is an area in your region where regenerating nature might be feasible? What synergies and trade-offs would you expect?

REFERENCES

Arnold, M. 2023. "ECB Flags Stark Economic Risks from Biodiversity Loss." *Financial Times*, June 7. https://www.ft.com/content/d83602d0-1296-4928-b58a-21cf2a6d2a0f.

Bashian, A., A. Mishra, D. Haynes, D. Bennell, K. Rudzinski, R. Raimond, S. Farley, S. Lang, T. Davids, and T. Crosby. 2024. *Financing for Regenerative Agriculture*. Pollination Group. Available at: https://www.rockefellerfoundation.org/wp-content/uploads/2024/06/Financing-for-Regenerative-Agriculture-Final.pdf.

Bhullar, G. S., D. Bautze, N. Adamtey, L. Armengot, H. Cicek, E. Goldmann, A. Riar, J. Rüegg, M. Schneider, and B. Huber. 2021. "What Is the Contribution of Organic Agriculture to Sustainable Development?" Available at: https://orgprints.org/id/eprint/39536/1/Syscom_Synthesis_Report.pdf.

Bioneers. 2024a. "Bridging the Gap Between Nature and Infrastructure: Insights from Leading Experts on Nature-Based Solutions." Available at: https://bioneers.org/bridging-the-gap-between-nature-and-infrastructure-insights-from-leading-experts-on-nature-based-solutions-ztvz2409/.

Bioneers. 2024b. "Erica Gies – The Slow Water Movement: How to Thrive in an Age of Drought and Deluge." Available at: https://www.youtube.com/watch?v=NMM3TBA2_HA.

Blanc, G. 2022. *The Invention of Green Colonialism*. Cambridge: Polity Press.

Cohen, P. 2024. "Africa's Debt Crisis Has 'Catastrophic Implications' for the World." *The New York Times*, August 24. https://www.nytimes.com/2024/08/28/business/african-debt-crisis.html.

de Oliveira, I. n.d. "Ernst Götsch: The Creator of the Real Green Revolution." *Believe Earth*. Available at: https://believe.earth/en/ernst-gotsch-the-creator-of-the-real-green-revolution/ (accessed February 14, 2024).

FAO. 2015a. "Healthy Soils for a Healthy Life." Available at: https://www.fao.org/soils-2015/en/.

FAO. 2015b. "Let's Talk About Soil." Available at: https://www.youtube.com/watch?v=invUp0SX49g.

FAO. 2015c. "Soil Is a Non-Renewable Resource." Available at: https://openknowledge.fao.org/server/api/core/bitstreams/c43bacdd-2fb6-46d2-b9d0-6731c566be91/content.

FAO. 2022. "Regional Workshop on Permaculture and Community Seed Production for ECA Region." Available at: https://www.fao.org/fileadmin/user_upload/reu/europe/documents/Events_2022/CNA15nov.pdf.

FAO, ITPS, GSBI, CBD, and EC. 2020. *State of Knowledge of Soil Biodiversity*. Available at: https://openknowledge.fao.org/items/24035f6b-eed4-40a8-a838-b45b866c35c2.

Gies, E. 2022. *Water Always Wins: Thriving in an Age of Drought and Deluge*. Chicago: University of Chicago Press.

Global Water Partnership Technical Advisory Committee. 2000. *Integrated Water Resources Management*. Available at: https://www.gwp.org/globalassets/global/toolbox/publications/background-papers/04-integrated-water-resources-management-2000-english.pdf.

Hailstone, J. 2024. "Group Aims to Scale Regenerative Farming in South America and Beyond." *Forbes*. https://www.forbes.com/sites/jamiehailstone/2024/10/02/group-aims-to-scale-regenerative-farming-in-south-america-and-beyond/.

Hancock, A., and A. Bounds. 2024. "US Urges EU to Delay Deforestation Law." *Financial Times*, June 19. https://www.ft.com/content/1b1c7541-92f8-478a-9e18-8c0419af7714.

Heinrich Böll Stiftung. 2023. *Mooratlas 2023*. Available at: https://www.boell.de/de/2022/11/17/mooratlas-2023.

High Seas Alliance. 2023. "Briefings: Deep Dives into the Provisions of the BBNJ Agreement." Available at: https://highseasalliance.org/resources/briefings-deep-dives-into-the-provisions-of-the-bbnj-agreement/.

Hu, T., S. L. Malone, C. Rumpel, and A. Chabbi. 2024. "Maximizing Soil Organic Carbon Stocks Through Optimal Ploughing and Renewal Strategies in (Ley) Grassland." *Communications Earth & Environment* 5 (1). https://doi.org/10.1038/s43247-024-01202-3.

Indigo Ag. 2024a. "Carbon Calculator." Available at: https://app.indigoag.com/programs/how-much-can-i-earn-carbon-farming?tillage=NO_CHANGE.

Indigo Ag. 2024b. "Connecting Sustainability and Profitability." Available at: https://www.indigoag.com/carbon-credits.

Indigo Ag. 2024c. "Indigo Ag's Carbon Program to Deliver Record Carbon Credits to Microsoft." Indigo Ag, Inc. Available at: https://www.indigoag.com/pages/news/indigo-ags-carbon-program-to-deliver-record-carbon-credits-to-microsoft (accessed February 14, 2025).

Ionova, A. 2024. "A Changing Climate Is Scorching the World's Biggest River." *The New York Times*, October 8. https://www.nytimes.com/2024/10/06/world/americas/amazon-river-climate-change-brazil.html.

IUCN. "About IUCN." Available at: https://iucn.org/about-iucn (accessed February 14, 2025).

Jabr, F. 2020. "The Social Life of Forests." *The New York Times*, December 12. https://www.nytimes.com/interactive/2020/12/02/magazine/tree-communication-mycorrhiza.html.

Jarvis, B. 2018. "The Insect Apocalypse Is Here." *The New York Times*, November 27. https://www.nytimes.com/2018/11/27/magazine/insect-apocalypse.html.

Jungcurt, S., E. Kosolapova, and E. Bonaccorsi. 2024. "Why We Shouldn't Give Up on the SDGs." IISD. Available at: https://www.iisd.org/articles/deep-dive/why-give-up-sdgs-summit-future-agenda-2030. Accessed February 14th, 2024.

Klein, J. 2017. "How Ants Figured Out Farming Millions of Years Before Humans." *The New York Times*, November 4. https://www.nytimes.com/2017/04/11/science/ant-fungus-farmers-evolution.html.

Krenak, A. 2020. *Ideas to Postpone the End of the World*. Trans. A. Doyle. Toronto: Anansi International.

Kynge, J. 2023. "China Grants Billions in Bailouts as Belt and Road Initiative Falters." *Financial Times*, March 27. https://www.ft.com/content/9b2cb53f-e6f0-479e-bb94-a2e0c8680e88.

Lal, R. 2004. "Soil Carbon Sequestration Impacts on Global Climate Change and Food Security." *Science (New York, N.Y.)* 304: 1623–1627.

LeZaks, D., and M. Ellerton. 2021. *The Regenerative Agriculture and Human Health Nexus: Insights from Field to Body*. Croatan Institute. Available at: https://croataninstitute.org/wp-content/uploads/2021/09/RegenAg_HumanHealth_2021.pdf.

Marin Carbon Project. 2025. "Discovering the Soil's Solution to Global Climate Change." Marin Carbon Project. Available at: https://marincarbonproject.org (accessed February 14, 2025).

Marston, J. 2023. "'No Comment' From Indigo Ag on Valuation Nose-Dive Report." Available at: https://agfundernews.com/no-comment-from-indigo-ag-on-valuation-nose-dive-report.

McDougall, M. 2023. "Poor Countries' Debt Costs to Hit 'Crisis' Levels, Says World Bank." *Financial Times*, December 13. https://www.ft.com/content/7b77b045-8b5b-4249-8fb9-551577559550.

McMahon, P. 2024. *Investing in Regenerative Agriculture*. SLM Partners.

Miralles-Wilhelm, F., and T. Iseman. 2021. *Nature-Based Solutions in Agriculture: The Case and Pathway for Adoption*. Rome: FAO and The Nature Conservancy.

Muthuri, C., C. Njoki, R. Kinuthia, A. Mukuralinda, W. Gebretsadik, C. Galabuzi, J. D. Ndayambaje, C. Okia, A. Kuria, J. Oduol, P. Smethurst, A. Derero, A. R. Cyamweshi, S. Tadesse, J. Buyinza, A. Toib, A. Malesu, J. Kung'u, A. Girmay, … F. Sinclair. 2024.

"Developing 'Farmer First', Locally Adapted Agroforestry in Eastern Africa." In D. Race and G. Wettenhall (Eds.), *Raising Trees and Livelihoods*. ACIAR.

Pro Natura Solothurn. n.d.-a "Chlepfibeerimoos." Available at: https://www.pronatura-so.ch/de/chlepfibeerimoos.

Pro Natura Solothurn. n.d.-b "Regenerations- und Aufwertungsprojekt Chlepfibeerimoos." Available at: https://www.pronatura-so.ch/de/regenerationsprojekt-chlepfibeerimoos.

Ramachandran, A. 2023. "Reaping What We Sow." UBS. Available at: https://www.ubs.com/us/en/assetmanagement/insights/thematic-viewpoints/sustainable-impact-investing/articles/investing-in-nature.html (accessed February 14, 2025).

Sacks, O. 2019. "Oliver Sacks: The Healing Power of Gardens." *The New York TImes*, April 18. https://www.nytimes.com/2019/04/18/opinion/sunday/oliver-sacks-gardens.html.

Savage, S. 2024a. "The Dubious Climate Gains of Turning Soil into a Carbon Sink." *Financial Times*, January 22. https://www.ft.com/content/91ed3d25-9d5c-4f46-8bd9-23cbb37e1288.

Savage, S. 2024b. "Investors Plough Record Amounts into US Farmland." *Financial Times*, February 20. https://www.ft.com/content/c691e279-710f-42aa-8894-4dd4bd331017.

Seguada, W., and A. Banerjee. 2024. *Harnessing Indigenous Knowledge Systems for Global Knowledge Cooperation*. Bonn: IDOS.

Simard, S. W., D. A. Perry, M. D. Jones, D. D. Myrold, D. M. Durall, and R. Molina. 1997. "Net Transfer of Carbon Between Ectomycorrhizal Tree Species in the Field." *Nature* 388 (6642): 579–582. https://doi.org/10.1038/41557.

Terzo, G. 2024. "Vietnam-Focused PE Firm Clinches Grant to Support Farmers in Lower Mekong Region." Available at: https://www.globalaginvesting.com/vietnam-focused-pe-firm-receives-grant-to-support-farmers-in-lower-mekong-region/.

The Mother Tree Network. 2024. "Mother Tree Network." Available at: https://www.mothertreenetwork.org.

UNCCD. n.d. "Why Sustainable Land Management Matters." Available at: https://www.unccd.int/land-and-life/land-management-and-restoration/why-sustainable-land-management-matters.

UNEP, and IUCN. 2021. "Nature-Based Solutions for Climate Change Mitigation." Available at: https://wedocs.unep.org/xmlui/bitstream/handle/20.500.11822/37318/NBSCCM.pdf.

UNFCCC. 2024. "COP29 UN Climate Conference Agrees to Triple Finance to Developing Countries, Protecting Lives and Livelihoods." Available at: https://unfccc.int/news/cop29-un-climate-conference-agrees-to-triple-finance-to-developing-countries-protecting-lives-and.

United Nations. 2024. *The Sustainable Development Goals Report 2024*. New York: United Nations.

Velasquez-Manoff, M. 2018. "Can Dirt Save the Earth?" *The New York Times*, April 18. https://www.nytimes.com/2018/04/18/magazine/dirt-save-earth-carbon-farming-climate-change.html.

Vo, M. 2024. "Mekong Capital Receives Convergence Grant for its Regenerative Agriculture and Sustainable Land Use Initiative in the Lower Mekong Region." Available at: https://www.mekongcapital.com/communications/press-releases/mekong-capital-receives-convergence-grant-for-its-regenerative-agriculture-and-sustainable-land-use-initiative-in-the-lower-mekong-region/.

Wang, Y., I. J. M. de Boer, U. M. Persson, R. Ripoll-Bosch, C. Cederberg, P. J. Gerber, P. Smith, and C. E. van Middelaar. 2023. "Risk to Rely on Soil Carbon Sequestration to Offset Global Ruminant Emissions." *Nature Communication* 14 (1): 7625. https://doi.org/10.1038/s41467-023-43452-3.

West, T. O., and W. M. Post. 2002. "Soil Organic Carbon Sequestration Rates by Tillage and Crop Rotation: A Global Data Analysis." *Soil Science Society of America Journal* 66 (6): 1930–1946.

Wilson, K. R., M. K. Hendrickson, and R. L. Myers. 2024. "A Buzzword, A 'Win-Win', or a Signal Towards the Future of Agriculture? A Critical Analysis of Regenerative Agriculture." *Agriculture and Human Values*. https://doi.org/10.1007/s10460-024-10603-1.

World Bank. 2023a. *Social Dimensions of Climate Change in Indonesia*. Washington, DC: International Bank for Reconstruction and Development/The World Bank.

World Bank. 2023b. *The World Bank Annual Report 2023*. Washington, DC: World Bank.

World Bank. 2025a. "Indonesia Sustainable Landscapes Management Program (SLMP)." World Bank Group. Available at: https://www.worldbank.org/en/programs/indonesia-sustainable-landscapes-management-program/overview (accessed February 14, 2025).

World Bank. 2025b. "Mangroves for Coastal Resilience Project." World Bank Group. Available at: https://projects.worldbank.org/en/projects-operations/project-detail/P178009. (accessed February 14, 2025).

WWF (World Wildlife Fund). 2025. "WWF Impact." World Wildlife Fund. Available at: https://www.worldwildlife.org/pages/wwf-impact-investing (accessed February 14, 2025).

Young, K. J. 2017. "Mimicking Nature: A Review of Successional Agroforestry Systems as an Analogue to Natural Regeneration of Secondary Forest Stands." In F. Montagnini (Ed.), *Integrating Landscapes: Agroforestry for Biodiversity Conservation and Food Sovereignty*. Cham: Springer International Publishing, pp. 179–209.

Part IV
Change is local

We could stop altogether here and get to work on our preferred hotspots (e.g., regenerating nature). But we would miss out on three crucial aspects: (1) how change supports transformation; (2) how gender inequality cuts across achieving sustainable development; and (3) what we mean by resilience, and how it translates into development policymaking and implementation. These three aspects are the focus of Chapters 13–15 and their overarching tenet that change ultimately is local. Upon completing Part IV, we will take a look at sustainable development beyond 2030, chart a course forward and discuss your role in achieving change in sustainable development.

DOI:10.4324/9781003400028-18

13 How change leads to transformation

CHANGE IS LOCAL

Local change matters for both the Global North and the Global South, as no country has yet demonstrated its ability to meet basic human needs without drawing on unsustainable resources (Independent Group of Scientists Appointed by the Secretary-General 2019, xx). As captured in Figure 13.1, Vietnam and Germany are closest to the shaded upper left-hand corner, where social achievements are most significant and the number of biophysical boundaries transgressed lowest.

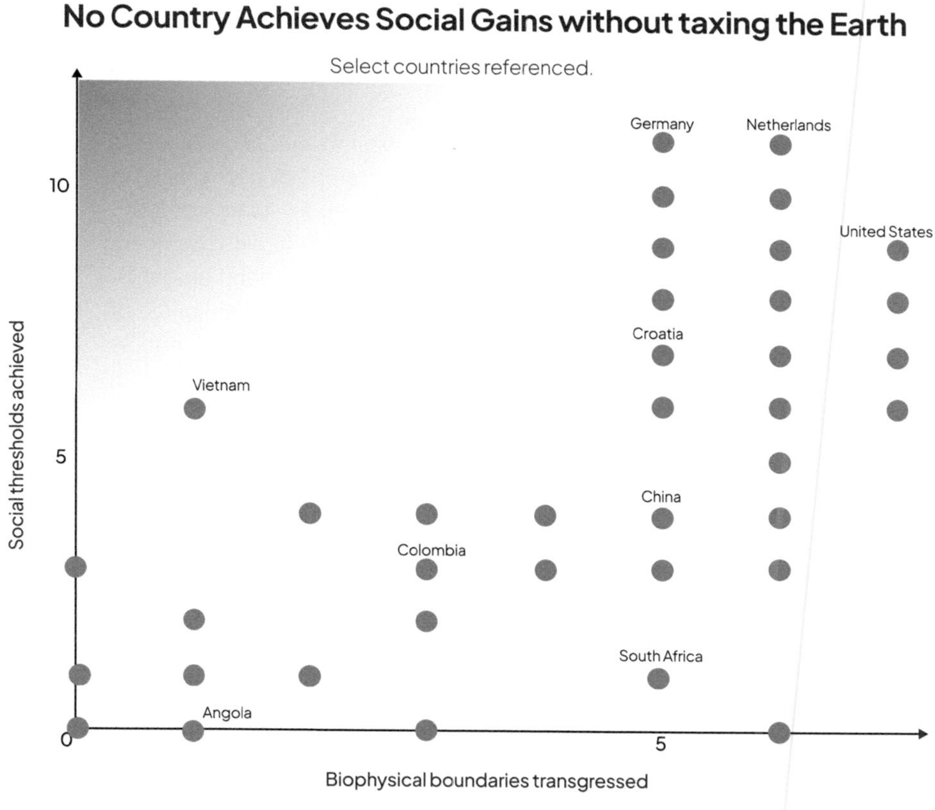

Figure 13.1 No country achieves social gains without taxing the Earth, adapted from Independent Group of Scientists Appointed by the Secretary-General (2019, xx)

DOI:10.4324/9781003400028-19

Considering this pattern, and combined with our success stories at sub-national levels, we get a sense of how change ultimately occurs locally.

Change also travels across nations. While observers are heralding the end of globalization, we are still living in an increasingly interconnected world. Trends in the flows of information, goods, capital, and people highlight the importance of coordinated action to bring about change. Exports in goods and services have grown from a few trillion US dollars in the 1960s to reach US$30 trillion by 2023 (WTO 2024). The number of mobile cellphone subscribers reached nearly 9 billion by 2023 (Taylor 2024). Some 4.5 billion people used airplanes across the globe, approaching a record post-pandemic (Statistica Research 2024). Even as USA-China trade has slowed down, approximately 21 percent of the value of all goods and services were traded internationally (Altman and Bastian 2024). As a whole, the global connectedness (including trade, capital, flow of information and people) has increased over the last two decades, not decreased.

A third thread of how change can lead to transformation is that context matters. Efforts at localization can play a role. Localizing the SDGs includes embedding the mantra to "leave no one behind" in human rights and their specific contexts (Mendelson 2023).

THE LOGIC OF CHANGE

As a development practitioner, we are fond of (or at least trained to do so) using a logical framework (short: logframe) that captures the steps of how change is supposed to happen (see Figure 13.2). Justifiably critiqued for its lack of inclusion,[1] flexibility, causing stress and greater proceduralization of aid (Chambers 2017), a logframe accounts to a donor for how inputs (e.g., cement) are transformed via activities (e.g., digging wells) into outputs (e.g., a reliable source of fresh water). A cumulation of outputs (usually at least two) lead to effects that are captured as outcomes and ultimately impacts. Outcomes capture both

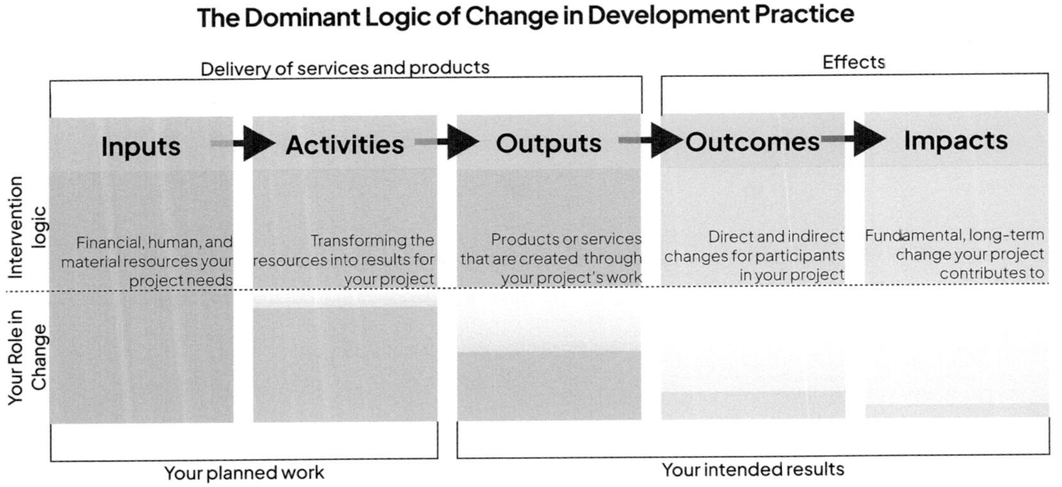

Figure 13.2 The dominant logic of change in development practice, adapted from SDC (2011)

the direct and the indirect effects on the beneficiaries (e.g., rural families use clean drinking water). Impacts describe longer-term changes to the society within which sustainable development occurs (e.g., improved health in families). Many other factors could contribute to the overarching impact on health, such as better public health care or a better diet. Commonly referred to as an attribution gap, this challenge impresses upon us the value of holistic interventions.

From a donor's perspective, you are held responsible for the ability to deliver successfully on the products and services under outputs, and how well your programming has effects on outcomes. As reflected in Figure 13.2, your sphere of influences reduces continuously as you move from inputs and activities toward impact.

For technical interventions, such as those including the construction of new water wells, a logical framework works well at your first glance to explain how change happens. The logframe, with its origins in engineering, reduces complexity and focuses on what steps you are doing to bring about the change using an 'if … then' logic: if I construct wells, then people will drink better water. In reality, attempts at improving sustainable development depend on individual and collective actions that include societal transformations.

What is the missing link between someone having a well built in their neighborhood and them using it? First, they need to have access to it, which could include a new set of social policies within the village. And, importantly, the well needs to be *functioning*. NGOs, such as Helvetas, which are leaders in such water-based interventions, have invested in training local water committees that both ensure equal access across the communities, but also know how to maintain the wells. As donor funding is time-bound and limited, these water committees will frequently collect small funds from the users to ensure that any repairs can be secured and financed in case problems arise.

If the intervention is more advanced – e.g., using a new toilet as opposed to a pit latrine – then you would also need a sensitization campaign to create awareness of the health benefits. You cannot force project participants to use newly built toilets. This is where you should notice the need for moving beyond *things* in development practice, and resharpen your focus on behavioral changes that are crucial to reach a desired outcome. Only when you translate the *things* into a change for the better for *people* can we speak of a positive impact (e.g., improved health).

YOU CANNOT IMPOSE CHANGE

Which brings me to how change happens: you and I, we cannot impose it. Change comes from the individuals themselves. The literature (and the filing cabinets of past development projects) are filled with stories of how top-down development interventions have failed.

One such example is *ujamaa* – the villagization campaign in Tanzania by popular post-independence President Julius Nyerere (Scott 1998). Farmers relocated to their allotted plots and villages subverted the efforts by continuing to farm their old lands, ultimately causing the demise of the entire venture. This same lesson would extend to giving a community member access to a new input (e.g., an improved bean variety) with no prior consultations or understanding of the cultural context. In the event that the seed grows successfully and increase yields, if there is no market nor any appetite to eat the beans, the benefits will be short-lived at best.

LISTENING AND QUALITY MATTERS

This has been the finding from an extensive World Bank-supported review of recipients of aid (Anderson, Brown, and Jean 2012). Participatory, reflexive approaches are especially important when considering projects that necessitate a behavioral change. Think changing gender norms: you cannot impose a change on society, but want to engage all actors (including those in the dominant gender role) to pave the way for a longer-term change.

In a similar vein, when engaging in projects that aim to build skills, such as literacy or apprenticeships, the quality of the work you are doing is essential. Having 100 students attend a course – especially if attendance is subsidized with meals or vouchers – is relatively easy. Ensuring that you are creating a benefit or change for participants during (and beyond) project activities, that is the tougher nut to crack.

HOW CHANGE CAN TRANSFORM

As one of the six evaluation standards proposed by the OECD-Development Co-Operation Directorate (DAC) to assess progress toward the Sustainable Development Goals (SDGs), sustainability captures "the extent to which the net benefits of the intervention continue or are likely to continue" (OECD 2021, 71). Given how the timeframe for how change happens outlined above regularly extends beyond the traditional three-year funding period, this criterion should gain greater urgency in development practice. Unfortunately, it is also the one least funded – studies to assess longer-term benefits are rare. Understanding how change leads to transformation in the long run is important.

Let us start with a review of how change is managed across the scales. We will work our way down: from the society, to the organization, and the individual.

Managing transformational change at the societal scale

Dr. Duncan Green hosted on his blog, "*From Poverty to Power,*" years of wisdom on advancing global humanitarian and development efforts. Synthesized for his book, *How Change Happens*, he identifies seven principles summarized below (Green 2016, 20–22) that are crucial to bring about change through advocacy work:

1. *Be flexible*: You need to be able to "shelve the current plan in response to emerging events and your organization's culture should thank the staff who alert it to signals of change."

2. *Seek fast and ongoing feedback*: You should set your work up to be able to "detect changes in real time, especially when the windows of opportunity around such changes are short lived."
3. *Success is often accidental*: "Surprising breakthroughs … are a recurring feature of innovation and change." Being able to obtain rapid feedback will help you "to spot and respond to accidental successes as early as possible."
4. *Undertake multiple parallel experiments*: To gain feedback on what works, Green suggests you should "pursue a 'lean start-up' based on best guesses about what will work, followed by a fast and frugal cycle of experimentation and adaptation until you find something that really does."
5. *Learn by doing (and failing)*: You need to get comfortable to discuss failure. Green recommends asking "'What have you learned?' During the course of any given effort, which covers the same ground in a less embarrassing fashion." You should include getting "feedback on the outside world," not just from your project, "not least to detect unintended consequences."
6. *Identify and discuss your rules of thumb*: Instead of developing a detailed best practices guide, identify what your rules of thumb are for engaging in your work. Green suggests you should "[m]ake them explicit and review them regularly."
7. *Convene and broker relationships*: Identify who needs to be part of your stakeholders to bring about change. Green argues for creating space for "[b]ringing dissimilar local players together to find their own solutions."

Green's list mirrors key lessons shared by Dr. Rajiv Shah (2023), current president of the Rockefeller Foundation, which include the importance of experimenting and pivoting as well as giving up control (in order to bring others on board). Scodanibbio et al. (2023) also show how knowledge brokers and their social networks can strengthen the use and co-production of knowledge for decision-making to address transformations.

Managing transformational change at an organizational scale

Green's insights are very valuable, given the important role of advocacy in sustainable development. To achieve collective change, we also need to understand best practices on how to mobilize action from an organizational setting.

Best practices in change management include a myriad of models and resource guides. Based on a summary from USAID (2015), two models stand out: Dr. John Kotter's (1995) eight-step process, shown in Table 13.1 and the ADKAR model developed by Prosci, shown in Table 13.2. Both these models are derived from a synthesis of research on how change happens within the private sector.

John Kotter's eight-step process

Kotter identified three focal areas: (1) setting the climate for change; (2) enabling and engaging in organization; and (3) implementing and sustaining new ways. Reflecting our earlier discussion, the steps include a shift away from top-down

Table 13.1 Dr. Kotter's eight-step process. Source: Kotter (1995) and USAID (2015)

Steps	Focus
1. Creating a sense of shared need and urgency	*Setting the climate for change*: Create an understanding of the rationale and opportunities
2. Developing and being part of a guiding coalition	
3. Developing a vision and strategy	
4. Communicating the change and mobilizing commitment	*Enabling and engaging in organization*: Prepare people to work in the new environment; demonstrate how the change will work; explain how the change will affect individuals.
5. Enabling employees to make the change	
6. Generating and celebrating short-term wins	
7. Consolidating gains and building on success	*Implementing and sustaining new ways*: Monitor updates and governance
8. Making it stick	

Table 13.2 From awareness to reinforcement: a model for change management. Source: Prosci and USAID (2015)

Change management model – ADKAR	Select factors influencing success
Awareness of the need for change	A person's view of the current state and problems; the credibility of the awareness messaging
Desire to support and participate in the change	The nature of the change, and a person's intrinsic motivation
Knowledge of how to change	Capabilities of the person, and resources available for education and training
Ability to implement required skills and behavior	Psychological barriers, available time to overcome skill gaps
Reinforcement to sustain the change	Demonstrated progress, absence of negative consequences, and accountability reinforcing the change

implementation toward establishing buy-in and crucially enabling people to make the change themselves.

While the missing Step 0 masks a more top-down approach, the eight-step process highlights how change cannot succeed without individuals making the change themselves, even in large, powerful corporations. It speaks to the individual's agency in these processes, and the importance of buy-in.

Presci's ADKAR model

We previously reviewed the AIDA model in Vignette 1, focused on consumer behavior and the three steps for individual change in the KAP model: knowledge, attitude and practice. The ADKAR model focuses on organizational change. Although focused on the private sector, the model complements Dr. Kotter's steps by showcasing how change can be sustained through investing in the ability for people to adapt and offering reinforcement.

VIGNETTE 1 AIDA AND COCA-COLA

The goal needs to be to empower people to bring about the change (that they are seeking) themselves. This logic of how change takes place is at the core of our own consumption patterns.

During my grammar school days in the 1990s in Switzerland, we were taught the marketing model of AIDA – *Attention, Interest, Desire, Action*. Attributed to late nineteenth-century advertisement pioneer St. Elmo Lewis, the model captures the steps we go through when we make purchasing decisions.

That new [insert smartphone of your choice] you got, you first noticed it through a peer or an ad. You developed an interest in learning more about its benefits (or even the company and its range of products). It stimulated a growing desire to the point where you ultimately purchase said gadget.

Even as we engage in conspicuous consumption, the argument holds that you ultimately made the decision to purchase the product. In development practice, the acronym KAP, standing for *knowledge, attitude*, and *practice* offers additional insights into how we need to first create awareness, support a shift in attitude, to then ultimately translate into a change in practice. Adapted from health care where doctors can prescribe but not order an individual to change, the combination of qualitative and quantitative methods yields deeper insights into how individuals decide.

To illustrate my next point, let us use the example of a Coca-Cola.[2] Ubiquitously available across East Africa, its price of around US 25 cents still represents a substantial cost for rural households that earn less than US$2 a day. Successful marketing gets customers' attention, interest, and translates their desire into subsequent action. Coca-Cola is excelling at the above, reaching nearly everywhere across the region.

When attempting to achieve sustainable transformation, we are tempted to "be-like-Coke" and focus our attention on reaching the largest number of beneficiaries via outreach or awareness campaigns. Getting people's attention is the easy part. Unlike Coca-Cola, to create a desire for change and take action will take months or years.

And if you create a desire, you had better ensure that your product or intervention is accessible to the individuals. I have worked on development projects that successfully generated a demand, but had no supply available locally (e.g., a new seed in a local shop). If that happens, you will have done more harm than good on your journey of supporting sustainable transformations. The same applies to skills-building programs that neglect to check whether there is an actual demand in the marketplace.

And this is the reason behind Coca-Cola's dominance: its supply chain is just as powerful as its marketing. Coca-Cola's success motivated USAID to leverage their supply trucks to improve access to medicine into the rural areas under the *Project Last Mile* (USAID n.d.).

Managing transformational change at an individual level

Change can never be imposed. In contrast to achieving transformational change at an organizational level, transformational development project implementation requires more bottom-up approaches. Implementation shifts from a "need" or "problem" to leveraging existing capabilities and a focus on how we can empower stakeholders to achieve their desired outcomes. Vignette 2 explores what development practice looks like if we shift from *things* to *people*, and *delivery* to *collaborations*.

VIGNETTE 2 DIFFERENCE BETWEEN BOTTOM-UP AND TOP-DOWN DEVELOPMENT[3]

Chuck Kleymeyer (1991, 38) describes grassroots or bottom-up development as a "process by which disadvantaged people organize themselves to overcome obstacles to their social, cultural and economic well-being." These efforts "normally involve small-scale, practical efforts to achieve change, and are carried out by organizations such as village or neighborhood associations, production or service cooperatives, cultural groups, workers' associations, ethnic coalitions or federations of such organizations" (Kleymeyer 1991, 38).

Policy innovation still happens from the top (Mosse 2005, 233), but local action makes the difference. As Kleymeyer (1991, 38) remarks, what distinguishes bottom-up from top-down development is that "potential beneficiaries themselves take the initiative to identify problems and possible solutions. With their own human and material resources, and whatever outside technical, financial, or political assistance is available, they then address those problems."[4]

Bottom-up development is an answer to the satirical Lord's Prayer written by Elnathan John (2019):

> Our donors who art abroad, hallowed be thy purse. Thy aid come in dollars and pounds. Thy will be done in our countries, as promoted by Bono. Give us this day our yearly funding. And lead us not into self-reliance. But deliver us from ourselves. For thine are dollars, the pounds and the euros, forever and ever. Amen.

Or as summarized by Pilling (2022), who reviewed the latest work by Stefan Dercon, former head of the UK's Department for International Development:

> [International aid] is never a determining factor. It can help countries, but only if they have gambled on development themselves. If not, then aid is at best a sticking plaster and at worst an enabler of a derelict elite. But if a country is on the right road, aid can increase the upside and reduce the risk of failure.

Dr. Robert Maguire developed a helpful typology to contrast top-down and bottom-up development for us. As captured in Table 13.3, bottom-up development shifts the focus to the communities and recipients of aid. It empowers them by orienting our development paradigm around the belief that local communities know how to undertake development and what to do.

As a practitioner of bottom-up development, your opening conversation is focused on questions around *what do you do?* instead of *what do you need?*. Rather than fixing, our role is to support and assist. To get us there, be responsive and listen, rather than prescriptive and deciding.

In the end, instead of delivering on projects, we aim to collaborate as partners, shifting the focus from a distant they/them to a collaborative we/us. To succeed, we subsequently seek avenues to support processes and activities, as opposed to pushing for outputs. Bottom-up development support hence focuses less on things and techniques, and more on empowering people and building relationships.

To expand on Maguire's typology, Chambers (2017) presents a more in-depth shift from a dominant view of development – focused on *things* – versus a *people*-centered approach to development. Maguire captured the juxtaposition of blueprint and process, planning and participation, a linear logic, and acknowledging complexity. In Table 13.4, Chambers highlights the differences in methods and approaches (e.g., a randomized controlled trial vs. a participatory rural appraisal), values (e.g., making work replicable vs. relevant), interactions with local participants (e.g., extractive vs. empowering), and how they are seen (e.g., informant vs. collaborators). Research outputs also shift from a focus on statistics that provide universal knowledge to case studies that focus on the particular.

Table 13.3 Bottom-up and top-down development typology. Courtesy of Robert 'Bob' Maguire

Top-down development	Bottom-up development
We know how	They know how
Projects	Partners
Deliver	Collaborate
Fix	Support/assist
Things/Techniques	People/Relationships
Prescriptive	Responsive
What do you need?	What do you do?
Targets	Outcomes
Outputs	Process/activities
Deciding	Listening
They/them	We/us

Table 13.4 Putting people at the center of development. Source: Chambers (2017)

Select points of departure/ reference	Things	People
Conceptualization	Top-down Blueprint Planning Linear	Bottom-up Process Participation Complex
Methods and approaches	Randomized controlled trials Logframes Questionnaires Standardized	Participatory rural appraisals Systems analysis Observations Context-dependent
Values	Precision Replicable	Insight Relevant
Interactions with local participants	Questioning Extractive	Facilitating Empowering
Local people seen as	Informants	Collaborators
Research outputs	Statistics Universal	Case studies Particularistic

HOW WE TRANSLATE BOTTOM-UP DEVELOPMENT INTO CHANGE

Translating the typology into the practice of sustainable development project implementation from a bottom-up perspective, our role is much more around empowering individuals and local organizations to bring about a change. Following an adapted version of Kotter's eight-step process or the ADKAR model could be a viable pathway for some, but this also has its shortcomings.

In the Global South, or marginalized communities, the ability of individuals and communities to take risks is limited. A Western corporation engaging in change management has sufficient capital that can act as safety nets in case the change turns out to be detrimental. A smallholder farmer, operating on half a hectare of land, or a recent urban migrant who has no excess income at the end of the week, is less likely to have the capacity to adapt to changes.

As a result, we are less focused on creating an awareness of the need for a change, but more on awareness building to establish the confidence that a change would have benefits. Failure to do so would create a lack of desire to support and take part in the change.

ON THE NEED TO ADAPT BY INCORPORATING PILOT PROJECTS AND FAILURE

We need a step 0 that focuses on co-defining and demonstrating the viability of the change. Usually, these take the form of participatory trials, pilots, or community-wide discussions. As a development practitioner, one avenue to support such endeavors is drawing on the practice of human-centered design (HCD) (Figure 13.3).

I tried my first step into HCD as I was on a journey to Brighton, in the south of England, doing a sprint for a week to start a new social enterprise – *farmbeter* (see Chapter 9). Our five-day retreat applied the steps of former Google Ventures design partners: Day 1, brainstorming around the target of the sprint; Day 2,

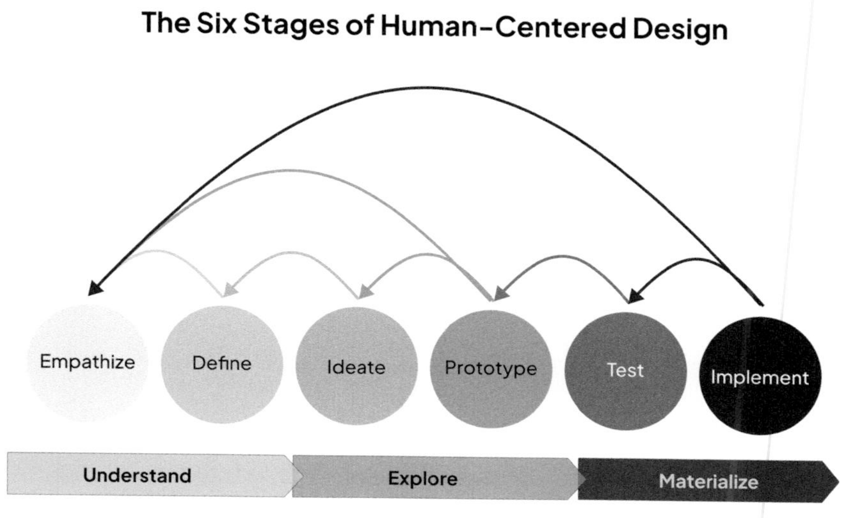

Figure 13.3
The six stages in human-centered design, adapted from Moran and Brown (2024)

sketching competing solutions; Day 3, deciding on the best; Day 4, building a prototype, and Day 5, testing with target customers (Knapp, Kowitz, and Zeratsky 2016).

I was formally introduced to HCD by Scott Stein,[5] who was spearheading my university's pitching competition. I brought him into our graduate course to showcase to development practitioners how they can apply concepts from the start-up world in their practice (and possibly launch their own venture as an outcome from it).

To take you through the cycle of HCD (Moran and Brown 2024), it starts with *Understanding*, followed by *Exploring* and *Materializing*.

The Understanding phase

As captured in Figure 13.3, understanding entails the process of *Empathize* and *Define*. Empathize is where you start your research to increase your understanding of your user (Moran and Brown 2024).

As represented by the *Empathy Framework* in Figure 13.4, it asks you to put the user at the core of your investigations, capturing: what they *say*, what they *think*, what they *feel*, and what they *do*. The framework demands that you draw on your observational skills and senses to get a more complete picture. While listening is important, you want to capture more holistically their expressions and reactions. In our practice, it allows us to stretch explicitly our focus beyond the voices from our interviews, surveys, or focus groups.

The *Empathize* step is followed by *Define*, where we "combine all your research and observe where your users' problems exist" (Moran and Brown 2024).

A sample problem statement – courtesy of Scott Stein – I used with my students is to capture the following:

> I observed that [person/group of people/user] has a [need/problem] which is causing [this adverse effect]. Surprisingly, I learned that [insight].

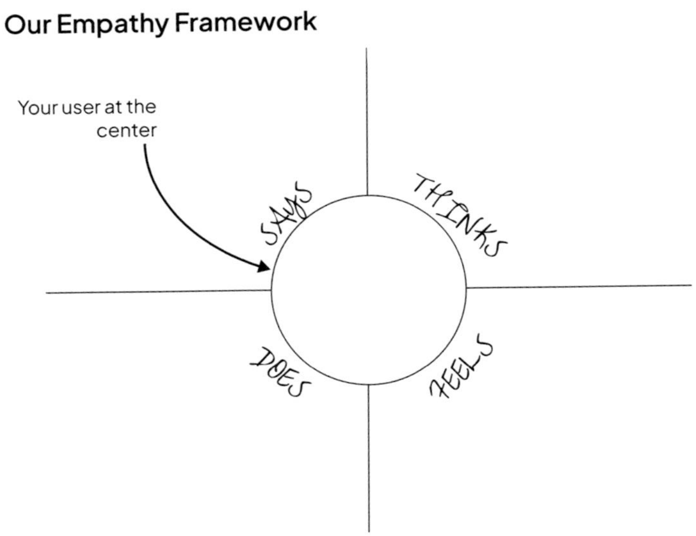

Figure 13.4 Our Empathy Framework, adapted from Scott Stein

My students, having worked earlier in the semester on identifying the problem, usually were quick to articulate the first sentence. The second one – around the insights – proved more challenging; from our classroom setting, we were often drawing on what we gained from the review of the literature. The resulting distance to *end users* (i.e., beneficiaries or participants in development projects) was a source of frustration for them.

Unfortunately, their situation might not be too different from a government worker tasked with managing development aid stationed at an embassy abroad with limited trips into the field. That gap subsequently illustrated to all of us the importance of scoping missions, listening to beneficiaries and stakeholders, or at least engaging in key informant interviews in order to move beyond the written text.

The Exploring phase

Upon completing the define stage, you move into the exploring phase where you *ideate* — "generate a range of crazy, creative ideas." I complete this stage together with my students using "How would..." prompts. Scott Stein introduced us to thought-provoking ones, such as trees or rain. He also offered prompts that were more out there, such as *Wonder Woman* (who had a lasso of truth that my students had to teach me about) or *NASCAR* (a US car racing series).

In the beginning, my students struggled with taking things too literally: for trees, they focused on reducing carbon emission. After some practice and guidance, the trees nearly always ended up being the most fruitful topic. Here are two illustrative examples: "A tree has a strong root systems that can withstand shocks, so our project on reducing gender-based violence needs to be deeply grounded in the community." Or "A tree exists in a larger forest ecosystem, so we need to ensure that our gender project is engaging across all genders and not only with women." In contrast to the problem statement, these prompts nearly always generated new ideas during a friendly brainstorming setting.

Timing is important, as recipients might be too far along into their development design to still incorporate them.

Once completed, you will move onto developing a *prototype*, which is a first "real, tactile representations for a range of your ideas" (Moran and Brown 2024). Ideo.org has implemented prototyping as a central part of their quest to create larger social impact for their stakeholders. Focused on designing for impact, their projects are filled with examples of prototypes that have been developed, and rapidly failed. In development projects, failure is not an option (or at least we try not to reveal them). In a human-centered design approach, it is built into the logic with the goal that the second or fifth prototype will emerge successfully to generate the longer-lasting impact we intended.

The Materializing phase

For user-centered design, you want to get rapid feedback, something I had to get comfortable with as an academic. In the *materialize* phase, you start by *testing* where you "return to your users for feedback" (Moran and Brown 2024). At this stage, you might either head toward *implementation*, the final stage "where you put your vision into effect" (Moran and Brown 2024), or decide you need to rework your prototype.

As alluded to above, HCD's unique part is planned and expected failure. It represents an iterative process (as captured in Figure 13.2). You might stop at prototyping, for example, and head back to ideating. You might even return to the beginning with empathizing. These quick actions circumvent the risk of taking too long to perfect your own *solution* (to where it might even become obsolete). The option to fail ensures we put the end user (or beneficiary) at the center of our work. HCD pushed me out of my comfort zone and showcased the importance of failing, revising, and improving to create a sustained impact.

MIXING UP THE MODELS FOR SUSTAINABLE DEVELOPMENT

For sustainable transformations, we can benefit from a model that is adapted to a context in the Global South and focuses on the change in the individual. Awareness about the project in the first place and its intervention is a necessary part of achieving change. Development projects, however, should focus most of their funding on building capacities for beneficiaries, participants, or stakeholders to bring about change. The existing practice of backstopping works similarly to providing reinforcement to sustain such change. After initial project activities have concluded, implementers step back and provide support and advisory services to partners.

CHANGE IN SUSTAINABLE DEVELOPMENT IMPLEMENTATION

Combining the above frameworks with my lessons learnt, I distilled the following five steps in Figure 13.5 on how change *LEADS* to sustainable development implementation and subsequent transformation: **L**isten, **E**xperiment, **A**dapt,

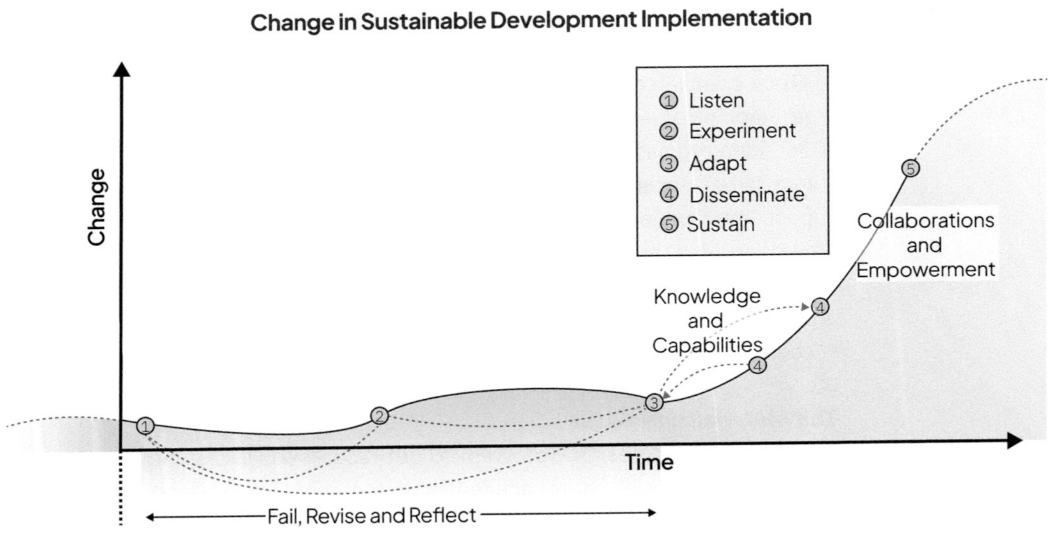

Figure 13.5 How change happens in sustainable development implementation

Disseminate, and **S**ustain. These steps are inspired by a focus on bottom-up development and strengthening capabilities. Figure 13.5 is motivated to reflect that we want to focus on creating sustained change that goes beyond any set project implementation period (usually determined by funding).

1 Listen

As opposed to deciding on what to implement, listen to your stakeholders. While we all have preconceived ideas on what could work, this first phase of listening focuses on ownership to create any sustained change. As reflected in Figure 13.5, this first step will entail what could be perceived as a lack of progress or even regressing in your goal of achieving change.

Listening should start even before you move toward implementation, by providing avenues for (less privileged) colleagues to take the lead in framing development activities and goals. Such a participatory approach stands in contrast to and offers to circumvent the inherent bias within the development funding landscape for funding organizations from the Global North to do work in the Global South.

2 Experiment

Development projects fail if they are not testing the assumptions early on. You can co-create small, rapid pilots to get feedback on what works and what does not. Given the discussed limitations to experiment by the poorest of the poor, these participatory pilots can build awareness of the intervention and provide sites for demonstrating its viability. Experimenting also allows you to fail, quickly, to the point of forcing you to step back and revert to listening again before you revise your small-scale pilots.

3 Adapt

Adoption is the name of the game when it comes to reporting development "success" to donors. The real key to success, however, is adaptation. Any results from your pilots – once successful – would expect to yield half as much once moved outside the more controlled (or biased) conditions. For example, even if you held trials at a local farmer, the farmer you are working with is likely better off than their peers, has more capacity, and greater capabilities. They also might have signed up as hosts, as they had the greatest desire for change. Instead of being prescriptive, to achieve change, you need to be responsive. Each location, each household, each individual is unique. Even in large-scale initiatives such as AFR100 (African Forest Landscape Restoration Initiative), involving 34 countries (Peel and Bryan 2024), its success comes down to context: "Ultimately, the right trees and the right number need to be planted in the right place" (Parr, Te Beest, and Stevens 2024). Context matters. And since your organization cannot be in every location at the same time, you should focus on supporting and assisting people, as opposed to trying to fix it for them. To sustain this change, you need to ultimately build knowledge and expand capabilities as you move into dissemination.

Adaptation marks another point where you reflect on what works and what does not. This reflection includes looking at effects beyond your project to account for unexpected synergies or negative externalities. These will become more visible as you move into dissemination and reach a larger number.

You might also decide at this stage that you need to revert to experimenting again, or even back to listen to your stakeholders if adaptation is too difficult. Allow yourself to even fail at this stage. Pushing through a change from the top-down will not only fail to yield sustained change, you might not even create meaningful adoption through a dissemination campaign.

Isabel Guerrero, who reflected on her work in helping development organizations increase their impact through scaling, captures the importance of focusing on essentials when scaling:

> Scaling up is really a long journey of discovery, failure, adaptation, and letting go. Sometimes, letting go involves parting with what is not essential, striving towards a minimum viable intervention. Unfortunately, one cannot scale an intervention with all its bells and whistles, and it is important to make the solution leaner as it transfers to other contexts.
>
> (Gokhale and Fahsbender 2024)

4 Disseminate

We understand that most dissemination efforts work best when undertaken in-person and with more intimate numbers of participants. We also know that if you have successfully experimented and adapted, it will be easy to generate greater awareness and interest in the change you are supporting with your development work.

Cost-effectiveness is important, as you can reach hundreds of thousands of listeners with a well-timed radio campaign, but these efforts might not translate into action. As change cannot be imposed and occurs at an individual level, the focus in this phase needs to be on ensuring you build the needed knowledge and capacities for people to act upon their interests and desires.

Adoption will happen through a behavioral change, as an attitude turns into a practice. Adaptation will happen through supporting capabilities to re-innovate and draw on new tools and resources. Avoid falling into the trap of delivering an intervention, and re-focus on building collaborations.

Self-sustained scaling is the dream of every development practitioner (including mine). A Swiss analogy is of an avalanche that picks up speed and size as dissemination continues. The more appropriate are viral videos that reach your social media channels. The reality is for every viral campaign, there are thousands that gained no traction.

5 Sustain

If change cannot be imposed, then empowerment is key. We need to invest in people and relationships. Investing in items or technologies alone will always be self-limiting, as we are operating with limited budgets.

To hedge toward a more sustained change, bring to the table diverse stakeholders from our four changemakers. You should be able to draw upon them as they have been engaged since your first step in listening. In development implementation, creating a sustainability plan would be a good step (PM4NGOs 2020). Loosely attributed to my former boss, we should put ourselves out of business by the end of a project. To get there, empowerment and ownership are indispencible. If neither of these exists, our change trajectory will regress to the starting point.

SUSTAINED CHANGE AND SHIFTING OUR DEVELOPMENT PARADIGM

The groundbreaking work by Anderson, Brown, and Jean (2012) in their book, *Time to Listen: Hearing People on the Receiving End of International Aid*, analyzed over 6,000 voices and the resultant insights from development stakeholders. Anderson et al. engaged with over 125 international and local development organizations across 20 countries that are aid recipients. Their insights – supported through funding by the World Bank – offer a unique window into how development can create a more sustained change.

One major phenomenon identified by aid recipients is *Projectitis* – projects are often too short-term (three years) and discreet (focused on silos) to create a more systemic change. Unfortunately, a development project – rather than creating a sustained change – is creating… another development project. This project cycle is captured in development speak by discussing project phases: a three-year project would try to extend existing or apply for new funding as it reaches its last year, to start another three-year project.

As captured by Anderson, Brown, and Jean (2012, 74):

> [P]rojects do not add up. In [aid recipients'] experience, one project simply leads to other projects; they are often piecemeal interventions that are not strategic and cannot, with such limited time commitments, support systemic change. The point here is that many recipients identify the proceduralization of the processes of the project cycle (assessments, proposal writing, reporting) as contributing to and reinforcing the piecemeal nature of assistance. They call for more holistic approaches and long-term thinking that, they believe, would support sustainable impacts.

To bring about more sustained change, we need longer durations for projects, shifting our focus from fixed plans to context-relevant and adaptive implementation. The goal should be to empower people to expand their range of available options to bring about a lasting change.

Anderson, Brown, and Jean (2012, 138) ultimately argue for a paradigmatic shift from an externally driven aid delivery system to one that is collaborative. As captured in Table 13.5, it reflects Robert Maguire and Chambers (2017) by integrating a greater focus on how incentives are placed. Staff could be rewarded for the quality of their relationships and based on an evaluation of their work's results from recipients. Projects need to move away from growth to making yourself no longer of use. And spending should be adapted to the realities on the ground, as opposed to what works best for the HQ.

Table 13.5 Elements of two paradigms: a comparison. Source: Anderson, Brown and Jean (2012)

Externally driven aid delivery system	Collaborative aid system
Local people seen as beneficiaries and aid recipients	Local people seen as colleagues and drivers of their own development
Focus on identifying needs	Focus on supporting/reinforcing capacities and identifying local priorities
Pre-planned/pre-determined programs	Context-relevant programs developed jointly by recipient communities and aid providers
Provider-driven decision-making	Collaborative decision-making
Focus on spending on a pre-determined schedule	Fit money and timing to strategy and realities on the ground
Staff evaluated and rewarded for managing projects on time and on budget	Staff evaluated and rewarded for quality of relationships and results that recipients say make lasting positive changes in their lives
Monitoring and evaluation by providers on project spending and delivery of planned assistance	Monitoring, evaluation, and follow-up by providers and recipients of results and long-term effects of assistance
Focus on growth	Planned draw down and mutually agreed exit/ end of assistance strategy

NOTES

1 I introduce my students to logframes as learning how to read a new language. Logframes are usually developed by a select few rather than a broader community of stakeholders. Stakeholders who have not been trained in logframes at all will struggle to secure funding if applications require them.

2 With thanks to my former colleague, Jimmy Pittchar, for alerting me to this issue when we were planning a new development project together.

3 The following section was based on my collaboration together with my predecessor, Dr. Robert "Bob" Maguire, with whom I have had the pleasure of co-teaching his "Bottom-up Development" course over the years. Before his career in academia, Bob was a long-time practitioner of bottom-up development at the Inter-American Foundation.

4 David Mosse (2005, 233), writing an ethnography of aid policy and practice, captures well that "even in bottom-up projects, policy innovation comes from the top."

5 Scott is currently the Principal Technical Advisor, Product Development at LMI.

REFERENCES

Altman, S. A., and C. R. Bastian, 2024. *DHL Global Connectedness Report*. Bonn: DHL.

Anderson, M. B., D. Brown, and I. Jean. 2012. *Time to Listen: Hearing People on the Receiving End of International Aid*. Cambridge, MA: CDA Collaborative Learning Projects.

Chambers, R. 2017. *Can We Know Better? Reflections for Development*. Rugby, UK: Practical Action Publishing Ltd.

Gokhale, S., and J. Fahsbender. 2024. "Scaling Up Development Impact: An Interview with Author Isabel Guerrero." https://www.hks.harvard.edu/centers/cid/voices/scaling-development-impact-interview-author-isabel-guerrero (accessed July 25, 2024).

Green, D. 2016. *How Change Happens*. Oxford: Oxford University Press.

ideo.org. "Work." Available at: https://www.ideo.org/work (accessed March 23, 2025).

Independent Group of Scientists Appointed by the Secretary-General. 2019. *Global Sustainable Development Report 2019: The Future is Now – Science for Achieving Sustainable Development*. New York: United Nations. Available at: https://sdgs.un.org/gsdr/gsdr2019.

John, E. 2019. *Be(com)ing Nigerian: A Guide*. Abuja: Cassava Republic.

Kleymeyer, C. 1991. "What is 'Grassroots Development'?" *Grassroots Development: Journal of the Inter-American Foundation* 15 (1).

Knapp, J., B. Kowitz, and J. Zeratsky. 2016. *Sprint: How to Solve Big Problems and Test New Ideas in Just Five Days*. New York: Simon & Schuster.

Kotter, J. P. 1995. "Leading Change: Why Transformation Efforts Fail." *Harvard Business Review* June. Available at: https://hbr.org/1995/05/leading-change-why-transformation-efforts-fail-2.

Mendelson, S. E. 2023. *Paradigm Shift: Creating More Just Societies Within the SDGs, Human Rights, and Innovations in Higher Education*. Washington, DC: Center for Sustainable Development at Brookings, Brookings Institution..

Moran, K., and M. Brown. 2024. "Design Thinking: Study Guide." NN/g. Available at: https://www.nngroup.com/articles/design-thinking-study-guide/.

Mosse, D. 2005. *Cultivating Development*. New York: Pluto Press.

OECD. 2021. *Applying Evaluation Criteria Thoughtfully*. Paris: OECD. Available at: https://doi.org/10.1787/543e84ed-en.

Parr, C. L., M. Te Beest, and N. Stevens. 2024. "Conflation of Reforestation with Restoration Is Widespread." *Science* 383 (6684): 698–701. https://doi.org/10.1126/science.adj0899.

Peel, M., and K. Bryan. 2024. "Global Tree-Planting Push Threatens African Grasslands, Warns Report." *Financial Times*, February 15. https://www.ft.com/content/4f8d91b9-c8b6-4dc5-8318-27d75c30e542.

Pilling, D. 2022. "Gambling on Development — How Economic Miracles Can Happen." *Financial Times*, June 6. https://www.ft.com/content/e1f203a4-2b6a-4fe3-b29b-5517d1207696.

PM4NGOs. 2020. *Project DPro: Project Management for Development Professionals Guide*. Available at: https://pm4ngos.org/methodologies-guides/project-dpro/.

Prosci. "The Prosci ADKAR Model." Available at: https://www.prosci.com/methodology/adkar (accessed March 23, 2025).

Scodanibbio, L., G. Cundill, L. McNamara, and M. du Toit. 2023. "Effective Climate Knowledge Brokering in a World of Urgent Transitions." *Development in Practice* 33 (7): 755–761. https://doi.org/10.1080/09614524.2022.2159932.

Scott, J. 1998. *Seeing Like a State*. New Haven, CT: Yale University Press.

SDC. 2011. *Module 2: Planning for Results*. SDC and ETH Zürich.

Shah, R. 2023. *Big Bets: How Large-Scale Change Really Happens*. New York: First Simon Element.

Statistica Research. 2024. "Global Air Traffic – Scheduled Passengers 2004–2024." Available at: https://www.statista.com/statistics/564717/airline-industry-passenger-traffic-globally/ (accessed March 23, 2025).

Taylor, P. 2024. "Number of Global Mobile Subscriptions 1993–2023." Available at: https://www.statista.com/statistics/262950/global-mobile-subscriptions-since-1993/ (accessed March 23, 2025).

USAID. 2015. *Change Management Best Practices Guide*. Washington, DC: USAID.

USAID. n.d. "Project Last Mile." Available at: https://www.usaid.gov/cii/project-last-mile.

WTO. 2024. "World Trade Statistics 2023." Available at: https://www.wto.org/english/res_e/statis_e/statistics2023_e.htm (accessed March 23, 2025).

14 Focus I

Gender inequality

THE DIFFERENT INEQUALITIES OF GENDER

We covered in Chapter 8 how women face the double burden of both productive and reproductive household labor. They also face a third burden of additional organizational labor, time-poverty, and inequalities at the intersections of gender, poverty, and marginalization (Kabeer 2015). The COVID-19 pandemic accentuated gender inequalities, with women being more severely impacted by job losses, domestic violence, and food insecurity (Béné et al. 2021). Women are also more exposed to climate change risks (FAO 2018). In the study site for my dissertation of Meatu District in Northern Tanzania, accusations of the practice of witchcraft is on the rise during droughts and other shocks. As shown by Dr. Edward Miguel, old women, accused of being witches, are more likely to be killed when climate shocks cause yield losses (Miguel 2005).

In summary, gender inequalities not only manifest at the intra-household level as additional burdens, but are often accompanied by increased forms of violence. As shown in the next section, addressing these inequalities offers one of the best pathways to achieve sustainable development.

GENDER AS A CROSS-CUTTING THEME IN ACHIEVING SUSTAINABLE DEVELOPMENT

Across our six hotspots of transformation, gender and its inequalities are omnipresent. Gender serves as an SDG that Nilsson, Griggs, and Visback (2016) score as *indivisible* (+3), as it is "inextricably linked to the achievement of another goal." Nilsson, Griggs, and Visback (2016, 321) use the example how "[e]nding all forms of discrimination against women and girls is indivisible from ensuring women's full and effective participation and equal opportunities for leadership."

Transformations toward sustainable, local food systems move us closer to ending poverty (SDG 1) and hunger (SDG 2), improved health and well-being (SDG 3), more responsible consumption (SDG 12), as well as a better life on land (SDG 15). According to UN Women and UN DESA (2023, 4), "if current trends continue, over 340 million women and girls will still live in extreme poverty by 2030 … [and] close to 1 in 4 women and girls are expected to be moderately or severely food insecure." Negative effects from climate change are also projected to affect women and girls harder: 158 more million of women and girls – in a worst-case scenario by 2050 – will experience extreme poverty

DOI:10.4324/9781003400028-20

due to climate change. Another 236 million more women and girls will become food insecure due to climate change (UN Women and UN DESA 2023).

Our transformation hotspot of more just economies needs to build upon gender equality, in conjunction with improving decent work and economic growth (SDG 8), reduced inequalities (SDG 10), and strengthening peace, justice and strong institutions (SDG 16). Clear synergies exist. Offering women the same opportunities as men would generate additional economic growth, as the gender gap in labor force participation is nearly 30 percent (UN Women and UN DESA 2023). These goals do not need to conflict with each other. Yet persistent inequalities continue to hold back our advances in sustainable development.

Gender inequality negatively impedes advances made

Investing in addressing gender inequality can address multiple challenges, as the synergies outweigh the trade-offs. Across the SDGs, in order to achieve gender equality and women's empowerment, additional funding of US$360 billion is needed (UN Women and UN DESA 2023).

Reducing inequality would pay for itself. As covered in Chapter 6, getting rid of the gender gap in uneven access to resources in agriculture would increase yields alone by 20 to 30 percent and reduce hunger globally by 12 to 17 percent (FAO 2011). According to UN Women and UN DESA (2023), global GDP would rise nearly US$1 trillion if the gender gaps in agrifood systems were to be closed.

You might argue that more growth is not what is needed – especially if it comes with increased environmental costs. Let us connect with our earlier discussion on building local, sustainable food systems. My argument was that business-as-usual is no longer an option. The limits to further intensification of production are becoming more visible as the marginal return of an additional investment in fertilizer, for example, is diminishing. The tragic reality from a gendered perspective is that the baseline from which women are to invest in agricultural production, including intensification, is significantly lower. With lower intensity of fertilizer or pesticide use, negative environmental impacts from run-offs are also much less likely as captured in just Earth-system boundaries.

Ultimately, the gendered gap is worse than starting with a 20 percent handicap on a climbing competition – the ceiling itself is lower. Cracking that ceiling needs to be the goal. But just as in the newly introduced Paris Olympics Speed Climbing competition, we do not have to all follow the same path. Shifting from conventional to sustainable agriculture can also create "spaces of empowerment for women farmers" (Trauger 2007, 290) by adopting a different development trajectory.

A gender lens allows us to review and identify sustainable development challenges and opportunities. Without it, we could falsely assume that any further investment in "business-as-usual" is counterproductive, when in reality some groups have been substantially marginalized: they have been excluded from "business" in the first place and are on a different sustainable development trajectory.

Inequalities extend beyond access to agricultural inputs

Useful social networks for learning new technologies are less accessible to women as well (Fisher et al. 2019). In Switzerland, for example, wives of farmers (including my oldest sister) receive training to manage their household and garden, while their husbands gain a farming education. Combined with their double-burden, women have less time to learn about the latest technology trends, such as digital agriculture (Hackfort 2021). Women more frequently lack ownership of land and hold smaller landholdings. Both are key factors that reduce the ability of women farmers to adopt, experiment, and adapt new technologies (van Campenhout 2021).

Lacking land (and the resultant lower abilities to experiment) reinforces that female farmers are less visible to extension agents (Ragasa 2014), further marginalizing their rural positioning. In Malawi, for example, women were less likely to be chosen as lead farmers for demonstrations (Duffy et al. 2021). In summary, inequalities extend well beyond the productivity domain in agriculture and magnify both the need and potential for a transformation.

A CLOSER LOOK AT GENDER AND DEVELOPMENT

Gender in development has come a long way from the 1970s when Ester Boserup's volume on "Women's Role in Economic Development" set off an entire field of inquiry and action within the development sector. Her work led to gender breaking out from its marginalization in development, as "the central role of women in the world economy as well as the development process has been established beyond doubt" (Pearson 2019, 158).

I draw here on Dr. Ruth Pearson's scholarship to highlight the progress made on gender as it relates to our changemaker of policymaking and governance. Development agencies stick to a technical or equality-based approach to gender. USAID, for example, by the 1980s focused on greater participation, benefits, and inclusion of women (Pearson 2019).

When it comes to translating policymaking into implementation, our sector prioritizes addressing more practical gender interests that focus on "changes and investments required to address women's disadvantage in terms of poverty and deprivation" (Pearson 2019, 161). More strategic gender interests that aim to reduce women's disadvantages in education, gender-based violence, or wage inequalities are less prominently featured (Pearson 2019, 161). Women's persistent fears of gender-based violence negatively impact their contributions to society. In least developed countries, nearly four out of ten women report having experienced domestic violence (Independent Group of Scientists Appointed by the Secretary-General 2019).

By now, it is well established that a gendered approach to development work should include men just as much as women to advance gender norms (IDS Bulletin 2000; Jackson 2000; cited in Pearson 2019, 166). Let me offer two examples from development programming.

First, a systematic review of female genital mutilation (FGM) found that "[s]ocial obligations and the lack of dialogue between men and women" were noted by men as barriers to abandoning FGM (Varol et al. 2015, 1). Advocacy efforts to counter the "silent culture" between men and women, including conversations on health, are expected to improve the effectiveness of interventions (Varol et al. 2015, 3).

Second, Dr. Rachel Bezner-Kerr and her colleagues have conducted longitudinal participatory research and ecological development in Northern Malawi (Bezner Kerr, Berti, and Shumba 2011). Over the years, they have observed the effects of gender inequalities. Women undertake the brunt of the work in the household and in the field. Inequalities also manifest around child malnutrition and food insecurity. Their aim was to get men more involved in household labor and nutrition. After trialing one-on-one approaches, they developed the *Recipe Days*: a public, collective engagement where men learnt about cooking recipes from women, and "[transgressed] a gender role as they pounded soya, sieved flour and stirred various dishes" (Patel et al. 2014, 35).

While starting small, the intervention was scaled throughout the region. As noted by a male respondent who was asked about any observed changes in gender relations after his participation in Recipe Days:

> They've changed so much. Like now I'm here at the clinic, maybe my wife is here at the garden, so when I go home, I cannot sit and wait for her. I can prepare a meal, eat my share, and when she gets back she can eat. I cook nice meals! My last-born son is 15 years, he also cooks and helps at home.
>
> (Patel et al. 2014, 35–36)

Gender mainstreaming has become pervasive across the UN system, NGOs, and other development stakeholders. A rights-based approach to development (RBA) affirmed the role of gender as stated in Article 6 of the UN Declaration on the *Right to Development*: "All states should co-operate with a view to promoting, encouraging and strengthening universal respect for and observance of all human rights and fundamental freedoms for all without any distinction was to race, sex, language or religion" (United Nations 1986).

While gender has been mainstreamed, accountability and compliance, however, are perceived to be still lacking (Pearson 2019, 167). Analogous to discussion on Diversity, Equity, and Inclusion (DEI) efforts, Pearson (2019, 174) fears that organizations are just paying lip-service to gender, with gender perceived as an entry-point for development, as opposed to a desirable end in itself. Pearson (2019, 174) provides the example of reproductive rights, where development projects facilitated access to contraceptives as opposed to actively improving "women's autonomy and choice in terms of reproduction and sexuality."

The World Bank already argued back in 2003 for linkages between gender equality, poverty, and achieving the MDGs (Pearson 2019, 168–169). I posit that the SDGs provide the strongest case yet to break the silos that have limited the reach of gender equality across our development hotspots.

Beyond breaking silos, we need to leverage gender across all four changemakers. Let me use the example of gender-responsive budgeting – an intervention leveraging *economy and finance* – first introduced to me by a former student of mine around a decade ago who worked for Oxfam. In Tanzania, for example, gender-responsive budgeting provides greater equality and improves women's standing as it relates to economic development. It receives support from the Constitution and supporting laws (*governance and policymaking*). Women's voices at local levels are frequently still not heard. As a result, women farmers fail to receive the resources they need. Local and national government officials also lack capacity around gender-responsive budgeting, pointing to the crucial role of civil society actors in stimulating effective gender-responsive budgeting (*individual and collective action*). To support other actors in the field, Oxfam produced a series of research briefs on the topic. Insights on what works and why can improve how budgetary processes address gender inequalities (*science and technology*) (Koda and Mtasingwa 2021).

GENDER INEQUALITY IN UNPAID CARE AND DOMESTIC WORK ACROSS THE GLOBAL NORTH AND THE GLOBAL SOUTH

Around 700 million women are unable to enter the workforce due to unpaid care responsibilities (ILO 2024). Demographic trends, such as an overall aging population, combined with expected increases in climate shocks and stresses on households, will increase demand for elderly care (Grown 2023). Populations, on average, are aging across Asia, Europe, and the Middle East, raising the risk that women will end up feeling greater pressures to leave their job to focus more on domestic care (Grown 2023). To address these challenges of unpaid care and domestic work, UN Women (2022) developed a toolkit – loaded with case studies and resources – that includes the proposed following five steps: *recognize, reduce and redistribute unpaid work, and reward and represent paid caregiving*.

Beyond the main avenue to establish new national policies for parental leave or long-term care, local municipalities are innovating with new approaches. Grown (2023) references the example of Bogota, Colombia, in adopting "a comprehensive care framework, centered around CARE blocks that currently operate in 10 areas of the city, providing a range of quality services for caregivers and recipients." These CARE blocks – similar to the 15-minute city featured in Chapter 10 – aim to offer professional training, wellness promotion, and income-generating activities within 15 to 20 minutes' walking distance. They include community spaces, such as urban gardens, spaces for women entrepreneurship, and public food courts (Rodríguez Franco 2022).

In addition to the public sector, our changemaker of *economy and finance* has a role to play. Employer-supported childcare has been shown to have an oversized impact, especially "in low-income and post-conflict contexts where fiscal space may be restricted and publicly provided services are limited" (Grown 2023, 11).

HOW BREAKING SILOS AND CHANGEMAKERS MATTER

Ultimately, to address gender inequalities, we need to break down silos and all four changemakers need to be engaged. On *economy and finance*, for example, increasing access to capital for women (including intra-household) is crucial, such as through self-help groups and capacity building (Peters et al. 2016). Improving women's relative standing in their own household and in the economy (including reducing wage disparities) will also contribute significantly to economic growth. Even private companies with greater gender equality and diversity in their leadership team have historically generated higher returns (Dixon-Fyle et al. 2023) and improved efficiencies in their organizations (Profeta 2017).

On *governance and policymaking*, new legislation can support dealing with some of the thornier issues, such as reducing pay gaps. Greater representation of women in decision-making will translate into better governance around gender inequalities and foster overall a more representative democracy. Their enfranchisement also reduces the likelihood of conflicts between democratic states (Barnhart et al. 2020).

On *collective and individual action*, creating social groups will increase empowerment – as Agnes Quisumbing (2017) puts it like this, "[i]t is not development agencies that empower women; women empower themselves, and often, each other." By including all genders, movements can emerge to address gender stereotypes.

On *science and technology*, women still face significant barriers in entering male-dominated fields (Llorens et al. 2021). Even where they become better represented, discrimination still persists (Begeny et al. 2020). Research (including in development) focuses more extensively on male research participants, with female voices or bodies less featured (Neville 2024). Gender research is also under-funded, especially for more inclusive North-South, academic-practitioner collaborations (Caywood and Darmstadt 2024). Closing these research gaps promises to fuel further shifts in policymaking.

REFERENCES

Barnhart, J. N., R. F. Trager, E. N. Saunders, and A. Dafoe. 2020. "The Suffragist Peace." *International Organization* 74 (4): 633–670. https://doi.org/10.1017/S0020818320000508.

Begeny, C. T., M. K. Ryan, C. A. Moss-Racusin, and G. Ravetz. 2020. "In Some Professions, Women Have Become Well Represented, Yet Gender Bias Persists—Perpetuated by Those Who Think It Is Not Happening." *Science Advances* 6 (26). https://doi.org/10.1126/sciadv.aba7814.

Béné, C., D. Bakker, M. Juliana Chavarro, B. Even, J. Melo, and A. Sonneveld. February 2021. *Impacts of COVID-19 on People's Food Security: Foundations for a More Resilient Food System*. CGIAR COVID-19 Hub.

Bezner Kerr, R, P. R. Berti, and L. Shumba. 2011. "Effects of a Participatory Agriculture and Nutrition.Education Project on Child Growth in Northern Malawi." *Public Health Nutrition* 14 (8): 1466–1472.

Caywood, K., and G. L. Darmstadt. 2024. "Gender Mainstreaming at 25 Years: Toward an Inclusive, Collaborative, and Structured Research Agenda." *Journal of Global Health* 14: 04011. https://doi.org/10.7189/jogh.14.04011.

Dixon-Fyle, S., C. Huber, M. del Mar Martínez Márquez, S. Prince, A. Thomas, and V. Hunt. 2023. "Diversity Matters Even More: The Case for Holistic Impact." McKinsey & Company. Available at: https://www.mckinsey.com/featured-insights/diversity-and-inclusion/diversity-matters-even-more-the-case-for-holistic-impact#/.

Duffy, C., G. Toth, J. Cullinan, U. Murray, and C. Spillane. 2021. "Climate Smart Agriculture Extension: Gender Disparities in Agroforestry Knowledge Acquisition." *Climate and Development* 13: 21–33. https://doi.org/10.1080/17565529.2020.1715912.

FAO. 2011. *The State of Food and Agriculture: Women in Agriculture, Closing the Gender Gap for Development*. Rome: FAO.

FAO. 2018. "Empowering Rural Women, Powering Agriculture." Available at: https://openknowledge.fao.org/server/api/core/bitstreams/e31ea1b4-3378-499e-81da-85552d69ceeb/content.

Fisher, M., E. Habte, W. Ekere, T. Abate, and P. A. Lewin. 2019. "Reducing Gender Gaps in the Awareness and Uptake of Drought-Tolerant Maize in Uganda: The Role of Education, Extension Services and Social Networks." *Journal of Gender, Agriculture and Food Security (Agri-Gender)* 04 (1): 38–50. https://doi.org/10.19268/JGAFS.412019.4.

Grown, C. 2023. "Caring About Care: An SDG 5 Priority." In Center for Sustainable Development at Brookings Institute (Ed.), *The SDG Second Half: Ideas for Doing Things Differently*. Washington, DC: Brookings Institution.

Hackfort, S. 2021. "Patterns of Inequalities in Digital Agriculture: A Systematic Literature Review." *Sustainability* 13 (22). https://doi.org/10.3390/su132212345.

ILO. 2024. "The Impact of Care Responsibilities on Women's Labour Force Participation." International Labour Organization. Available at: https://www.ilo.org/sites/default/files/2024-10/GEDI-STAT%20brief_formatted_28.10.24_final.pdf.

Independent Group of Scientists Appointed by the Secretary-General. 2019. *Global Sustainable Development Report 2019: The Future is Now – Science for Achieving Sustainable Development*. New York: United Nations. Available at: https://sdgs.un.org/gsdr/gsdr2019.

Kabeer, N. 2015. "Gender, Poverty, and Inequality: A Brief History of Feminist Contributions in the Field of International Development." *Gender and Development* 23: 189–205. https://doi.org/10.1080/13552074.2015.1062300.

Koda, B. O., and L. V. Mtasingwa. 2021. *Gender-Responsive Budgeting in Tanzania*. https://oxfamilibrary.openrepository.com/bitstream/handle/10546/621342/rr-gender-responsive-budgeting-in-tanzania-291221-en.pdf.

Llorens, A., A. Tzovara, L. Bellier, I. Bhaya-Grossman, A. Bidet-Caulet, W. K. Chang, Z. R. Cross, R. Dominguez-Faus, A. Flinker, Y. Fonken, M. A. Gorenstein, C. Holdgraf, C. W. Hoy, M. V. Ivanova, R. T. Jimenez, S. Jun, J. W. Y. Kam, C. Kidd, E. Marcelle, … N. F. Dronkers. 2021. "Gender Bias in Academia: A Lifetime Problem That Needs Solutions." *Neuron* 109 (13): 2047–2074. https://doi.org/10.1016/j.neuron.2021.06.002.

Miguel, E. 2005. "Poverty and Witch Killing." *Review of Economic Studies* 72 (4): 1153–1172. https://doi.org/10.1111/0034-6527.00365.

Neville, S. 2024. "How Medical Research Is Failing Women." *Financial Times*, August 1. https://www.ft.com/content/ce9895f9-8ead-4b43-b473-874aebabc0e6.

Nilsson, M., D. Griggs, and M. Visback. 2016. "Map the Interactions Between Sustainable Development Goals." *Nature* 534: 320–322. https://doi.org/10.1038/534320a.

Patel, R., R. Bezner Kerr, L. Shumba, and L. Dakishoni. 2014. "Cook, Eat, Man, Woman: Understanding the New Alliance for Food Security and Nutrition, Nutritionism and its Alternatives from Malawi." *The Journal of Peasant Studies* 42: 21–44. https://doi.org/10.1080/03066150.2014.971767.

Pearson, R. 2019. "The Rise and Rise of Gender and Development." In U. Kothari (Ed.), *A Radical History of Development Studies*. London: Bloomsbury Publishing.

Peters, M. D. J., C. Lockwood, Z. Munn, S. Moola, and R. K. Mishra. 2016. *People's Views and Experiences of Participating in Microfinance Programmes: A Systematic Review of Qualitative Evidence*. London: IOE.

Profeta, P. 2017. "Gender Equality in Decision-Making Positions: The Efficiency Gains." *Intereconomics* 52 (1): 34–37. https://doi.org/10.1007/s10272-017-0640-4.

Quisumbing, A. 2017. "How Development Organizations Can Help Women's Empowerment." *IFPRI*. Available at: https://www.ifpri.org/blog/how-development-organizations-can-help-womens-empowerment/ (accessed March 23, 2025).

Ragasa, C. 2014. "Improving Gender Responsiveness of Agricultural Extension." In A. R. Quisumbing, R. Meinzen-Dick, T. Raney, A. Croppenstedt, J. A. Behrman, and A. Peterman (Eds.), *Gender in Agriculture*. Cham: Springer, pp. 411–430.

Rodríguez Franco, D. 2022. *The Bogotá CARE System: How a City Is Reorganizing Itself For Women*. Washington, DC: Center for Sustainable Development at Brookings.

Trauger, A. 2007. "'Because They Can Do the Work': Women Farmers in Sustainable Agriculture in Pennsylvania, USA." *Gender, Place & Culture* 11 (2): 289–307. https://doi.org/10.1080/0966369042000218491.

UN Women. 2022. *A Toolkit on Paid and Unpaid Care Work: From 3Rs to 5Rs*. Available at: https://www.unwomen.org/en/digital-library/publications/2022/07/a-toolkit-on-paid-and-unpaid-care-work.

UN Women, and UN DESA. 2023. *Progress on the Sustainable Development Goals: The Gender Snapshot 2023*. Available at: https://www.unwomen.org/en/digital-library/publications/2023/09/progress-on-the-sustainable-development-goals-the-gender-snapshot-2023.

van Campenhout, B. 2021. "The Role of Information in Agricultural Technology Adoption: Experimental Evidence from Rice Farmers in Uganda." *Economic Development and Cultural Change* 69: 1239–1272. https://doi.org/10.1086/703868.

Varol, N., S. Turkmani, K. Black, J. Hall, and A. Dawson. 2015. "The Role of Men in Abandonment of Female Genital Mutilation: A Systematic Review." *BMC Public Health* 15: 1034. https://doi.org/10.1186/s12889-015-2373-2.

15 Focus II

Resilience

INTRODUCING RESILIENCE

Resilience has been around for several decades. You will find resilience guidance in development organizations dating back decades, although frequently privileging or limited to climate resilience (World Bank 2013). To get a better sense of what resilience means today for sustainable development, I offer below a breakdown of its definition, followed by two examples of how prominent actors have integrated it into their own work.

A START: A DEFINITION

The UN defines resilience as:

> the ability of individuals, households, communities, cities, institutions, systems and societies to prevent, resist, absorb, adapt, respond and recover positively, efficiently and effectively when faced with a wide range of risks, while maintaining an acceptable level of functioning without compromising long-term prospects for sustainable development, peace and security, human rights and well-being for all.
>
> (United Nations Sustainable Development Group 2020, 3)

Take a deep breath. And now let us dissect the above definition together:

"Resilience is the ability of individuals, households, communities, cities, institutions, systems and societies … ": while we might have individuals and households at the top of our minds, resilience scales across multiple levels up to systems and societies. To use the example of the COVID-19 pandemic, the shock not only resulted in individual illness and death, it wrecked havoc across our supply chains. For example, it highlighted the vulnerabilities to shocks and stresses of our global food system. Effects from social distancing also caused negative societal impacts.

"to prevent, resist, absorb, adapt, respond and recover positively, efficiently and effectively …": based on the UN's definition, we are not just focused on preventing the risk from occurring or resisting it. We need to manage the risk and build adaptive capacities. For example, global warming will continue to increase the likelihood of risks from intense hurricanes in the Atlantic (Colbert 2022). We cannot prevent the risk itself, but we can improve how we handle it:

DOI:10.4324/9781003400028-21

building early warning systems will improve our response to it; creating more mangroves would better absorb the risk (especially from flooding). We can adapt to the risk by improving our housing structures. Improved infrastructure – such as underground electricity cables or better stormwater runoff – will also allow us to recover sooner.

The focus on efficiency and effectiveness provides some boundaries as to what is reasonable. In my Swiss childhood home, for example, every house that was built needed to have a safety shelter in case of a nuclear fallout. While potentially effective in absorbing the risk, it is a high-cost endeavor that would be better undertaken at a communal level. In my family's case, it served as the storage area for old skiing equipment and binders.

"when faced with a wide range of risks ...": We have moved well beyond just resilience to climate change. We face shocks from disease outbreaks, earthquakes, or cybersecurity. In July 2024, one of the world's largest IT outages occurred. A faulty software update by the security software provider CrowdStrike caused 8.5 million Microsoft Windows devices to feature the infamous blue screen of death (The Editorial Board 2024). As three companies dominate two-thirds of the cloud market, and CrowdStrike provides cyber security services for one-fifth, it represents a highly vulnerable system. This time it was only a faulty software update; a malicious cyber attack brings much greater risks.

"while maintaining an acceptable level of functioning without compromising long-term prospects for sustainable development, peace and security, human rights and well-being for all": Nearly every definition of resilience will emphasize the ability (or lack thereof) to recover from the shock to a prior set stage. As shown in Figure 15.1, if we get hit by a shock without building our resilience, we

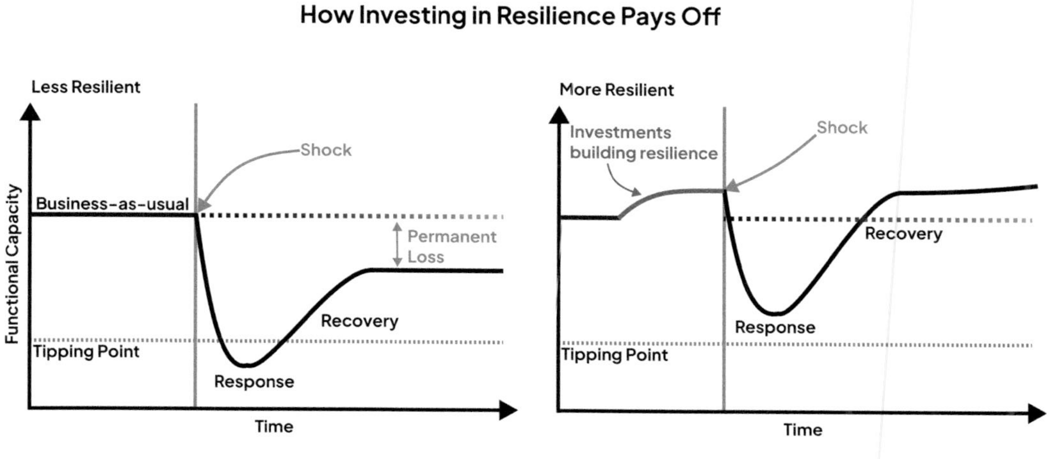

Figure 15.1 How investing in resilience pays off, adapted from NOAA (2024)

might dip below a tipping point. The shock creates a permanent loss from which we cannot fully recover.

In an ideal case, we have invested prior to the shock in building our resilience, and can recover to a new stable state through our continued action. The slogan of *Build Back Better* had that spirit at its core. Importantly, the UN's definition captures that reaching this new steady state cannot come at the cost of jeopardizing future efforts in sustainable development, nor fundamental rights or increases in inequality.

A SHIFT TOWARD SYSTEMS THINKING ECHOING HOW CHANGE LEADS TO TRANSFORMATION

As resilience has become increasingly embedded within sustainable development practice, Reyers et al. (2022) identify six shifts in one of the most dominant schools of resilience – social-ecological systems (SES) – as it relates to sustainable development. Table 15.1 captures the shift, its description, and an example of a development intervention. These shifts – from capital to capacities, or outcomes to processes – connect with our earlier findings on how change happens.

Table 15.1 Shifts in resilience and sustainable development. Source: Reyers et al. (2022)

Shift	Description	Example
Capitals to capacities	From a focus on natural and anthropogenic capitals to an understanding of the dynamic capacities that shape responses to change	Investing in "capacities at multiple scales for coordinated decision-making, collective action, innovation and experimentation"
Objects to relations	From a focus on distinct, independent objects to the dynamic relationships between social and ecological objects	Address "ecological and social processes and interactions that shape resilience of family farms"
Closed to open systems	From determined and static boundaries to porous, dynamic and fluid boundaries	Include additional stakeholders outside the project's sphere of influence
Outcomes to processes	From a focus on outcomes as fixed endpoints to outcomes as shaped by and shaping the processes of sustainable development	Apply "a co-evolutionary perspective of the interplay between social and ecological systems and resilience capacities in agro-ecological systems"
Generic to context-sensitive interventions	From generic interventions to the complex interplay between intervention and context	Integrate nonlinear behavior to better understand how change occurs through an intervention
Linear to complex causality	From assumptions of linear cause–effect relationships to complex causality	Move toward "nonlinear theory of change for development programming [that accounts for] the complexity of resilience causality"

Barriers remain to implement these shifts. We overemphasize the resilience of assets and capital. Limitations on how projects are funded or evaluated have effects. The need to create scalable solutions in a short time frame begs us to ignore complexity and the context sensitivity that are essential in resilience. The dominant model of linear theories of change or logframes also prevents a shift to moving from outcomes to processes and accounting for these complexities more adequately (Reyers et al. 2022, 659). Masking the role of power dynamics and knowledge (Vink, Dewulf, and Termeer 2013) also limits our understanding on how to address more complex challenges.

To better understand how resilience translates into organizations, we next review how the UN and USAID have integrated it into their programming.

THE MULTILATERAL AGENCY

The United Nations, in its *Common Guidance on Helping Build Resilient Societies*, offers a blueprint for UN teams to improve their work across international, national and subnational levels in furthering resilience (United Nations Sustainable Development Group 2020).

It recognizes four key elements to build resilience (United Nations Sustainable Development Group 2020, 4), connecting to several points raised by Reyers et al. (2022):

1. "[Understanding] the context and the multiple and interconnected dimensions of risks at play."
2. "[Recognizing] how systems are interconnected ... across global, regional, national, subnational, and local scales, [including] social, governance, economic, ecological and physical systems."
3. "[Including] multiple stakeholders[, inclusive of gender and a diverse] range of perspectives on risks [to ensure broad coverage of] needs, including those of the most vulnerable, are addressed."
4. Identifying presence of resilience capacities, including "absorptive, adaptive, anticipate, preventative and transformative capacities and resources to cope with, withstand and bounce back from shocks."

Chapter 4 introduced the dimensions of resilience integrated by the UN FAO for their assessment of smallholder and pastoralist resilience. The SHARP approach was one of several developed.

To better link-up UN programming, the following steps should be considered: (1) an assessment to reach a common understanding of the risks and context; (2) aligning planning to generate a joint outcome and results; and (3) advancing joint implementation and monitoring efforts. To get us there, coordination, partnerships, and financing for resilience are needed (United Nations Sustainable Development Group 2020).

THE DEVELOPMENT DONOR

In the aftermath of widespread famine in 2011 in East Africa, USAID released its inaugural resilience policy in 2012. In 2024, USAID updated this policy to account for new challenges from polycrises, including "pandemics, accelerated democratic backsliding, and rising conflict and violence, as well as the intensifying impacts of climate change" (USAID 2024, 2). Reflecting a more holistic approach, the new policy covered risks beyond food security, spreading across "climate, health, peace education, democracy and governance, poverty alleviation and economic growth" (USAID 2024, 2).

While the initial goal for resilience was to identify avenues with a view to "reducing humanitarian need in areas of recurrent crisis" (USAID 2024, 2), USAID explicitly integrated the Humanitarian-Development-Peace (HDP) approach (see Chapter 8). Beyond stressing how HDP can successfully reduce resilience, USAID's Resilience Policy was also reflected in USAID's 2022–2030 Climate Strategy, 2022 Local Capacity Strengthening Policy, and the US Government's 2022–2026 Global Food Security Strategy (USAID 2024, 2).

In programming, USAID identified five shocks and stresses – natural hazards, economic shocks, chronic marginalization, conflict, and disease – that might impede well-being outcomes. It aimed to focus on major sources of resilience to achieve the desired outcomes (USAID 2024, 6): human capital (e.g., "education, women's empowerment, good health, and psychosocial factors such as aspirations for the future and self-efficacy" (USAID 2024, 8)); diversified livelihoods, relationships and networks (e.g., "social capital, social cohesion, and exposure to information through early warning systems" (USAID 2024, 8)), and effective systems (e.g., "shock-responsive social protection and safety nets, local leadership, resilient market systems, and sustainable ecosystems and natural resource management" (USAID 2024, 8)).

USAID wanted to embrace a more holistic, cross-sectoral approach. The new policy cited the example of development work in Nicaragua, where a donor

> combined cash transfers with either vocational training or a grant for investment in an economic livelihood [... which] enabled internal migration in response to weather shocks as well as increased wage work, the investment grant expanded income, and both diversified household income and protected households from weather shocks two years after the intervention.
>
> (USAID 2024, 9)

More shock-responsive programming (USAID 2017) aimed to ensure that the flexibility exists through contingency planning to mobilize additional resources. The policy referenced the example of Somalia during the El Niño in 2015, where increased budget flexibility was leveraged to purchase sandbags for vulnerable villages facing flooding. According to USAID, this low-cost activity saved US$28 in emergency aid and US$91 in future income for every dollar invested (USAID 2024, 10). Examples like these highlight the high return on investments to resilience funding if we avoid reaching the tipping point in Figure 15.1.

RESILIENCE IN PRACTICE

Resilience is best understood when put into praxis. To investigate resilience in development practice, I reviewed two case studies: Case Study 15.1 on insurance in Africa, and Case Study 15.2 on resilience building efforts with a focus on Nepal. In addition, our online website sustainabledevelopment.me also summarizes five resilience vignettes at the intersection of two different transformation hotspots, showcasing the cross-cutting relevance of resilience. The cases are drawn from across the globe, featuring efforts in the Maldives, Yemen, Brazil, India, and Indonesia.

CASE STUDY 15.1 ACRE AFRICA INSURANCE

When I teach my students about how to assess and mitigate risks in sustainable development implementation, there are four categories of actions we apply: (1) *avoiding* a risk by not doing it or doing something differently (e.g., drop an activity); (2) *transferring* a risk, by shifting it to another party or project area (e.g., insurance); (3) *reducing* the risk through mitigation by taking measures that lower the impact of the risk (e.g., security); or (4) *acknowledging* the risk and signing up to keep monitoring it (PM4NGOs 2020).

Out of those, there is always one action that is the most difficult to conceptualize in the Global South: *transferring* the risk using the example of insurance. In the Global North, the ability to obtain insurance is at the core of wealth creation and sustaining livelihoods through shocks and stresses. In the Global South, insurance outside of mandatory coverage (e.g., for cars) is largely absent. Therefore, social safety nets and government support are increasing in importance in providing stability as climate shocks become more frequent.

Over the last two decades, innovative approaches have been spearheaded by the insurance industry to capture growth opportunities for their own market, and in the process extend insurance to customers in rural areas of the Global South. I present here the case of *Agriculture and Climate Risk Enterprise (ACRE) Africa*, an intermediary that offers "risk management solutions to mitigate agricultural and climate-related risks in Africa" (ACRE Africa 2024). Incorporated in 2014 with initial funding from the foundation of the crop giant Syngenta (Global Index Insurance Facility 2017), the company was a pioneer in offering micro insurance to African farmers who lacked access to formal insurance before. The following text captures their main insurance offering, with more details on our companion website.

Crop insurance includes weather index or picture-based insurance to overcome the challenge of monitoring remote farmers where a physical visit is not economically feasible. The same applies for index-based livestock insurance using satellite data. ACRE also offers a complete *farmer protection cover*, which combines insurances for crops, livestock, personal accident, health, and farming assets (ACRE Africa 2024).

ACRE's first venture dates back to 2008, when a project team piloted and subsequently rolled out an insurance called *Kilimo Salama* (translated from Kiswahili as *safe* or *peaceful agriculture*). In partnership with Syngenta's foundation and the main phone provider Safaricom, farmers could purchase the add-on insurance when they bought their (Syngenta) chemicals, fertilizers, and hybrid seeds each season.

Payout took place based on weather station reports (Kilimo Salama 2010a). At a premium of around 5 percent, it only covered the inputs which you paid for, however, and not the actual harvest you lost (Kilimo Salama 2010b). Both sign-ups and payouts were initially limited. In Rwanda, only 1,600 out of a total 13,000 insured farmers received an average payout of US$6.50 in 2012 (Kilimo Salama 2016).

Limits to insurances in East Africa

Dr. Leigh Johnson has dedicated the last decade of her career to better understanding the opportunities and challenges of insurances in rural development (Johnson 2020, 2021; Johnson et al. 2023). Contrary to expectations that the extension of weather insurance to smallholder farmers in the Global South represents a further financialization of agrarian production, the "market has proved far less successful than advocates hoped or critics feared" (Johnson 2021, 1224). Drawing on extensive interviews, ethnographic data and surveys, Johnson et al. (2023) find that the underlying assumption of insurance transferring the risk for an individual pastoralist in Northern Kenya runs counter to the local *Boraan* pastoralists' moral economy where animals do not belong to individuals but to the larger community:

> An individual who loses animals may request to be restocked by the community, but if they have not previously "invested" in others' livestock by giving in times of need, elders declare "loon ka uufi d'abee hin hirbinaafi" (they lost their own livestock, do not restock them). Such a person is said not to know that livestock belongs to all *Boraan* and is consequently dismissed as selfish and excluded from *hiriba* (restocking).
>
> (Johnson et al. 2023, 1939)

Mechanisms of reciprocity and redistribution are reaching their limits, and "poorer pastoralists may be excluded, forced to migrate or take up non-pastoral livelihoods" (Johnson et al. 2023, 1939). Johnson et al. (2023, 1940) also note an increase in social stratification and "class formation

[has] emerged from a process of commoditization and growing market orientation, leading to the breakdown of redistributive practices and growing individualization of economic relations."

Formal insurance is just one of many hybrid strategies deployed by pastoralists:

> Insurance is always understood from within local moral economies and repertoires of exchange, in which it figures as just one amongst a diverse set of relational tools for confronting uncertainty. These include networked household relations, friendship, kin and clan relations, obligations via social and faith-based organizations, and reliance on the spirit world ...
>
> (Johnson et al. 2023, 1944)

As farmers and agropastroalists are more reluctant to sign up for insurance than anticipated and unwilling to pay the market price needed to sustain insurance operators, Johnson (2021) questions the logic in providing donor and government subsidies to sustain a profit-driven insurance market. Instead of these companies pocketing development funds, she argues for establishing "more equitable, systematic risk-sharing infrastructures to replace the current fragmentary archipelagos of weather insurance" (Johnson 2021, 1224).

Or in the words of an insurance advisor:

> Having an agricultural insurance market is great for the insurance industry, but it's not great of the farmers ... Extension services, marketing boards, revising trade policies ... We have ample opportunity to change the conditions for smallholder farmers without then spending subsidies that end up at the Lake of Zurich at Swiss Re [headquarters].
>
> (Johnson 2021, 1236)

Insurances have played a crucial role in the economic growth across the Global North, with micro insurance offerings becoming increasingly popular (Lester 2024). As illustrated in the case of rural households in East Africa, formal insurances represent one of several risk-sharing approaches in use. Just as insurance companies trying to break into these new markets learn, so do we as development practitioners need to understand the complexity of how risks are managed locally. And even if we were to excel in offering formal insurance across unserved regions that improves resilience, we can already see its limits in the Global North and the Global South.

CASE STUDY 15.2 IDE GLOBAL

BRACED – **B**uilding **R**esilience and **A**daptation to **C**limate **E**xtremes and **D**isasters – was one of the UK government's largest resilience programming to date. In its initial phase, it awarded three-year grants to 15 projects across South and Southeast Asia, as well as the African Sahel and neighboring countries (BRACED, n.d.-a).

To illustrate its workings, let me introduce to you one of its flagship projects: Anukulan in Nepal. Including a consortium of local to international partners, implementation was overseen by the non-profit iDE Global (2024) and its country office in Nepal. The project, Anukulan, was funded for 4.5 years between 2015 to 2019, with the first phase from 2015 to 2017, and the second phase from 2018 to 2019. Its overall goal was "to improve the well-being of rural poor, especially women and children, coping with climate change-related shocks and stresses" (BRACED, n.d.-b). By creating public-private partnerships, it reached 600,000 poor people living in Western Nepal who were affected by climate extremes and disaster.

To illustrate its workings, the project combined a complex combination of interventions: conservation agriculture, improved nutrition, and women's empowerment. It successfully raised the percentage of people above the poverty line from 38 to 57 percent, increased dietary diversity by around 50 percent, and improved water access, saving four hours for women each day (Tye and Suarez 2021).

To accelerate and sustain the intervention, it created "commercial pockets focused on climate-smart economic opportunities in agriculture, water resource management, and community forestry" (Ambrosino et al. 2018, 3). Planning for disaster risk reduction was supported and integrated with strategies for climate change adaptation. Its gender focus included empowering women and marginalized populations "to take leading roles in rural institutions and contribute to economic opportunities" (BRACED n.d.-b.).

I chose Anakulan as a case study to illustrate resilience-building development projects that activate all changemakers. The project engaged local economies to ensure that any yield increases from conservation agriculture translated into additional incomes.[1] Its "commercial pocket" approach[2] – activating the *economy and finance* – organizes producers together to have greater negotiating power in the marketplace. The project leveraged *science and technology* by supporting, for example, the adoption by over 23,000 households of micro-irrigation technologies, such as drip or sprinkler systems. Solar power was used to pump water to marginalized communities (BRACED n.d.-b.).

The project also strengthened local *governance* and *individual and collection action* by establishing 41 local action plans of adaptation (LAPA) in collaboration with local communities, government, and district agencies. These plans were used "to allocate and implement investments for climate adaptation" (BRACED n.d.-b.).

Establishing these market system interventions increases the resilience (Ambrosino et al. 2018). The project also rigorously implemented resilience monitoring for its learning. It developed a composite index that tracked five determinants across household levels, with the corresponding resilience capacity in parentheses: (1) income (*absorptive capacity*), (2) dietary diversity (*absorptive capacity*), (3) access to climate-smart services and technologies (*adaptive capacity*), (4) access to improved water sources (*adaptive capacity*), and (5) access to an early warning system (*anticipatory capacity*) (Ambrosino et al. 2018). Based on a survey of 600 households (300 treatment and 300 control), the project doubled a household's resilience score, increasing the share of resilient households by 61 percent. Households that engaged most in market activities through the commercial pocket approaches increased their resilience more (Ambrosino et al. 2018).

The theory of change in Figure 15.2 captures the second phase of the project (2018–2019) as it shifted toward capacity building and policy change. As illustrated in Figure 15.2, capacities were built for both vulnerable households to access and benefit resilience-enhancing products and services, as well as the private sector to offer these products in the first place.

The theory of change also included capacity building of relevant institutions, undertaking complementary research that can inform how resilience building works, and steps to translate the findings into disaster risk reduction planning and climate change adaptation strategies. Capacity building is assumed to translate into reduced vulnerabilities for poorer households, improving their resilience and well-being by strengthening their ability to absorb, anticipate, and adapt to shocks and stresses (see *Outcome* in Figure 15.2).

Anukulan's implementation focused on locally-led climate adaptation. The case links back to the importance of gender and our LEADS model for change. A review of over 70 evaluations by Tye and Suarez (2021, 1) found that such adaptation efforts where "local actors have decision-making power in planning, implementing, and monitoring and evaluation, can play an essential role in achieving successful and sustainable adaptation." Tye and Suarez identified Anukulan as a project that successfully avoided the pitfalls of male leaders dominating priority setting, as women represented the majority in the farmers' groups and were encouraged to hold leadership positions. These efforts increased "women's confidence to speak out in meetings and participate in public affairs … creating conditions for women's leadership" (Tye and Suarez 2021, 8).

Mirroring our earlier discussion of how change happens, Tye and Suarez (2021, 1) recommend the following actions to improve the effectiveness of locally-led adaptation action:

> encouraging flexible and iterative approaches to program design and implementation; including local actors in decision-making from start to finish; increasing technical capacity building; working with existing governance and finance mechanisms and networks to achieve scale and reduce transaction costs; and building the evidence base on locally led adaptation.

Buisson, Clement, and Leder (2022) found that women in the Anukulan project with higher levels of critical consciousness (cf. Freire 1972) were more willing to gain agency, emphasizing the importance of "how critical consciousness influences women's will to change the status quo" (Buisson, Clement, and Leder 2022, 128). Women with higher critical consciousness exhibit more will to have better access to capital, both productive and unproductive (Buisson, Clement, and Leder 2022, 133–134).

iDE's own transformation journey in evaluating success

Beyond project implementation, iDE invested into rigorous monitoring and evaluations, and ultimately learning. Attributed to Rick Davies and Jess Dart (2015), iDE applied the "*Most Significant Change*" methodology to empower local communities to collect stories of change among themselves. These stories are based on their lived experiences as participants of a project, reviewed and aggregated by diverse stakeholders of the project, and then shared back with the community.

For iDE, applying this methodology culminated in their *Inclusive Strategy Report*, recording changes across nine of their project countries: Bangladesh, Cambodia, Ethiopia, Ghana, Honduras, Mozambique, Nepal, Vietnam, and Zambia (iDE 2024).

By both listening to project beneficiaries and empowering them to undertake their own peer evaluations, they found that respondents valued more than just an improved income, which historically was the primary focus of iDE evaluations. Respondents mentioned personal and family well-being (e.g., improved nutrition or health); assets (e.g., improved housing or livestock); community infrastructure (e.g., access to electricity or schools); business and entrepreneurship (e.g., fair markets or skills training); climate resilience (e.g., sustainable agricultural practices or better preparedness for climate shocks and stressors); and gender equity and social inclusion (e.g., increased decision-making power or more equal access to resources and opportunities) (iDE 2024).

This case captures what happens when an NGO – anchored around entrepreneurship and monitoring financial improvements – transforms itself.

The findings not only reinforce the value of piloting more participatory methodologies, but highlight an urgency to move beyond a specific programmatic silo to address the sustainable development challenges at heart. It also shifts our focus from the individual to the community, as shared by the following respondent from Cabo Delgado province in Mozambique:

> I would like to see women in my district engaged in business. These changes are important because they support people in my community in covering their children's educational expenses and improve the lives of [local district] residents. With these changes, I would feel prosperous. When I say prosperous, I mean that I would feel happy and like a successful woman.
>
> (iDE 2024)

I am convinced if we collectively were to listen to our project partners, empower them to take an active role in shaping projects and learning, and allowing the findings to shake up our own programming, we would join iDE in doing a better job at strengthening resilience and contributing to achieving sustainable development for all.

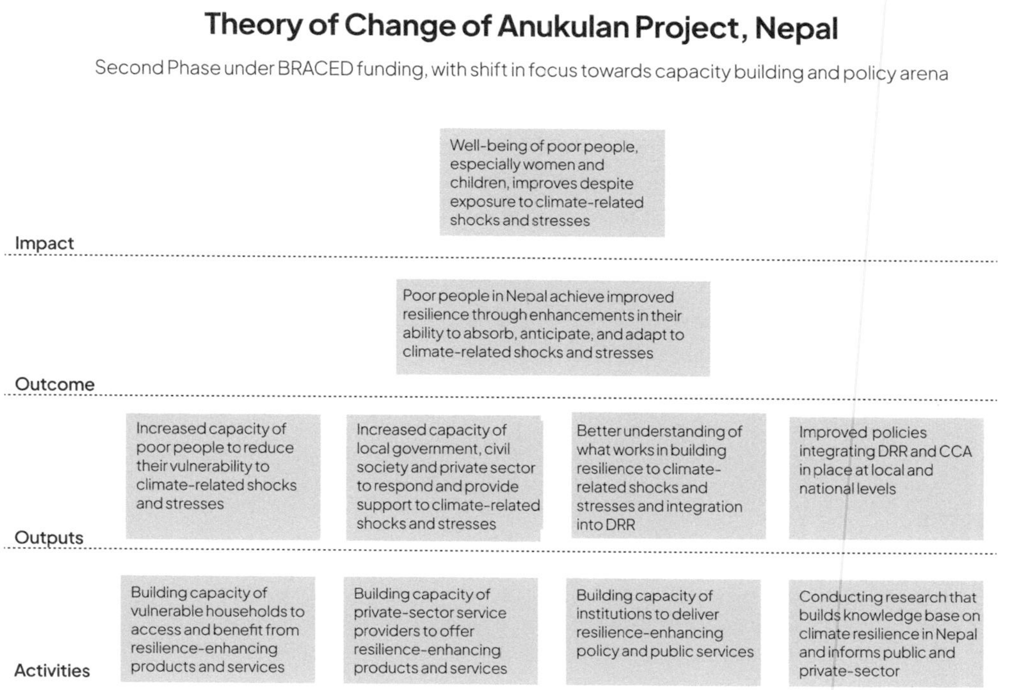

Notes: DRR: Disaster Risk Reduction; CCA: Climate Change Adaptation.

Figure 15.2 Theory of change of Anakulan Project, Nepal, adapted from iDE Global

WHEN INSURANCE IS NOT ENOUGH: THE CASE OF LOW-LYING COMMUNITIES

The US news more frequently features beach homes that overcome to the force of coastal erosion, such as in the Outer Banks in North Carolina. Stresses from climate are especially severe in such coastal areas that become no longer habitable. Insurance costs, as a result, have risen astronomically to the point of no longer being affordable. In the Outer Banks, an annual premium increase of around 60 percent is expected (Saunders 2024). Millions of homeowners across the world face a similar fate as climate risks increase (Smith, Mooney, and Williams 2024).

Where the risk becomes too large to insure against, ensuring human mobility is subsequently one solution left to increase the resilience of inhabitants and counteract future forced displacement. Fiji implemented a voluntary relocation of communities – "a painful but necessary trade-off" – to reduce their future exposure to sea-level rise and coastal erosion (United Nations 2023).

In the process, all four changemakers were leveraged: first, the Fijian government analyzed current and projected climate impacts (*science and technology*). Next, it engaged individuals and communities through a local consultation process (*individual and collective action*). To reach consensus on a policy decision, a threshold of 90 percent was established across all community stakeholders, including youth and people with disabilities. This consensus needed to be reached also within the receiving communities where relocation would occur (*governance and policymaking*) (United Nations 2023).

In the end, 42 villages were identified, with six having moved as of 2023. A living document – *Planned Relocation Guidelines: A Framework to Undertake Climate Change Related Relocation* (Ministry of Economy 2018) – was established to continuously integrate lessons learnt and improve efforts made. Relocation aimed to ensure that it resulted in minimal disruptions of livelihoods. It coincided with the establishment of service infrastructure and ensuring that schooling would be continuous throughout the process (*economy and finance*) (United Nations 2023). These actions reinforced how achieving solutions to sustainable development needs to break free from short project cycles and beyond building infrastructure, focus on building trust and strengthening relationships (Baleinakorodawa and Boege 2024, 13).

NOTES

1 Beyond income, it also links a nutritional component to support mothers with at-risk children through training (BRACED, n.d.a).

2 Commercial pocket approaches date back to the "*Making market systems work for the poor approach" (M4P)* that rose in popularity in development in the early 2000s (Ambrosino et al. 2018).

REFERENCES

ACRE Africa. 2024. "Who We Are." Available at: https://acreafrica.com (accessed March 23, 2025).

Ambrosino, C., R. Rose, M. Pariyar, T. Bui, R. Kothari, S. Phillips, M. Hernandez Lagana, and L. Savino. 2018. "Effect of Market System Interventions on Household Resilience: Comparison of Two Methodologies." Paper presented at *Resilience Measurement, Evidence and Learning Conference*, New Orleans.

Baleinakorodawa, P., and V. Boege. 2024. *Climate Change-Induced Community Relocation in Fiji: Challenges and Ways Forward.* Toda Peace Institute.

BRACED. n.d.-a. "About the Projects." Available at: http://www.braced.org/about/about-the-projects/.

BRACED. n.d.-b. Anukulan - BRACED Project. Available at: https://s3.amazonaws.com/www.ideglobal.org/files/protected/Anukulan-Factsheet.pdf?mtime=20190828174314.

Buisson, M.-C.., F. Clement, and S. Leder. 2022. "Women's Empowerment and the Will to Change: Evidence from Nepal." *Journal of Rural Studies* 94: 128–139. https://doi.org/10.1016/j.jrurstud.2022.06.005.

Colbert, A. 2022. "A Force of Nature: Hurricanes in a Changing Climate." NASA. Available at: https://science.nasa.gov/earth/climate-change/a-force-of-nature-hurricanes-in-a-changing-climate/ (accessed March 23, 2025).

Davies, R., and J. Dart. 2015. "The 'Most Significant Change' (MSC) Technique: A Guide to Its Use." Available at: https://realkm.com/wp-content/uploads/2020/11/The_Most_Significant_Change_Technique_A_Guide_to_Its_Use.pdf.

Freire, P. 1972. *Cultural Action for Freedom.*. Harmondsworth: Penguin.

Global Index Insurance Facility. 2017. "Agriculture and Climate Risk Enterprise (ACRE)." Washington, DC: World Bank Group.

iDE. 2024. *Inclusive Strategy Report.* Available at: https://sites.google.com/ideglobal.org/inclusive-strategy/home?authuser=0.

iDE Global. 2024. "Powering Enterpreneurs to End Poverty." Available at: https://www.ideglobal.org (accessed March 23, 2025).

Johnson, L. 2020. "Sharing Risks or Proliferating Uncertainties? Insurance, Disaster and Development." In I. Scoones and A. Stirling (Eds.), *The Politics of Uncertainty: Challenges of Transformation.* New York: Routledge.

Johnson, L. 2021. "Rents, Experiments, and the Perpetual Presence of Concessionary Weather Insurance." *Annals of the American Association of Geographers* 112 (5): 1224–1242. https://doi.org/10.1080/24694452.2021.1966294.

Johnson, L., T. S. Mohamed, I. Scoones, and M. Taye. 2023. "Uncertainty in the Drylands: Rethinking In/formal Insurance from Pastoral East Africa." *Environment and Planning A: Economy and Space* 55 (8): 1928–1950. https://doi.org/10.1177/0308518x231168396.

Lester, R. 2024. *Insurance and Inclusive Growth.* Washington, DC: The World Bank.

Ministry of Economy. 2018. *Planned Relocation Guidelines: A framework to undertake climate change related relocation.* Fiji.

NOAA. 2024. "Resilience Graphs." U.S. Climate Resilience Toolkit. Available at: https://toolkit.climate.gov/image/3144 (accessed March 23, 2025).

PM4NGOs. 2020. *Project DPro: Project Management for Development Professionals Guide.* Available at: https://pm4ngos.org/methodologies-guides/project-dpro/.

Reyers, B., M-L. Moore, L. J. Haider, and M. Schlüter. 2022. "The Contributions of Resilience to Reshaping Sustainable Development." *Nature Sustainability* 5 (8): 657–664. https://doi.org/10.1038/s41893-022-00889-6.

Salama, Kilimo. 2010a. "About Kilimo Salama." Available at: https://kilimosalama.wordpress.com/about/ (accessed March 23, 2025).

Salama, Kilimo. 2010b. "Protect Your Farming Investment Against Extreme Weather Events." Available at: https://kilimosalama.wordpress.com/wp-content/uploads/2010/02/kilimosalama_v031.pdf.

Salama, Kilimo. 2016. "Fact Sheet on Kilimo Salama 'Safe Agriculture'." Available at: https://www.syngentafoundation.org/sites/g/files/kgtney976/files/media/document/2016/11/23/factsheet_rwanda_payout_dec12.pdf.

Saunders, C. 2024. "As Outer Banks Faces Major Insurance Hikes, Commissioner Addresses Residents' Concerns." *The Virginia-Pilot*. Available at: https://www.pilotonline.com/2024/03/22/commissioner-addresses-residents-concerns-as-outer-banks-faces-major-insurance-hikes/.

Smith, I., A. Mooney, and A. Williams. 2024. "The Uninsurable World: What Climate Change is Costing Homeowners." *Financial Times*, February 12. https://www.ft.com/content/ed3a1bb9-e329-4e18-89de-9db90eaadc0b.

The Editorial Board. 2024. "Lessons from the Global IT Outage." *Financial Times*, July 23.

Tye, S., and I. Suarez. 2021. "Locally Led Climate Adaptation: What Is Needed to Accelerate Action and Support?" World Resources Institute. Available at: https://doi.org/10.46830/wriwp.20.00039.

United Nations. 2023. *GAR Special Report 2023: Mapping Resilience for the Sustainable Development Goals*. New York: United Nations General Assembly.

United Nations Sustainable Development Group. 2020. *UN Common Guidance on Helping Build Resilient Societies*. New York: United Nations.

USAID. 2017. *Shock Responsive Programming and Adaptive Mechanisms*. Available at: https://pages.usaid.gov/sites/default/files/shock_responsive_programming_and_adaptive_mechanisms.pdf.

USAID. 2024. *Resilience Policy*. Available from the author.

Vink, M. J., A. Dewulf, and C. Termeer. 2013. "The Role of Knowledge and Power in Climate Change Adaptation Governance: A Systematic Literature Review." *Ecology and Society* 18 (4). https://doi.org/10.5751/es-05897-180446.

World Bank. 2013. *Building Resilience: Integrating Climate and Disaster Risk into Development*. Available at: https://documents.worldbank.org/en/publication/documents-reports/documentdetail/762871468148506173/main-report.

Part V
Outlook beyond 2030

The SDGs are aspirational goals. They will be succeeded come 2030 by another set of shared goals and targets. Given their pathbreaking nature, we need to acknowledge how far we've come and look forward. Chapter 16 offers a perspective on Where we do go from here?. Chapter 17 concludes our book by capturing your role in achieving sustainable development. We started our journey on how to achieve sustainable development in Chapter 1 by capturing how it requires breaking down both institutional and our own personal silos to embrace integrated, holistic approaches. The SDGs and what follows beyond 2030 present a unique opportunity for us to build on the interconnections among social, environmental, and economic dimensions of sustainability in order to create transformative change.

DOI:10.4324/9781003400028-22

16 Where do we go from here?

A RECAP TO MOVE US FORWARD

We started our journey on how to achieve sustainable development in Chapter 1 by capturing how it requires breaking down institutional and our personal silos to embrace integrated, holistic approaches. The SDGs and what follows beyond 2030 present a unique opportunity for us to build on the interconnections among the social, environmental, and economic dimensions of sustainability in order to create transformative change.

In Chapter 2, we encountered how humanity is approaching environmental and social tipping points that demand transformative actions rather than business-as-usual approaches. Climate change, inequality, and our own footprints necessitate that we approach sustainable development with both local and global actions.

In Part II, our changemakers, we emphasized the limits of "silver bullets." We need to leverage multiple changemakers (Independent Group of Scientists Appointed by the Secretary-General 2019) – from governance and policymaking, economy and finance, individual and collective action, to science and technology – to drive sustained change. If we take complexity seriously, we can harness co-benefits and manage trade-offs. And more systematic, holistic rural development pays off, as shown in my research on multi-dimensional interventions in rural Ethiopia.

For our first changemaker – Chapter 3 governance and policymaking – we discussed the importance of stakeholders, including civil society and the private sector. Governance is essential given the need to address transboundary issues such as climate change or migration. Case studies highlighted how poor governance feeds food insecurity, as famines are man-made. On the flip side, efforts to address food insecurity in Brazil have withstood changes in administration because of their popularity and effectiveness. We also investigated the ripple effects from the Ukraine War on global food security and governance, including the importance of diplomacy and a more inclusive approach to sustainable development.

In Chapter 4, we identified how financial flows and interventions are crucial levers for change to address existing funding gap. Global financial flows (including remittances) play an important role in stimulating local development, with government policies creating the right incentives (including pricing of environmental externalities or realigning expenditures from harmful subsidies).

DOI:10.4324/9781003400028-23

To overcome funding gaps, the private sector clearly needs to step up – and to a large extent it has. However, transparency and accountability are important in minimizing greenwashing concerns. Innovative approaches to financing, such as patient capital for biodiversity or sustainable agriculture, can drive sustainable development.

In Chapter 5, we investigated how actions at the local level can contribute to achieving the SDGs, especially with a strong civil society and empowered individuals. Top-down approaches are rarely successful, as citizen engagement and participation are crucial to overcome major challenges, such as water insecurity, climate change, or gender inequality. Local communities are tackling sustainable development through investments in renewable energy, sustainable transportation, or local food systems. Given how race, gender, and socioeconomic status affect sustainability challenges, we can consider an intersectional approach in order not to further marginalization or exclusion.

In Chapter 6, we established the importance of R&D to achieve the SDGs and the need to address existing gaps. Greater collaboration, knowledge diversity, and moving beyond traditional Global North-Global South dynamics are major contributors to achieve sustainable development. Current unevenness in access to R&D funding and inequalities around who can access technologies continue to slow down progress in achieving the SDGs. Closing R&D gaps demands boosting research capacity in the Global South and emphasizing adaptation capacities. Integrating indigenous knowledge and moving toward co-production of knowledge are important steps to ensure that scientific advances are translated into sustained development.

Part III introduced a mapping of our six transformation hotspots (adapted from Independent Group of Scientists Appointed by the Secretary-General 2019), capturing which SDGs they address and how they interact with the four changemakers.

The first transformation hotspot – Chapter 7 – covered how strengthening sustainable food systems can reduce poverty (SDG 1), hunger (SDG 2), promote good health and well-being (SDG 3), reduce food waste and provide more sustainable production (SDG 12), and reverse land degradation and biodiversity loss (SDG 15). Agriculture is the epitome of how local actions can contribute to global change, as agriculture is part of the problem and plays an integral role as part of the solution.

The second transformation hotspot – Chapter 8 – highlights avenues to build social equity locally to remain within our own planetary boundaries. Transformation can contribute to reducing the number of working poor (SDG 1), gender and income inequality (SDGs 5 and 10), improving labor rights for economic growth (SDG 8), and addressing corruption (SDG 16).

The third transformation hotspot – Chapter 9 – centers empowerment of individuals and communities as the foundation for achieving the SDGs. Transformative education post-COVID not only reduces learning gaps (SDG 4) but also poverty (SDG 1). Investments in education and improved access to healthcare (SDG 3) and water (SDG 6) also further building of resilience. Transforming human capabilities also involves strengthening local agency, empowering an active citizenry, and investing in collective leadership.

In the fourth transformation hotspot – Chapter 10 – we examined the critical role of urban areas in achieving sustainable development, focusing on addressing challenges of uneven urbanization and creating more resilient and inclusive urban planning. By promoting integrated urban-rural development, leveraging information and incentives, and investing in green infrastructure and risk management strategies, we identified how local actions taken by city regions contribute to building sustainable cities and communities (SDG 11). We also demonstrated their connections to clean energy access (SDG 7), responsible consumption and production (SDG 12), climate change and biodiversity (SDGs 13 and 15).

In the fifth transformation hotspot – Chapter 11 – we explored how improving access to clean, affordable energy (SDG 7) also has synergies with increasing industrial growth (SDG 9), reducing inequality (SDG 10), addressing climate action (SDG 13), and improving life below water (SDG 14). Case Study 11.1 in Chapter 11 on South Africa's JETP underscores both the challenges and the great potential of inclusive energy transformation to address global inequality and environmental impacts. Addressing disparities between the Global North and the Global South and prioritizing governance, innovation and inclusion of all stakeholders are pivotal mechanisms for catalyzing transformative change.

In the sixth transformation hotspot – Chapter 12 – we reviewed how integrated approaches, such as sustainable land management, can address multiple SDGs, including SDGs 6, 13, 14, 15, and 17. Local efforts reviewed include regeneration of bogland and regenerative agriculture, which offer to contribute to environmental sustainability and climate mitigation. Building on our discussion around changemakers, integrating indigenous knowledge and moving away from extractive economic models are important as we revalue our ecosystems.

In Part IV we inspected how change happens in the first place, and the role of gender and resilience as they relate to sustainable development.

Chapter 13 argued for a paradigm shift from top-down, project-centric interventions to bottom-up, people-centered approaches that empower local communities as agents of change. For change to occur, we need to foster ownership and build local capacity for long-term sustainability. These are not novel concepts – but we need to look outside of our comfort zones. Approaches include human-centered design that explicitly embeds failure into any project design. Based on the above, I presented my LEADS model, as change happens when we *L*isten, *E*xperiment, *A*dapt, *D*isseminate, and *S*ustain.

In Chapter 14, we review how gender inequalities act as a cross-cutting theme impacting all aspects of sustainable development. Local actions are needed to tackle these disparities. Addressing gender inequalities can unlock progress across multiple SDGs, particularly poverty (SDG 1), hunger (SDG 2), economic growth (SDG 8) and reduced inequalities (SDGs 5 and 10).

Examples such as gender-responsive budgeting or improved care initiatives show how locally implemented policies and programs can empower women, promote gender equality, and contribute to more just and sustainable development outcomes.

As gender inequality is exacerbated by climate change and other shocks, we reviewed in Chapter 15 the importance of building capacities among individuals and societies to withstand and recover from shocks while maintaining long-term development goals. Actors prominently embedded resilience into their programming. Systems thinking and local capacity strengthening are two central aspects. We reviewed the potential to transfer risks through insurance in East Africa, and holistic development to strengthen resilience in Nepal.

Throughout the chapters, four themes have emerged: (1) we need to take complexity seriously to achieve sustainable development, embracing local knowledge and actors from the Global South; (2) we need to work toward systematic change, including a shift in the development paradigm toward leveraging all changemakers; (3) to achieve such sustained change, we need to empower others as change cannot be imposed from the top-down; and (4) while we are not on track to achieve the ambitious SDGs by 2030, the UN 2030 Agenda has offered new avenues to break down silos and advance change from the local to the global level beyond 2030. We pick up on this last point in our next section on why we should not give up.

WHY WE SHOULD NOT GIVE UP

Jungcurt, Kosolapova, and Bonaccorsi (2024) argue that even though only 17 percent of SDGs are on track to be achieved by 2030, the "'assessment by averages' can hide important success stories." They cite achievements in education, with girls surpassing boys in school completion rates, in health, with reduction in HIV infections and new all-time lows for child mortality. The aftermath of COVID-19 has created new challenges in inequalities for us to tackle, including a decline in students' reading skills or an overall reduction in life expectancy (Jungcurt, Kosolapova, and Bonaccorsi 2024).

Given the aspirational nature of the SDGs, Jungcurt, Kosolapova, and Bonaccorsi highlight a shift from a focus on targets to fostering more inclusive development that brings marginalized groups into the decision-making process. As the discussion moved from equality to equity, they argue that: "The key to achieving equity is giving agency to those with lived experience in poverty and marginalization. They should decide how marginalization is defined and measured so that their voices are heard and counted in decision making" (Jungcurt, Kosolapova, and Bonaccorsi 2024).

With the pendulum swinging back away from equity in the Global North, a focus on inequality and how it relates to post-pandemic growth and security is expected to return in the center. I argue that the aspirational, optimist spirit of the SDGs is more relevant than ever. As captured by Henry Mance, drawing on the work of Yascha Mounk, the activism over the last decade "was built on a rejection of the civil rights movement's optimistic pursuit of equality and racial integration" (Mance 2025). With its focus on leaving no one behind, the SDGs represent an inclusive pathway. As noted by a leading HR professional, "[w]e were beginning to focus too much on all these areas of difference, versus saying this is fundamentally about inclusion and respecting difference ..." (Mance 2025).

The retreat of the USA from the global stage of development at the beginning of the second Trump administration – commemorated in Figure 16.1 – weakens domestic and international efforts in sustainable development. The vacuum left behind could be filled by other actors, such as China or Europe. As China is stepping up, European countries cut their own (development) budgets to account for increased security spending in light of the Ukraine war. Beyond these state actors, localization of development has taken hold (Jungcurt, Kosolapova, and Bonaccorsi 2024). Trans-national, sub-national, and local actors have played an increasingly larger role in achieving sustainable development, enabled by the 2030 Agenda. Monitoring their progress is important to create greater visibility of how change happens – even if political tremors obfuscate it. As of December 2024, over 230 cities, municipalities or regions have completed their own local voluntary reviews (see Figure 16.2).

They reflect a foundational tenet of the 2030 Agenda: these are our goals. Not just the Global South or the Global North. They are our shared global goals and have local relevance, from Agadir in Morocco (Ville d'Agadir 2023) to Zapopan in Mexico (Gobierno Municipal de Zapopan 2024).

Figure 16.1 Messages from the first day USAID got shut down, with permission from Jason Gooljar (jasongooljar.org)

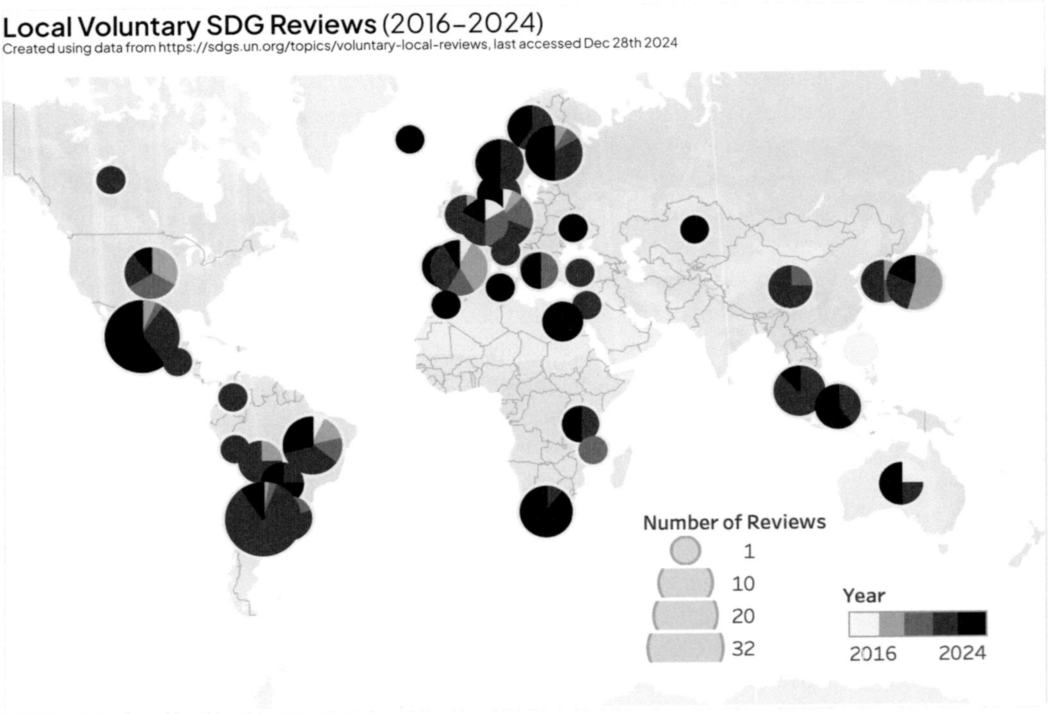

Figure 16.2 *Local voluntary SDG reviews (2016–2024)*, from https//sdgs.un.org/topics/voluntary-local-reviews

Across the globe, significant investments have also been made in a more systematic tracking for improved learning moving forward. These efforts will continue to pay off as the 2027 SDG Summit shapes the *Beyond 2030 SDG Framework* (Jungcurt, Kosolapova, and Bonaccorsi 2024).

In its run-up, the UN adopted the *Pact for the Future* (United Nations 2024), which includes 12 specific actions on sustainable development and financing for development (see Table 16). With its heavy focus on international peace and security, as well as global governance, UN Secretary-General Guterres emphasized that "we are here to bring multilateralism back from the brink" (IISD 2024, 1). The Pact included a new *Global Digital Compact*, focused on closing the digital divides, greater inclusion, and the role of artificial intelligence. It also includes a *Declaration on Future Generations*, covering themes around the importance of peace, security, empowerment, equality, inclusion, and intergenerational solidarity, as part of a "groundbreaking commitment by governments to listen to young people and involve them in decision making at all levels" (IISD 2024, 1).

The Pact offers a runway for future action until 2030 (with a review planned in 2028 by the UN General Assembly). Crucial to its success is our ability to deal with security and peace challenges that slow down the momentum for other relevant causes to future generations, such as climate change (IISD 2024).

Table 16.1 Actions under the Pact for the Future. Source: United Nations (2024)

Action	Description
1	We will take bold, ambitious, accelerated, just and transformative actions to implement the 2030 Agenda, achieve the Sustainable Development Goals and leave no one behind
2	We will place the eradication of poverty at the centre of our efforts to achieve the 2030 Agenda
3	We will end hunger and eliminate food insecurity and all forms of malnutrition
4	We will close the Sustainable Development Goal financing gap in developing countries
5	We will ensure that the multilateral trading system continues to be an engine for sustainable development
6	We will invest in people to end poverty and strengthen trust and social cohesion
7	We will strengthen our efforts to build peaceful, just and inclusive societies for sustainable development, provide access to justice for all and build effective, accountable and inclusive institutions at all levels and uphold human rights and fundamental freedoms
8	We will achieve gender equality and the empowerment of all women and girls as a crucial contribution to progress across all the Sustainable Development Goals and targets
9	We will strengthen our actions to address climate change
10	We will accelerate our efforts to restore, protect, conserve and sustainably use the environment
11	We will protect and promote culture and sport as integral components of sustainable development
12	We will plan for the future and strengthen our collective efforts to turbocharge the full implementation of the 2030 Agenda for Sustainable Development by 2030 and beyond

TIME TO TAKE COMPLEXITY SERIOUSLY

I encourage you to translate our reading on the importance of context, complexity, and empowerment into your own practice or studies. Adapted from a final assignment I conduct with my graduate students, apply the lens of sustainable development as you review existing or proposed technological innovations, interventions, or development projects. This approach applies when your organization has fully embraced the SDGs (possibly even using them excessively without a clear understanding of their critical contributions). If your organization has never heard of them, use them at least as inspiration. For example, the indicators used for measuring SDG targets are relevant for most monitoring or evaluation efforts.

In your analysis of how to achieve sustainable development, start by assessing the ability for the innovation, intervention or project to address the targets for the main SDG goal it relates to. Review any existing data or literature on demonstrated impact to understand what the knowledge gaps are. Based on our introduction in Chapter 1, assess your intervention on its ability to break down silos and identify relevant pathways that can address targets beyond the main SDG goal. Building on our book's map, identify how the intervention can support transformations and which changemakers have been already engaged.

As you identify its relationship to a transformation, consider how you can leverage the framework from Nilsson, Griggs, and Visback (2016) around synergies and trade-offs, and the extent to which there are constraining, counteracting or canceling interactions. Based on your review, consider how a more holistic, local impact could be strengthened. Assess the extent that the intervention can be sustained, and the degree it focuses on empowerment.

Thinking beyond the planned or existing programming, review how the intervention could be scaled out or scaled up (nationally or regionally). Consider the importance (and subsequent limitations) imposed by the contexts you are operating in, and how changemakers could be embedded more. Based on this review, formulate your recommendations on how achieving sustainable development can be strengthened – at the local or global level.

MAKING THE INVISIBLE, VISIBLE

In your practice, I encourage you to focus on trying to move beyond the surface to gain a deeper understanding of the drivers for change. I want to highlight two tools for you.

The first model stems from my former training as a political ecologist (by political ecologists). One of its memorable tenets was the differentiation between proximate versus ultimate causes. To use an example from my mentor, the proximate cause for degraded soils in Mali might be cotton farming practices, but the ultimate cause are national policies that drive exploitations of soils for export production (Moseley 2017). Applying this concept has helped me to get unstuck from the most visible, immediate answer when assessing a sustainability challenge, by looking at hidden, structural forces that drive the unsustainable practices.

The second model is used by colleagues who apply systems thinking modeling. By their work they contribute to our understanding of the root causes by mapping out all the drivers of a complex system. As captured by Untools.co, a website that compiles a range of useful tools, the tip of the iceberg represents events – what is happening right now (Untools.co 2025). Below the iceberg are patterns, structures, and mental models. The patterns capture the trends (usually over time and space). To use the example of soil degradation in Mali, we would investigate their trend. The structures would aim to uncover what influences the patterns and their connections, such as government policies. The most challenging aspect to capture is the mental model, which encompasses the underlying values, beliefs, or assumptions that govern the system, such as export-led development focused on maximizing production (Untools.co 2025).

As you apply these approaches, your continued learning can support change. Inspired by Bill Mollison (1988, 140), the co-founder of permaculture described in Chapter 12, in the process of making the invisible visible, let the complex system you are trying to understand also act as your teacher. As you learn, remain aware of your positioning. As captured by Desmond (2023, 44), "Complexity is the refuge of the powerful .... Hungry people want bread. The rich convene a panel of experts." So don't fall into the trap of retreating from development practice for the sake of a never-ending sabbatical researching complexity.

FACING THE REALITY AND KEEPING THE MOMENTUM GOING

Krenak challenges us to think about our role as we find that the gift-wrapped world we have received is not what we hoped to receive:

> What sort of world are you boxing and wrapping for future generations? You keep talking about another world, but have you asked the generations of tomorrow if the world you're building is the world they want? Most of us won't be here when the package arrives. It's your great-grandchildren and grandchildren, your elderly sons and daughters, who will have to sign for it. If each of us were to imagine a world, there'd be billions of worlds and they be delivered to specification in billions of places. But what world, and what mode of delivery, are you asking for? There's something deeply insane about the way we gather to grouse about the world our ancestors envisioned for us; we are more than a little bratty in our contention that, if it had been up to us, we'd have done a far better job.
>
> (Krenak 2020, 66–67)

We encounter a backlash against progressive policies across the globe. Journalist Pilita Clark called it *greenlash* (Clark 2024). Our momentum for sustainable transformations generated through individual and collective action – including through the voting booths – are clashing with populist (in)action.

As the world looks to the USA for its next steps, the EU is expected to play a greater leadership role in sustaining momentum. With highly visible farmer protests and inflation pressuring the pocketbooks, European lawmakers have already scaled back their most ambitious farming (Hancock and Bounds 2024) or energy transformation agendas (Hancock 2024). Creating an enabling environment for producers of green technologies to thrive – as they compete for subsidies and with trade tariffs – will be a crucial test.

As the political pendulum swings, the reality does not change: Europe's rise in temperature has been twice the global average, and a failure to stay within 1.5° Celsius will be intimately felt each year (Hancock 2024). Leveraging all changemakers – from policymakers, to citizens' actions, to the private sector and R&D – will be fundamental to ensure that we will do a better job. Avoiding a Western-centric view of the world helps to see who is filling the vacuums left behind by others. China continues to invest not only heavily in green technologies (Hancock 2024), but also is advancing sustainable development through its Global Development Initiative (UN DESA n.d.).

SUB-NATIONAL DIPLOMACY IN A CHANGING WORLD

The latest US presidential elections have caused shockwaves across the world. Levitsky and Way (2025) expect a breakdown of democracy and a move toward competitive authoritarianism. Civil society is regrouping, hoping that the silver lining of the election translates into increased activism (and donations) to ensure we do not steer too far from our transformation pathways. The question comes up whether transformations can succeed if we only activate three out of the four changemakers (i.e., dropping governance and policymaking). The short answer is no. The longer answer is below.

As shown throughout this book, transformations succeed if they are too popular or effective to cancel by an incoming administration. Much like strong roots support the survival of trees, the popular support fueled the success in Belo Horizonte in Brazil. Belo Horizonte also highlights the importance of sub-national governance and policymaking. The national government in Brazil has veered from one to the other extreme of the political spectrum over the course of two decades.

The program in Belo Horizonte sustaining itself reflects the increasing role of cities and regions in steering sub-national action. US state governors and city mayors, for example, are making plans to keep their environmental agenda intact (Hutchinson 2024). Strengthening alliances (including sub-national ones) across the globe was at the core of the COP29 climate change negotiations in Baku, Azerbaijan. Demands by Cities and Regions in the UNFCCC Process, the Global Taskforce of Local and Regional Leaders, and ICLEI (Local Governments for Sustainability) (2024, 1) included to "accelerate inclusive, multilevel, and gender-responsive climate action by incorporating local and subnational governments into new Nationally Determined Contributions (NDCs) and Adaptation Plans, and other national goals."

AVOID MAKING THE SAME MISTAKES AGAIN

On our path to create global change from local actions, I compiled three simple mistakes that we should not repeat in our pathway forward.

Invest in the boring

It is important that with the transition, we do not repeat the mistake of underfunding maintenance. For example, as the wind turbines are aging in Europe, we need to invest in their upgrades as they reach the end of their lifecycle (Jopson 2024). We need to invest in projects that do not make great PR or fundraising. We should resist chasing the latest fad and, in the process, abandoning proven solutions to sustainable development.

A classic example is the use of biological pest control in Africa, which over decades of application yielded at least US$200 of benefits for every single dollar invested (Zeddies et al. 2001). Biological pest control has been overshadowed by more exciting approaches, such as hopes for pest-resistant varieties through genetic engineering. While headline-grabbing in the 2000s, what captures our attention today are artificial intelligence-powered approaches. Although still in their infancy, they already come with real trade-offs: creating an AI image uses as much power as fully charging your phone (Heikkilä 2023). More promising are applications of AI to monitor climate change emissions or improve modeling of risks or early warning systems (Radwan 2024).

Support organizations, not projects

We also need to invest in improving how we engage in the process of achieving sustainable development, aligned with the underrated SDG 17: Partnerships for the Goals. The most valuable financing you can have as a development actor is not project funding, but so-called core contributions (or unrestricted funds). This type of funding provides an organization the opportunity to invest the funds as they usually see best fit – within reason. A separate opportunity is to increase overhead rates for local partners to provide more flexibility in funding (Inter-Agency Standing Committee 2024).

I have witnessed numerous colleagues experiencing stresses from managing short-funding cycles, upon which their salaries usually depend. In a three-year project, you will need to seek funding by the end of the second year. With organizations that have demonstrated consistent impact, they deserve to receive core contributions. They provide stability in implementation and improve the mental health of the staff. Unrestricted from a logical framework or fixed implementation plan, those funds allow adapting to change, taking advantage of new opportunities, and subsequently bringing about more sustained impact.

Don't force the change, but build capacity

On the notion of change, a common tenet in our book has been that you cannot force change upon someone. Ultimately, it is always the individual who will bring about change themselves. Closely related to the above point, we need to avoid repeating the mistake to engage in top-down development processes that subsidize a transformation for the few (given the limited funding). Instead, we need to identify avenues for empowering people to bring about the change themselves through bottom-up processes.

The prevalence of climatic (and other) shocks over the recent years demands more inclusive development interventions that focus on building capacity to adapt to the changes. In the past, societies successfully adapted to shocks and stresses "by revitalizing a common will to overcome adversity, drawing from both old experience and new information to revise or develop collective strategies for survival" (Butzer and Endfield 2012, 3630). Efforts to build individual and societal capacities also create a more sustained change in the long-run, well past any donor funding has subsided.

REFERENCES

Butzer, K. W., and G. H. Endfield. 2012. "Critical perspectives on historical collapse." *Proceedings of the National Academy of Science U S A* 109 (10): 3628–3631. https://doi.org/10.1073/pnas.1114772109. https://www.ncbi.nlm.nih.gov/pubmed/22371580.

Cities and Regions in the UNFCCC Process, Global Taskforce of Local and Regional Leaders, and ICLEI (Local Governments for Sustainability), 2024, "COP29: Local and Subnational Leaders Demand Urgent Multilevel Action and Urban Climate Finance." Available at: https://www.cities-and-regions.org/wp-content/uploads/cop29_lgma_-press_conference_media_release.pdf.

Clark, P. 2024. "Year in a Word: Greenlash." *Financial Times*, December 27.

Desmond, M. 2023. *Poverty, by America*. New York: Crown.

Gobierno Municipal de Zapopan. 2024. "Revisón Local Voluntaria: Contribución a la Agenda 2030 para el desarrallo sostenible. " Available at: https://sdgs.un.org/sites/default/files/vlrs/2024-08/revision_local_voluntaria_gobierno_municipal_de_zapopan.pdf.

Hancock, A. 2024. "EU Hits Roadblocks in Reaching Green Milestone as Elections Loom." *Financial Times*, March 27. https://www.ft.com/content/b9e7e292-8ac6-45c9-8098-8318ddb83256.

Hancock, A., and A. Bounds. 2024. "Brussels Bows to Farmers' Protests by Slashing Environmental Targets." *Financial Times*, February 6. https://www.ft.com/content/86073030-84c1-471e-93a0-4b696116030a.

Heikkilä, M. 2023. "Making an Image with Generative AI Uses as Much Energy as Charging Your Phone." *MIT Technology Review*. https://www.technologyreview.com/2023/12/01/1084189/making-an-image-with-generative-ai-uses-as-much-energy-as-charging-your-phone/ (accessed March 23, 2025).

Hutchinson, D. 2024. "States Embrace Climate Leadership as Trump Heads to White House." *Bloomberg Law*. https://news.bloomberglaw.com/environment-and-energy/states-embrace-climate-leadership-as-trump-heads-to-white-house.

IISD. 2024. *Summit of the Future Highlights: Sunday, 22 September 2024*. Available at: https://enb.iisd.org/summit-future-22sep2024.

Independent Group of Scientists Appointed by the Secretary-General. 2019. *Global Sustainable Development Report 2019: The Future is Now – Science for Achieving Sustainable Development*. New York: United Nations. Available at: https://sdgs.un.org/gsdr/gsdr2019.

Inter-Agency Standing Committee. 2024. "Grand Bargain 3.0 Implementation Agenda." Available at: https://interagencystandingcommittee.org/sites/default/files/2024-10/3-%20GB%20AM%202024%20-%20Background%20Doc%20-%20The%20GB%203.0%20Implementation%20Agenda.pdf.

Jopson, B. 2024. "The Problem with Europe's Ageing Wind Farms." *Financial Times*, February 21. https://www.ft.com/content/7f742d23-673b-47d3-9ce9-64fa5d322abe.

Jungcurt, S., E. Kosolapova, and E. Bonaccorsi. 2024. "Why We Shouldn't Give Up on the SDGs." IISD. Available at: https://www.iisd.org/articles/deep-dive/why-give-up-sdgs-summit-future-agenda-2030 (accessed February 14, 2024).

Krenak, A. 2020. *Ideas to Postpone the End of the World*. Trans. A. Doyle. Toronto: Anansi International.

Levitsky, S., and L. A. Way. 2025. "The Path to American Authoritarianism." *Foreign Affairs* March/April. https://www.foreignaffairs.com/united-states/path-american-authoritarianism-trump.

Mance, H. 2025. "Where 'Woke' Went Wrong." *Financial Times*, February 1.

Mollison, B.C. 1988. *Permaculture: A Designer's Manual*. New Delhi: Tagari Publications.

Moseley, W. G. 2017. "The New Green Revolution for Africa: A Political Ecology Critique." *Brown Journal of World Affairs* 223(2): 177–190.

Nilsson, M., D. Griggs, and M. Visback. 2016. "Map the Interactions between Sustainable Development Goals." *Nature* 534: 320–322. https://doi.org/10.1038/534320a.

Radwan, G. 2024. "Harnessing AI to Accelerate the Sustainable Development Goals." Available at: https://sustainability.google/operating-sustainably/stories/harnessing-ai-to-accelerate-sdgs/ (accessed March 23, 2025).

UN DESA. n.d. *Global Development Initiative – Building on 2030 SDGs for Stronger, Greener and Healthier Global Development*. Available at: https://sdgs.un.org/partnerships/global-development-initiative-building-2030-sdgs-stronger-greener-and-healthier-global.

United Nations. 2024. *Pact for the Future, Global Digital Compact, and Declaration on Future Generations*. New York: United Nations.

Untools.co. 2025. "Iceberg Model." Available at: https://untools.co/iceberg-model/ (accessed March 23, 2025).

Ville d'Agadir. 2023. "Voluntary Local Review: Municipality of Agadir, Morocco." Available at: https://sdgs.un.org/sites/default/files/vlrs/2024-01/agadir_vlr-english.pdf.

Zeddies, J., R. P. Schaab, P. Neuenschwander, and H. R. Herren. 2001. "Economics of Biological Control of Cassava Mealybug in Africa." *Agricultural Economics* 24.

17 Your role in achieving sustainable development

YOU HAVE MORE AGENCY THAN YOU MIGHT IMAGINE

Even if the reader dreams of working for a large development agency, they might have more agency than they imagine. This chapter covers key insights on how one can shape one's role in achieving sustainable development. It includes a focus on imagining change and resisting standardization. In order to take complexity seriously, the right questions need to be asked as well as listening skills need to be improved. To effectively contribute to sustainable development, one also needs to focus on self-care, learning how to say "no," yet working collaboratively across traditional boundaries. It closes by reviewing the tools introduced in the book and how they support readers on their paths to become silo breakers and development changemakers, beyond 2030.

Changes in development spending by governments impact the make-up of jobs in the sector. At times of budget cuts, I recommend to my students to consider working for smaller, local NGOs, foundations less dependent on government funding, development actors in the Global South, or sustainable development efforts within the private sector. Especially within local NGOs and foundations, usually these positions offer more opportunities to learn (and engage) beyond your job description. If you currently dream of working for a large development organization in the mid-term future, you might think that you will have less agency there: however, the opposite can be true. The larger development organizations become, the less they have an overview of what is happening across the organization. Mosse (2005) uses the analogy of an hourglass, with significant power held by the lower-level staff. The reality is, even in large organizations, you can have leeway in how to undertake and reshape your development work, as you have more agency than you might imagine.

As Mosse captures in his review of development, "[j]unior staff withhold or reveal information strategically in order to secure reputations, conceal poor performance or to negotiate position in the organization or with outsiders (donors, villagers)" (2005, 12). During participatory development intervention, he found senior staff also failed to have a complete command over fieldworkers: "Bottom-up principles, competing agendas and accountabilities, and the constant interference of outsiders gave fieldworkers *de facto* autonomy to develop their own operational interpretation of both villager needs and project goals, and their own strategies of intervention in both areas" (Mosse 2005, 108–109). Senior staff needed field staffs' "knowledge, contacts and actions in

DOI:10.4324/9781003400028-24

order to secure the outputs for which they would, in turn, be held accountable" (Mosse 2005, 109). He concludes by arguing that "[u]ltimately it is not policy consensus, rational planning or bureaucratic procedures that make projects … run. It is personalities, brokering skills and the channels of influence of individual editors, buffers and filters" (Mosse 2005, 125).

While we are at the mercy of policy trends in development, if you manage to catch the waves, you can use your agency to surf toward more inclusive, transformative development.

KEEP DREAMING AND RESISTING STANDARDIZATION

> Where do we go to design parachutes? We go to that place beyond this hard earth: the land of dreams.
>
> (Krenak 2020, 64)

Krenak (2020, 64) instills the importance of dreams as "transcendental experience in which the human chrysalis cracks open onto unlimited new visions of life." He urges us to learn how to reach this world. As referenced throughout the book, expanding what counts as knowledge and whose voice is included will play a large role in whether we achieve a more inclusive development.

We also need to resist standardization. As Mosse (2005, 116) finds, during project implementation, the implementation logic – and all its plans and structures – pushed development practice toward standardization: "it was virtually impossible to ensure that 'participatory planning' involved local problem solving or even choosing between alternatives." Implementation was rushed toward the familiar and conventional (over the complex or risky), toward easily deliverable capital items (over collective actions), over work that offered quick, "dramatic and visible results" as opposed to longer-term efforts (Mosse 2005, 116).

MAKE SURE YOU ASK (THE RIGHT) QUESTIONS

If you are in the business of taking complexity seriously, you will have to invest in moving beyond a single indicator. To get you there, it is important that you draw from a pool of both qualitative and quantitative methods (see Appendices A and C for toolkits, including on participatory methods).

No matter the method you will deploy, you need to overcome, as Iarossi (2006) put it, "*[h]ow easy it is to ask the wrong questions.*" While his study was written for survey design, its primary focus on relevance and accuracy remains valid. Relevance captures that we obtain information that is "appropriate for the purpose of the study" (Iarossi 2006, 27). Accuracy ensures that we "[collect] the information sought in a reliable and valid manner" (Iarossi 2006, 27).

The examples used by Iarossi and frequently encountered during my fieldwork relate to getting recall estimates on household income or expenditures. Ask yourself, would you be able to offer an answer on your income or expenditures with no records to support you? Could you recall the number of climate shocks you have experienced over the last five years? If you have trouble, so will your participants. So do not allow complexity to trickle into making your research tools overly complicated. Dedicate more time to listening, learning, and being aware of your own assumptions. And in the end (or the beginning and middle), continue sharing and giving back what you find with the stakeholders you engage with.

SELF-CARE AND LEARNING HOW TO SAY "NO"

To effectively contribute to sustainable development, you need to focus on self-care. This includes not obsessing over negative thoughts or news, but learning how to accept them. Not silencing them, but acknowledging that they exist in a non-judgmental matter (Alderman 2017). While I struggle with it, I try not to let the bad news of the week (and, yes, there can be a lot) result in me throwing up my hands. Rather, I try to identify how I can make a difference. As shown by Dr. Frederickson's research, quoted by Jane Brody (2017): "taking time to learn the skills [such as compassion or meditation] to self-generate positive emotions can help us become healthier, more social, more resilient versions of ourselves." As suggested by Krenak (2020, 21), "[w]hen you feel the sky is caving in on you, just give it a push and breathe."

One of my mentors during my career in development urged me to say "no" more often. I ignored the advice for nearly too long, as I was always offering myself to step up when others failed to do so. Related to our next section, "*You can't do it all by yourself (and you shouldn't)*," what happened in the process is that I became overloaded with work.

A memorable example was when I took a summer vacation, but felt like I still needed to keep checking my work email every day. My mentor led by example: they did not check their phones during vacation. Or as they put it: "You are not so important that you are not replaceable." That hurt to hear, but it was the truth. It took me several years to put on an "away message" and stick to it. This small step represented a major achievement towards saying "no."

The next step was to actively turn down new opportunities as they came up in the daily grind. While the stakes of failing to follow through in the office seemed lower, the promises and commitments you make in the field have larger consequences. It was during my first fieldwork in 2006 in Uganda that I learned not to overpromise in development work.

You will encounter reasonable requests for support beyond what your project entails. It can range from the more mundane exchange of WhatsApp phone numbers to stay in touch with project beneficiaries, to the more elaborate ones regarding providing (personal) financial or technical support on top of what was planned. It was tempting to commit to all of them, as my position of privilege permits me more easily to make a difference in an individual life. Even as hard as it sounds, learning how to say "no" (or "I don't know" as an academic), was often the best course of action.

When traveling across southwestern Ethiopia for project trips, I would frequently stop at the roadside to visit farmers and engage in a conversation on their practices and challenges. Figure 17.1 depicts a common scene, as I was greeted by children on their way from school and family members nearby. Faced with deprivation, you will inevitably experience an urge to help, even when not faced with any explicit requests. Through my mentors, I learned that I would focus my efforts on what I can contribute to the most. In my development practice, it meant ensuring that we are implementing projects that are inclusive and most effective in bringing about a sustained change across the area.

This challenge was especially acute as I am a social scientist operating across the field of agronomy. Farmers frequently requested advice regarding a specific pest, plant, or livestock challenge. Respecting their livelihoods and expertise required me to acknowledge what I do not know. Failing to do so could likely violate the overarching tenet guiding development work of doing no harm.

It was through this experience that I ended up preparing my field trips by identifying other actors engaging in relevant topics and sourcing relevant knowledge-platforms that are freely available. From a beneficiary's perspective, my field visits must have seemed very myopic, as they failed to cover the complexity of needs and challenges present. My hope was offering a resource for

Figure 17.1 A snapshot on the way to Tolay, Ethiopia, Peter Lüthi/Biovision Foundation

building knowledge for those I encounter represents a viable compromise that improves upon the experience of how development is done.

For further reading, I wholeheartedly recommend the book, *Development Fieldwork: A Practical Guide* (Scheyvens 2014). While largely written for academic researchers planning to undertake fieldwork, its lessons on personal (Chapter 7) and ethical issues (Chapter 9) in the field, how you work with marginalized, vulnerable, or privileged groups (Chapter 10), as well as the politics and practicalities of leaving the field (Chapter 11) are highly useful in development work itself.

YOU CAN'T DO IT ALL YOURSELF (AND YOU SHOULDN'T)

Krenak, reflecting on his role in building networks with other peoples and governments to build solidarity with other Indigenous peoples, emphasizes the importance of collaborations across the traditional boundaries:

> We feel today as though we've been cut adrift in a cosmos devoid of meaning and shorn of any shared ethic. It weighs upon our lives, and at every turn we are reminded of the consequences of these choices. If we could only heed some vision beyond the blindness that has descended upon the world, perhaps it would open our minds to some form of cooperation among peoples, not so we can save others, but so can save ourselves … What I have learned over [the] decades is that we all need to wake up, because whereas before it was just us, the Indigenous peoples, who were facing a loss of meaning in our lives, today everyone is at risk, without exception.
>
> (Krenak 2020, 47)

I needed to learn in my own practice that I cannot do it all myself. Part of it was learning how to delegate, no matter how fast you think you are in doing it yourself. We are not seeking perfection. If you are striving for it each and every time, you are likely not lasting long in a team-based work environment. A precondition to delegating is having said "no" in the first place, so you have the space to mentor and follow up.

Delegating for me entailed becoming a better communicator. Reversing our work from human-centered design, it means sharing what you are seeing, what you are hearing, what you are feeling, and what you are thinking. You also might need to become a better listener (e.g., Stetson et al. 2007). The work introduced earlier by Anderson, Brown, and Jean (2012) – *Time to Listen* – captures well the spirit that people at the receiving end want to become partners, they want to be heard. They do not call for development to end. They call for a transformation of how development is done. Shockingly, one out of two partners never receive the final proposal submitted to a donor (Unlock Aid 2024). We can and need to do better.

To address transformations, this includes empowering knowledge brokers who can feed insights gained into their social networks and policymaking (Scodanibbio et al. 2023). For your own practice, it might start by improving how we build partnerships. These steps include understanding shared visions or values (Stetson et al. 2007). To formalize a partnership, you move toward addressing power differences and risk, negotiating an agreement (usually a non-legally binding Memorandum of Understanding (MoU)), that can form the basis for undertaking work and learning together (Stetson et al. 2007). Developing trust and respect rest at the core of your work. Treasure them, as they are difficult to rebuild (Stetson et al. 2007).

REUSE AND RECYCLE

You can do a myriad of personal activities to bring about a local change within your own household toward sustainable development: think reducing food waste, recycling your containers, replacing single-use plastic bags, bike and use trains. You can see a list of them for inspiration in Appendix A*, which offers a guide to tools for further action.*

Reuse or recycling is increasingly a big business. The US market for second-hand clothes has nearly doubled over three years (Milne 2024). Besides offerings like Goodwill, which have existed for decades, US companies like ThredUp, are offering services for you to part with your clothes and make some extra cash from their sales. Lego, the largest producers of plastic toys, is piloting how to reuse and recycle its bricks (Milne 2024). These efforts are in addition to its push to replace fossil fuels to make its bricks (Ziady 2024). A global effort to phase out single-use plastics was close to reaching an agreement late 2024, but momentarily failed because of objections from oil-rich nations such as Russia and Saudi Arabia (Speed and Chu 2024).

At work, you want to avoid creating everything from scratch. To give you the example from my work as a university professor, some graduate students are not thrilled to undertake a literature review. It can be perceived as unnecessary busy work standing between us and the most interesting, original part of our work. My argument is that failing to understand what has worked (or failed) before, risks not only duplicating work, but most likely investing your time (and your employer's money) into something that will be infinitely worse than if you'd have reused and recycled (or *upcycled*) existing scholarship and lessons learnt.

Just as you are leveraging larger networks in your daily practice at home, you want to engage and learn from partners and non-partners alike (including development actors you might traditionally perceive as competitors). In development, it is still rare that failures get published. One of the best way to learn about them is to pick up the phone or have an informal meeting to see how you could avoid repeating the same errors twice. And bring these colleagues on as stakeholders in your project – a well-chosen steering committee can provide priceless insights.

In Chapter 9, we encountered the work by the International Initiative for Impact Evaluation, 3ie. Its *Development Evidence Portal* has captured over 17,500 impact evaluations, over 1,400 systematic reviews, and evidence maps that identify where knowledge gaps persist and where sufficient evidence is available (2024). The site serves as a prime example where someone has already done the legwork, allowing you to assess whether you need to invest US$300,000 in understanding the benefits of school lunch programs on educational progress. If the existing evidence is strong, those limited funds can be best spent on reaching more children. And if you need to collect data yourself, don't go overboard.

On methods, Acumen (2015) was an early mover in developing a *Lean Data* methodology. Focused on social enterprises, its guide offers a refreshingly short approach to get data, quickly. The team has subsequently developed more focused guides, including leveraging lean data for *Understanding Gender Impact* (Acumen and Unilever 2018). During COVID-19, 60decibels (2024) – founded by previous Acumen staff – has emerged as a pioneer in getting social impact data, offering industry-wide insights that can guide programming. Leveraging the above will open up greater capacity for you to dive deeper, listen more, and co-learn in the process.

WALK THE TALK: MONITORING AND EVALUATING YOURSELF

Your work in sustainable development might come with a task to assess project performance – or at least hire an external evaluator. I previously introduced the OECD-DAC guidelines (OECD 2021), a likely framework to evaluate your development work. The guidelines are focused around the six criteria of relevance, coherence, effectiveness, efficiency, impact, and sustainability. On sustainability, the question is "Will the benefits last?". On relevance, the question is "Is the intervention doing the right things?". On impact, "What difference does the intervention make?". As you reflect on your own work, a good starting point can be for you to assess yourself using these questions. It allows you to more holistically capture your own contributions, and serve as a basis for your personal development to create a more sustained impact with your time and efforts.

What is your role in bringing about a sustainable transformation? Where should I start? Where should it end? Before answering the questions, let me highlight that Appendix B offers some career guidance for you, and our companion website – sustainabledevelopment.me – has additional resources for you.

We started our journey by exploring how business-as-usual is no longer possible. We share the realization that we are lagging in reaching the aspirational SDGs by 2030. We are united in working beyond 2030 to create sustained transformations that can withstand political, economic and climatic shocks.

ON WHERE TO START

Given the gap between progress made and the shared aspirations across the globe, a reasonable point of departure is anywhere you feel passionate about and have the competence in. No dimension of sustainable development is off-topic. Some fields of work will require more technical training to get into, such as SDG 3 dealing with health.

You identified early in our book an entry point of an SDG that resonates with you. As we aim to bring about a sustainable transformation, you need to move beyond that singular entry point and commit to making a difference that leverages the identified synergies and tackles the trade-offs across sustainable development.

My motivation for writing this book in the first place is the argument that we need new expertise and new learnings to get us there. For transformation to occur, we will need a new generation of leaders to emerge who can break free from the current development system and match their actions to the aspirations as framed by the UN 2030 Agenda. Given the levers of change – from policymaking and governance, to finance and the economy, to individual and collective action, and science and technology – there are a myriad of entry points. You can choose how you want to position yourself. Or you might succeed positioning yourself at the intersection of two (or possibly three) of these.

ON WHERE IT SHOULD END

Yes, development work is successful if you put yourself out of businesses. On paper, this could mean when we achieve an SDG target – with its political and methodological maneuvering – you can move on. In reality, the (country-level) averages will mask regional unevenness. Even though the UN video referenced in Chapter 1 promoted the achievements by the next generation, I do not expect development to go out of business soon.

The power you hold in dedicating your career to sustainable development applies to both the Global North and the Global South. Or using a different image, your work can start in your neighborhood and end anywhere along the road you will take from there. Depending on your entry point, your work might have a larger impact if you focus on the Global North (e.g., greenhouse gas emissions). Or your work might be more rewarding, if you get to build coalitions and connections locally to bring about a change in your neighborhood.

To advance our last point further, let me use the example of intersectional environmentalism and how it translates into our daily lives and communities. Based on discussions with my students, the following five threads emerged, ordered from the self to the collaborative. I have adapted them to capture how these set of actions can apply across any of the concepts and themes that inspired you in the book.

First, you can build your own knowledge base around environmental justice, past and current movements. Second, you can embody the learnings into your own daily actions and decision-making, such as in your consumption patterns. Third, you can raise awareness or share your knowledge and decision-making with friends and families. Fourth, you can support or join existing movements by donating your time (i.e., volunteering) or money (e.g., including leveraging others who are in a more privileged position to donate). And, fifth, you can integrate the findings in your own profession, and across our levers of policymaking, science, finance, or the non-profit sector.

Independent of the SDGs and what follows them, the greatest contribution you can make to sustainable development in the long run is to support a paradigm shift from the externally driven aid delivery system captured by Anderson, Brown, and Jean (2012) toward what they call the collaborative aid system (introduced in Chapter 13).

To support you, I have captured for each of their steps a practical action. Our companion website – sustainabledevelopment.me – features resources for each of those steps.

- *Local people seen as colleagues and drivers of their own development*: Build your own capacity in participatory, bottom-up development.
- *Focus on supporting/reinforcing capacities and identifying local priorities*: Spend your field visits less on demonstrations and performances, and more on observations and experiences.
- *Context-relevant programs developed jointly by recipient communities and aid providers*: Revise how you are soliciting development proposals in the first place, reducing barriers around language (e.g., English) and technical (e.g., logical frameworks) requirements.
- *Collaborative decision-making*: Use your agency to empower those closest in the field to propose pathways forward, to overcome challenges together.
- *Fit money and timing to strategy and realities on the ground*: Provide longer, more consistent funding cycles that fit the projected change trajectory.
- *Staff evaluated and rewarded for quality of relationships and results that recipients say make lasting positive changes in their lives*: Support holistic, bottom-up interventions that actively leverage multiple livelihood dimensions and levers of change.
- *Monitoring, evaluation, and follow-up by providers and recipients of results and long-term effects of assistance*: Encourage peer-to-peer monitoring and evaluations, and sharing the learnings beyond your organization.
- *Planned draw down and mutually agreed exit/end of assistance strategy*: Embed the sustainability of your work at the beginning, and not as an afterthought.

OUR SHARED WORLD BEYOND 2030

The SDGs represent a jumping-off point to gain momentum in our shared quest to create a more just and ecologically sound world. They have made a difference in their first decade. The 2030 Agenda and its aspirational language represented a big leap forward toward shifting development to face complexity, breaking down silos, and accounting for nature and its contributions to holistic development. Localization initiatives are aligned with a paradigmatic shift toward bottom-up development and the limitations of business-as-usual showcase the importance of context.

The world faces multiple crises at this moment that need to be urgently addressed – no matter which framework we operate within. That said, as sustainable development represents a decades-old concept, I urge you not to judge its success or failure based on where we stand by 2030. Use the momentum of progress made, and the painful shortcomings, as drivers for change for future generations.

Just as we need to shift our own expectations of what impact can be generated over a short period of time, we also need to ensure we keep the perspective well beyond 2030. In its global foresight report, the United Nations Environment Programme (2024) mapped out eight shifts or phenomena that will pre-occupy us in 2050, including misinformation, resource scarcity, AI, widening inequalities, and mass displacement. Both the framework we have adapted, as well as the lessons learned from our transformation hotspots, will keep their relevance over the next two decades.

As a trained geographer, it is rare that I quote a historian. But Fernand Braudel, who coined the term the *longue durée* – looking for changes over longer periods of time – left a lasting impression on me in my formative years (Braudel 1995). Braudel is quoted by Guldi and Armitage (2014, 15) stating, "[i]n truth, the historian can never get away from the question of time in history: time sticks to his thinking like soil to a gardener's spade." Translated into our domains, I argue that moving forward, a silo breaker can never get away from the question of sustainable development: sustainability sticks to our thinking like soil to a gardener's spade. And the ambitious, optimistic, and inclusive nature of the SDGs are a useful tool to plant and nourish the seeds for a better future. A future that is resilient to the shocks we face today and tomorrow. I hope this book has done its part to support you on your path to becoming a silo breaker and development changemaker, through all the ups and downs. We look for your leadership for generations to come.

REFERENCES

60decibels. 2024. "Insights." Available at: https://60decibels.com/insights/ (accessed March 23, 2025).

Acumen. 2015. "The Lean Data Field Guide." Available at: https://ssir.org/pdf/Acumen_Lean_Data_Field_Guide.pdf.

Acumen, and Unilever. 2018. "A Lean Data How-To Guide: Understanding Gender Impact." Available at: https://acumen.org/app/uploads/2024/03/Acumen-Understanding-Gender-Impact.pdf.

Alderman, L. 2017. "The Year of Conquering Negative Thinking." *The New York Times*, January 3. https://www.nytimes.com/2017/01/03/well/mind/the-year-of-conquering-negative-thinking.html.

Anderson, M. B., D. Brown, and I. Jean. 2012. *Time to Listen: Hearing People on the Receiving End of International Aid*. Cambridge, MA: CDA Collaborative Learning Projects.

Braudel, F. 1995. *A History of Civilizations*. New York: Penguin Books.

Brody, J. E. 2017. "Turning Negative Thinkers into Positive Ones." *The New York Times*, April 3. https://www.nytimes.com/2017/04/03/well/live/turning-negative-thinkers-into-positive-ones.html.

Guldi, J., and D. Armitage. 2014. "Going Forward by Looking Back: The Rise of the Longue Durée." In J. Guldi, and D. Armitage (Eds.), *The History Manifesto*. Cambridge: Cambridge University Press, pp. 14–37.

Iarossi, G. 2006. *The Power of Survey Design: A User's Guide for Managing Surveys, Interpreting Results, and Influencing Respondents*. Washington, DC: World Bank.

International Initiative for Impact Evaluation. 2024. "3ie Development Evidence Portal." Available at: https://developmentevidence.3ieimpact.org (accessed March 23, 2025).

Krenak, A. 2020. *Ideas to Postpone the End of the World*. Trans. A. Doyle. Toronto: Anansi International.

Milne, R. 2024. "The Relentless Rise of Second-hand." *Financial Times*, August 30. https://www.ft.com/content/9b07bad3-af81-4cd2-a98a-a750fcee9d2e.

Mosse, D. 2005. *Cultivating Development*. New York: Pluto Press.

OECD. 2021. *Applying Evaluation Criteria Thoughtfully*. Paris: OECD. https://doi.org/10.1787/543e84ed-en.

Scheyvens, R. 2014. *Development Fieldwork: A Practical Guide*. 2nd edn. London: SAGE Publications Ltd.

Scodanibbio, L., G. Cundill, L. McNamara, and M. du Toit. 2023. "Effective Climate Knowledge Brokering in a World of Urgent Transitions." *Development in Practice* 33 (7): 755–761. https://doi.org/10.1080/09614524.2022.2159932.

Speed, M., and A. Chu. 2024. "UN Plastics Treaty Talks Fail After Oil Producers Block Output Limits." *Financial Times*, December 1. https://www.ft.com/content/737b609d-d660-4f91-a016-740d9c476642.

Stetson, V., S. Hahn, D. Leege, D. Reynolds, and G. Sharrock. 2007. "*ProPack II: The CRS Project Package*." Baltimore, MD: CRS.

United Nations Environment Programme. 2024. *Navigating New Horizons: A Global Foresight Report on Planetary Health and Human Wellbeing*. Nairobi: United Nations Environment Programme.

Unlock Aid. 2024. "Glassdoor for Primes: Increasing Transparency and Accountability in Global Development Subcontracting." Available at: https://cdn.prod.website-files.com/64230dc9afdfc901ff425554/672142e2133864b5b7d9f5b0_Glassdoor%20for%20Primes.pdf.

Ziady, H. 2024. "Lego Plans to Ditch Oil in Its Bricks for Pricier Renewable Plastic as Profits Surge." *CNN*, August 28. https://www.cnn.com/2024/08/28/business/lego-bricks-renewable-plastic/index.html.

Glossary

Adaptive capacity ability to respond proactively to a shock or stress and engage in recovery

Agroecology the science, research, and practice of ecological agriculture, usually focused on more holistic, local production

Artificial intelligence leveraging advanced computing with the aim of solving human tasks in a more efficient fashion

Biodiversity life within ecosystems, contributing to sustainable development

Blended finance mix of public and philanthropic capital to mobilize additional private investments for development

Bottom-up development approach leveraging grassroots efforts with a focus on empowering local actors to bring about change; usually in contrast to top-down development

Capacity building developing and strengthening skills, processes, and resources to increase the abilities of an individual or organization

Changemakers actors or fields that drive sustainable development; in our book focused on governance and policymaking, economy and finance, individual and collective action, and science and technology

Circular economy system aimed at reusing, recycling, or upcycling products to eliminate waste

Civil society collection of non-governmental organizations, community groups, or individuals, usually engaging in organizing efforts or advocacy for change

Climate change long-term shifts in temperature or weather patterns primarily attributed to human activities

Climate shocks sudden climate events, such as flooding or storms, impacting ecosystems and society at-large

Co-benefits additional benefits arising from addressing a challenge

Collective actions collaborative efforts by individuals or across groups sharing similar goals for greater impact

Distributional justice fair allocation of resources and opportunities

Earth system relationship between air (atmosphere), land (geosphere), life (biosphere) and water (hydrosphere) on our planet, and their interactions

Ecosystem services benefits obtained from the environment, such as pollination services increasing yield

Energy transition shift to renewable energy from fossil fuels

Environmental externalities (unintended) impacts of usually economic activities that are not reflected in the cost of goods or services

Environmental justice a fair distribution of environmental benefits and burdens across society

Financial flows movement of funds, including investments, grants, but also remittances

Food justice framework for equitable access and the right to food, its production, and consumption

Food security availability, access, and utilization of sufficient, safe, and nutritious food

Food system framework capturing the interconnections of food production with food processing, distribution, access, consumption, and resource or waste recovery, looping back into production

Fossil fuels finite energy sources, such as coal, oil, or gas, from ancient organic matter that are known to contribute to greenhouse warming

Gender inequalities unequal access to resources, opportunities, or rights based on gender

Global North countries in the Northern Hemisphere, historically categorized as higher income

Global South countries in the Southern Hemisphere, historically categorized as lower income and with a legacy of colonialism

Governance use of political or administrative authority to manage institutions or countries

Green Revolution period of increased agricultural productivity from the 1960s onward through adoption of hybrid seeds, fertilizers, pesticides, and irrigation across the Global South

Holistic as in holistic development, considering interrelated aspects to improve well-being for all

Hotspots areas of substantial impact and potential for leveraging synergies to achieve sustainable development

Human capabilities individuals' abilities to make choices and improve their livelihood

Human-centered design approach primarily applied in product development, putting the user at the center and including rapid prototyping

Impact investing investments with intention to generate a positive social or environmental impact in addition to any financial returns

Indigenous knowledge traditional wisdom, practices, and skills developed over generations, supporting sustainable development

Intersectionality framework capturing how social identities overlap

Just economies economic systems that prioritize fairness, equity, and well-being of all

Just energy equitable access to clean, affordable energy for all

Just transition usually an environmental transition that accounts for impacts on workers and communities

Land degradation decline in land quality over time caused by human or natural activities

Leapfrogging bypassing intermediary stages of development or technology, for example, skipping landlines and adopting mobile technologies

Localization efforts at strengthening the role of local actors in development, including through capacity building and a greater share of total funding received

Logical framework tool adapted from engineering to define goals, outcomes, results, and activities of a development project

Marginalized communities economically, socially, or politically excluded groups that historically have lacked their fair share of access to resources and opportunities

Planetary boundaries safe operating space for humanity without tipping critical Earth systems

Procedural justice fair and transparent decision-making process, with rules applied in a consistent and inclusive fashion

Recognition justice to identify and address differences across groups, identities, or cultures

Regenerative agriculture farming practice focused on soil health, specifically carbon, using sustainable land management practices

Resilience ability for individuals, communities, or systems to absorb, adapt, and recover from shocks and stresses

Scaling out expansion of an existing development effort to reach larger numbers of people, organizations. Usually in new geographical areas

Scaling up integrating a successful intervention into policies, laws, or budgets at sub-national, national, or international levels

Silo breakers those who overcome barriers to collaborate across fields, sectors, or disciplines

Sustainable development development that "meets the needs of the present without compromising the ability of future generations to meet their own needs," based on the UN Brundtland Commission and operationalized in the UN Sustainable Development Goals

Tipping point threshold where irreversible change takes place

Trade-offs additional drawbacks arising from addressing one challenge

Traditional knowledge indigenous or local knowledge of relevance, but usually outside the Western scientific inquiry

Transformation a systemic change that creates long-term impacts across multiple development challenges

Appendices: Overview

The three appendices offer a starting points for your further development. As they feature substantial number of links to website, check our companion website – sustainabledevelopment.me – for any updated links and new references.

Appendix A: Guide for tools for further action offers a menu for change, with a sub-section on your investments, inspirations for you to become a silo breaker, and a note on teaching or mentoring others around sustainable development.

Appendix B: Career guide in sustainable development provides my perspective on jobs in the field, including investing in your professional development.

Appendix C: Guide to companion website: sustainabledevelopment.me offers a preview of what you can expect with additional resources, deeper dives, and videos and podcasts for your exploration and inspiration.

Appendix A Guide for tools for further action

YOUR MENU FOR CHANGE

The UN campaign to inspire citizen action for achieving sustainable development is called *Act Now*. You can access their main page at: https://www.un.org/en/actnow. The UN has also a dedicated app for their campaign, available for the iPhone and Android platforms at https://aworld.org. Google selected it as one of the best apps of 2023. The app allows you to log your actions, track your change, and add your voice to the movement for change.

The UN Act Now website offers concrete actions you can take in four domains overlapping with our transformation hotspots: *a healthy planet*; *a just society*; *a peaceful world*; and *a better economy* (United Nations 2025a).

Speaking up is a large part for bringing about change. UNEP has an excellent website on how to engage in advocacy for change to "[a]ppeal to world leaders, urge your city, your bank and your employer to take urgent action toward net-zero emissions" (UNEP 2025). An example includes choosing a bank with a positive impact.

The UN Act Now campaign also has a dedicated page on actions to speak up in your personal life (e.g., amplify diverse voices), in your community (e.g., be an ally), with your vote (e.g., encourage others to vote), with your wallet (e.g., choose fair trade and climate-friendly products), and with your time (e.g., stand up for rights) (United Nations 2025b). For climate activism, see also global movements like 350.org with its mission to reduce CO_2 emissions to 350 ppm.[1]

I have listed 40 actions curated by the UN for each domain for inspiration (United Nations 2025a). For a deeper dive, download the app or visit www.un.org/en/actnow.

Actions for a healthy planet

1. Save energy at home.
2. Change your home's source of energy.
3. Walk, bike. or take public transportation.
4. Switch to an electric vehicle.
5. Consider your travel.
6. Reduce, reuse, repair, and recycle.
7. Eat more vegetables.
8. Throw away less food.
9. Plant native species.

10. Clean up your environment.
11. Make your money count.
12. Speak up.

Actions for a just society

1. Teach children about equality.
2. Sponsor a scholarship.
3. Fundraise for schools.
4. Support a food bank.
5. Model self-care.
6. Advocate for quality health care.
7. Understand and help reduce mental health stigma.
8. Champion accessibility and inclusion.
9. Know your rights.

Actions for a peaceful world

1. Educate yourself about peace.
2. Say no to violence.
3. Practice understanding and solidarity.
4. Report bullying and harassment.[2]
5. Protest inequality and discrimination.[3]
6. Embrace inclusion and diversity.
7. Support peace advocates.
8. Follow reliable news sources.
9. Post with care on social media.

Actions for a better economy

1. Donate to charities.
2. Mentor young people.
3. Advocate for better work.
4. Pay fair tips and wages.
5. Buy from employee-friendly businesses.
6. Purchase fair-trade products.
7. Green your tourism.
8. Join the circular economy.
9. Use green building materials.
10. Do your fair share at home.

PICK WHAT MATTERS MOST TO YOU

You can use apps like the above to track the impact of your action. You might decide to start with the one that is most meaningful to you, or the low-hanging fruit that is easiest picked first. In the end, our cumulative local actions will matter for bringing about the global change.

If you'd like to dive deeper, here are tools specific to your sustainable consumption, climate change actions, and assessing your business or organization.

First, for more sustainable consumption, you can use websites and their apps like *Commons: Sustainability made simple* (https://www.thecommons.earth/) for shopping or *TooGoodtoGo*, the world's largest surplus food market (https://www.toogoodtogo.com/en-us).

Second, for a deeper dive into climate actions, Project Drawdown has curated 93 technologies and practices covering energy but also sustainable food systems available at https://drawdown.org/solutions. Apps like *Earth Hero: Climate Change* allows you to discover more tips for reducing your carbon emissions (https://www.earthhero.org/). Platforms like *We Don't Have Time* connect business leaders, policymakers, scientists and individuals united in climate action (including an overview of events) (https://app.wedonthavetime.org/). I've used the *CoolClimate Network* calculator to see how my emissions compare to the average and take actions to reduce them (https://coolclimate.berkeley.edu/calculator).

Third, if you are looking to assess and improve how your business contributes to the SDGs, UN Global Compact together with the B Corp community has developed an SDG Action Manager for businesses. You can set up a free account to take the assessment here at https://unglobalcompact.org/take-action/sdg-action-manager.

Your investments

If you are like me, you don't have excessive capital to invest. You might, however, have retirement or pension funds that are usually passively invested. You might also try to build up a nest-egg for your children. We discussed challenges around ESG in *Chapter* 4. To empower you to make your own decisions, I recommend two sources for you to review the potential environmental and social dimensions of your investments. Usually such information is only available to high net-worth individuals. I have hence tried to provide you here with freely available online resources.

For individual stocks, Robeco has developed their own metrics to score companies against their contributions to the SDGs. You can access it at https://www.robeco.com/en-us/sustainable-investing/how-do-companies-and-countries-score-on-sustainability. Several other providers allow public access to look up individual companies' ESG scores. One of them is MSCI, which has its ESG Ratings and Climate Search Tool accessible at https://www.msci.com/our-solutions/esg-investing/esg-ratings-climate-search-tool. It captures a company's commitment to reduce carbon emissions, if their emission goals are enough to keep us at 1.5 degree warming, how the company scores across the ESG, any controversies (e.g., labor right violations), and how well it is aligned to individual SDGs.

For your retirement portfolio, you are likely not invested in individual stocks but rather in mutual funds, ETFs, or other retirement funds. The non-profit *As You Sow* offers free access to their *Invest Your Values* tool. Available at https://www.asyousow.org/invest-your-values, it breaks down a fund's holding to reveal its exposure to issues such as fossil fuels, deforestation, guns, prisons, weapons, or tobacco. It also allows you to use positive screening (e.g., for gender equality). It provides a summary letter grade (with A being the best and F the lowest) and a sustainability report card capturing the above issues.

BREAK YOUR SILOS

Consider joining the silo breakers. Or gain inspirations from them. As you might have your specific area of expertise or focal point at work, I've tried to create a curated list below for you to peruse with a focus on examples across our changemakers and the Global North and the Global South:

SDG 1: No Poverty

Global Citizen is an advocacy platform that has at its core the mission to end extreme poverty. It engaged over a million people annually toward that cause, although questions have been raised regarding the attribution gap between new aid commitments and their advocacy efforts (Chadwick and Root 2021). Its current work focuses on tackling poverty by working on climate justice. Their work is able to reach young audiences through an engaging communication strategy, which includes an iPhone and Android app as well as star-studded celebrity ambassadors. You can join their movement and take action at www.globalcitizen.org.

SDG 2: Zero Hunger

Full disclosure: I am a current board member of Groundswell International headquartered in the USA. The reason I dedicate my time to its work is that the organization effectively supports grassroots partners across Latin America, Africa, and Asia who are innovators in holistic, sustainable agricultural development. Focusing on bottom-up efforts, they facilitate networking across the local NGOs putting a systems approach into action. You can watch inspiring submissions for their annual youth film festival capturing the beauty of diversity and resilience at https://www.groundswellinternational.org/youth-storytellers/.

SDG 3: Good Health and Well-being

Founded in 1972 in Bangladesh, *BRAC* (originally Bangladesh Rehabilitation Assistance Committee) has grown to become one the best-known global NGOs dedicated to holistic approaches to improve well-being. Working across the Global South, BRAC is also a pioneer in integrating the SDGs into their reporting, publishing their own voluntary reviews to showcase how their work contributes to achieving the goals in Bangladesh and beyond. You can read and get inspired from their latest reporting at https://www.brac.net/SDGs/.

SDG 4: Quality Education

Education Cannot Wait is a global fund for education in emergency settings in coordination with the UN. The fund links with gender and peace building, among others. Explore their programming and impact at https://www.educationcannotwait.org/.

SDG 5: Gender Equality

> Climate finance must not leave Indigenous, Afro-descendants and local community women and girls behind.
>
> (RRI 2022)

The *Rights and Resources Initiative (RRI)*, a coalition of over 200 organizations, aims to improve marginalized communities' access to their ecosystems. Working across 64 countries, its strategic review identified the need to empower women's advocacy in climate finance. Launched in 2022, it founded the *Women in Global South Alliance for tenure and climate (WiGSA)* to amplify their voices. Learn more about recent initiatives, including resources on women's rights and land, at www.rightsandresources.org.

SDG 6: Clean Water and Sanitation

Calvert Impact Capital – based in Bethesda, USA – developed the first Environmental Impact Bond in the USA to raise US$25 million in capital for water infrastructure projects in the nation's capital, Washington, DC (Rosser 2022). By successfully controlling pollution from stormwater runoff and improving water quality, the intervention leveraged private capital to invest in a previously neglected public good. Green infrastructure included creation of permeable pavements or rain gardens. To learn more how Calvert Impact Capital pioneers investing in small businesses or communities to create a social and environmental impact, and their lessons learnt, visit them at calvertimpact.org. You can also invest in their community notes starting with US$20.

SDG 7: Affordable and Clean Energy

Practical Action – originally co-founded back in 1966 by economist and philosopher E.F. Schumacher who authored the 1973 book, *Small Is Beautiful: Economics as if People Mattered* – stands out as an NGO that has integrated a holistic, systemic approach at its core. Learning is a crucial part for Practical Action, as they even have their own publishing house in addition to school kits and knowledge banks. In their work on improving renewable energy, they effectively break down silos and integrate for example rural energy access with agricultural livelihoods. You can find out more at https://practicalaction.org/.

SDG 8: Decent Work and Economic Growth

Patagonia stands out from the apparel industry as the leader in environmental sustainability and labor rights. Yvon Chouinard, its founder, was a pioneer in pushing for Fair Trade certification, organic cotton, and offering customers ways to reuse or repair their products. Chouinard and his family also put earth at the core of their business, transferring all of their Patagonia stock into a trust and the Holdfast Collective to support combating climate change and regenerating nature. You can read his letter titled, "Earth is now our only shareholder" at https://www.patagonia.com/ownership/.

SDG 9: Industry, Innovation, and Infrastructure

The UN *Sustainable Development Solutions Network* has over 2,000 member institutions (many of them universities) to "identify and develop global and local solutions for action on the world's most critical sustainable development challenges." Explore or join at https://www.unsdsn.org/.

SDG 10: Reduced Inequality

The GIZ-funded *Global Alliances for Social Protection* leverages a multi-stakeholder approach rich with webinars and resources (including job opportunities, publications, and databases): You can find a wealth of resources at https://socialprotection.org/.

SDG 11: Sustainable Cities and Communities

ICLEI – Local Governments for Sustainability (featured in Chapter 17) encompasses more than 2,500 local and regional governments committed to sustainable development across more than 125 countries. If you work in local government, consider becoming a member. Their website also features a valuable database of case studies or podcasts featuring sustainable planning that addresses energy, resilience and inclusivity. Learn more at https://iclei.org/.

SDG 12: Responsible Consumption and Production

The *Ellen MacArthur Foundation* is the world's leading non-profit dedicated to building a circular economy. Putting people and nature at the center of their work, the organization is fully committed to systems solutions and works across all changemakers we covered. You will find latest publications, podcasts (e.g., *The Circular Economy Show* with more than 170 episodes), and access to their network at https://www.ellenmacarthurfoundation.org/.

SDG 13: Climate Action

Co-founded by former Vice President Al Gore, *The Climate Reality Project* has built up a network of over 3 million individuals to ensure that just climate solutions don't fall off the radar of policymakers. Besides offering grants for climate justice projects, its programming includes an explicit focus on building the capacity of youth and future leaders. You can join a regional or local chapter to get involved, or apply for joining the next *Climate Reality Leadership Corps* training at https://www.climaterealityproject.org/.

SDG 14: Life below Water

The Global Mangrove Alliance launched at the World Ocean Summit in 2018 has grown into a global movement including all changemakers to restore mangroves and support livelihoods, with a focus on marginalized groups. Its knowledge hubs has a range of tools and resources, and you can join the alliance if you work for a government or a non-governmental organization. Learn more at https://www.mangrovealliance.org/.

SDG 15: Life on Land

Founded by late Nobel Laureate Wangari Maathai, the *Green Belt Movement* has grown from its Kenyan origins into a global movement that embraces local actions, advocacy, and awareness campaigns. Using tree planting as an entry point, the organization deploys a holistic approach by tackling improvements of socio-economic livelihoods, gender equality, and climate change. Get involved at http://www.greenbeltmovement.org/.

SDG 16: Peace, Justice, and Strong Institutions

The *United Network of Young Peacebuilders* is a global network of 130 grassroots organizations led by young people across 70 countries. They connect peacebuilding with education and gender inequality, providing thought leadership on topics like well-being and peacebuilding. Their website offers access to training manuals, advocacy tools, and you can join if you work for a youth-based organization at https://unoy.org/.

SDG 17: Partnerships for the Goals

Recalling our earlier discussion on *Invest in Boring* from Chapter 16, whose voice counts matters. The *Global Partnership for Sustainable Development Data* is actively engaged in addressing persistent inequalities in how data is collected and used. As part of its work, it has, for example, created a Data Values Manifesto that aims to break down these barriers, and includes extensive primers ranging from AI to advancing inclusive data collection. Your organization can join their growing global network of more than 600 organizations. Explore more at https://www.data4sdgs.org/.

YOUR TEACHING AND MENTORING

Some of you will go on to serve as teachers after graduation – be it in a domestic context or abroad. Some of you, like myself, will continue your path of education and might end up teaching five to ten years down the road. Here's a database of solutions that have been tested by teachers across the globe to improve how we educate on climate change: https://teachersfortheplanet.org/solutions/. If you are looking for engaging UN material for children, you can find their *Climate Action Super Hero* missions at https://www.un.org/sustainabledevelopment/climate-action-superheroes/.

Finally, Table A.1 captures eight themes, questions, and actions (UNPRPD and UN Women 2022) on how to translate intersectionality into your own practice.

Table A.1 Translating intersectionality into practice. Source: UNPRPD and UN Women (2022)

Themes	Questions	Actions
Reflexivity	Do I critically reflect on how my biases, attitudes, and beliefs influence my opinions and actions? How does my privilege directly or indirectly disadvantage others? What can I do to address this?	Uncover and interrupt your own unconscious biases and proactively seek the feedback of those experiencing intersectional discrimination Listen to others and be conscious of how your position/status may inhibit others from speaking up
Dignity, choice, and autonomy	Who has independence and who doesn't? Who shares their perspectives and who doesn't? Who has full control over how they live their life and who doesn't?	Be sensitive to people's situations and right to inherent dignity. Respect all opinions, be careful not to make assumptions or rely on proxies.
Accessibility and universal design	Have you asked people what they need to participate? Have you removed physical, transportation, information and communication barriers or provided reasonable alternatives? Have you addressed attitudinal, environmental, and institutional barriers?	Allocate resources (including budget) to ensure meaningful participation of all marginalized persons with disabilities including spoken, local and sign language interpretation, captioning, audio description, Braille language, plain language, easy-read formats
Diverse knowledges	How do we know what we think we know? Who told us? Who has not been consulted?	Actively engage with people who have intersecting experiences of discrimination at all stages of analysis Determine an analysis framework that privileges and learns from diverse forms of knowledge Dedicate resources (including time and budget) to seek out diverse knowledges, recognizing different cultures and ways of communicating
Intersecting identities	What are the intersecting identities of the people we engage with? Who is missing?	Explore how identities interact to create unique social effects and inequalities
Relational power	Who holds power and in what circumstances? Who makes decisions? How are they accountable?	Identify how power varies from one person to another and in what circumstances Explore how systems and attitudes influence power dynamics
Time and space	Does privilege look different in this location? Across different generations? Does discrimination look different in this location? Across different generations?	Consider how inequality and discrimination vary according to time (e.g., intergenerational change) and location (e.g., rural to urban, coastal, migrant, between countries)
Transformative and rights-based	Are we changing the way that resources are produced and/or distributed? Are we changing the way relationships are produced and/or distributed?	Identify gaps in broader formal and informal systems based on analysis of impacts on intersecting identities Analyze how social norms, roles and relations impact on those with intersecting experiences of discrimination

NOTES

1 Other notable movements are the infamous Fridays for Future, the disruptive Extinction Rebellion, or the symbolic Earth Hour.
2 See also UNICEF's campaign and resources on the topic at https://www.unicef.org/end-violence/how-to-stop-cyberbullying.
3 For a complete list of international days and weeks, see https://www.un.org/en/observances/list-days-weeks.

REFERENCES

Chadwick, V., and R. L. Root. 2021. "Donors Question Global Citizen's Claims to Impact." DevEx. Available at: https://www.devex.com/news/donors-question-global-citizen-s-claims-to-impact-98725?utm_source=chatgpt.com (accessed March 23, 2025).

Rosser, C. 2022. "Success Story: DC Water." Available at: https://calvertimpact.org/resources/success-story-dc-water.

RRI. 2022. *Climate Finance Must Not Leave Indigenous, Afro-descendant and Local Community Women and Girls Behind!* Available at: https://rightsandresources.org/publication/our-call-to-action/.

UNEP. 2025. "Act Now: Speak Up!". Available at: https://www.unep.org/interactives/things-you-can-do-climate-emergency/ (accessed March 23d, 2025).

United Nations. 2025a. "Act Now." Available at: https://www.un.org/en/actnow.

United Nations. 2025b. "Speak Up for Our Common Future." Available at: https://www.un.org/en/actnow/speakup.

UNPRPD, and UN Women. 2022. "Intersectionality Resource Guide and Toolkit." Available at: https://www.unwomen.org/sites/default/files/2022-01/Intersectionality-resource-guide-and-toolkit-en.pdf.

Appendix B Career guide in sustainable development

When thinking of your career in international development, consider working backwards. We usually start advising our first year M.A. students by asking them "What does your dream job look like after graduation?" If students have a good sense as to where they want to work, you can leverage websites like unjobs.org (not only for UN jobs) to search by duty station and review existing job openings. Based on the requirements listed, you would then want to plan accordingly your studies, your internships or work experience, as well as any additional ventures or travels you are undertaking.

If you are reading this book and you have a relevant M.A. or M.Sc. degree, you are likely well qualified (academically) for the position. My colleagues have written an in-depth book on exploring your career in international development (Brinkerhoff and Brinkerhoff 2005).

If you are an undergraduate or early-to mid-career professional who holds a Bachelor's degree, you inevitably will hit a ceiling as to how far you can go up the job ladder in most organizations. Consider me biased, but this is one of the main reasons that I feel passionate that investing in a graduate degree still pays off.

The other is that, too frequently, we believe that we can *just do good* no matter our training. In development, there's still the risk that you will end up doing more harm than good. Out of 100 job applications reviewed, your motivation to do good won't make you stand out, as it is shared by your peers. That's why I believe in investing in a professional training or graduate education to ensure you are not wasting only your time or money in development, but also those of the intended beneficiaries of your work.

My own story on how I got into development has shaped me considerably. I understood late during my undergraduate studies (triple-majoring in Political Science, International Studies, and Geography) that I was interested in working in development. The reality was, after my Bachelor's degrees, I was still underqualified. So I invested in a Master's degree in Geography, focusing on development in sub-Saharan Africa. While I loved the subject matter, on the job market, the Geography degree still did not position me well for the actual practice of development. Determined, I undertook a PhD in Economic Geography, working on a sustainable development intervention in Tanzania. My first job interview after graduation, I was told "you are overqualified." I still got the job and started my career in development management and implementation. But in retrospect it taught me that degrees, such as the M.A. in International Development Studies at George Washington University, where I work, are invaluable, as they not only cover the theory and policy of development (which you might get out of most social studies degrees), but especially the practice of development. For these reasons, make sure to also pick a strategic location for your studies where you can take advantage of internships and job opportunities in the development sector to build your resume.

As most of my students aim to get a part-time or full-time job during their first semester, I have compiled the following useful list of job searching tools:

- *LinkedIn*: Not only increasingly used to advertise jobs, but also the best source for you to get your foot in the door by doing a 15-minute informational interview with someone you have a second degree connection with. I encourage all my students to connect with me, so they can draw upon our alumni and their peers in job searches. Do the same with all your peers and professors.
- *Subscribe to newsletters*: I have four recommendations for newsletters for you that focus on jobs. Impact Pool (www.impactpool.org) covers UN jobs, as well as opportunities in the EU and among NGOs. Reliefweb (https://reliefweb.int/jobs) has a focus on humanitarian and development jobs. DevNetJobs (http://devnetjobs.org) offers a paid service that includes so-called premium jobs as well as consultancies. DevEx (www.devex.com/jobs) not only offers a large number of jobs, but provides helpful advice for building your resume and navigating changing funding environment (www.devex.com/career-center).
- *Get a good overview of the major actors in international cooperation*: The Swiss organization – cinfo – has as its mission to build the capacity for Swiss candidates for UN (or other multilateral) job opportunities. Your country might have a similar clearinghouse. The tools cinfo developed are readily accessible to you, irrespective of your residence, such as this overview of major development actors and what they are looking for (https://www.cinfo.ch/en/individuals/get-informed/overview-of-actors-in-international-cooperation?panorama_markets=44).
- If you are looking for opportunities outside North America: for academic positions, for Europe and Asia, I recommend https://academicpositions.com/. For searching for jobs by duty station, I recommend unjobs.org. My personal websites features some additional funding opportunities and tips for how to apply for jobs under my Frequently Requested Resources page (https://samueltledermann.me/faq).

REFERENCE

Brinkerhoff, D. W., and J. M. Brinkerhoff. 2005. *Working for Change: Making a Career in International Public Service*. Bloomfield, CT: Kumarian Press.

Appendix C Guide to companion website: sustainabledevelopment.me

Our companion website is available at www.sustainabledevelopment.me. Thank you to Pritam Garcha for helping me with the initial conceptualization of the site.

The main page is organized according to the book, with the Chapter Overview featuring for each chapter an opportunity to dive deeper with additional readings and resources. Select chapters include videos captured under the "To see more and get you talking" section, including relevant discussion questions. If relevant, we also included podcasts (e.g., "If you are on your daily commute" section), or social media accounts to follow (e.g., under the "To be inspired" section).

As printed links in the book will inevitably break at some point, I will also aim to capture any corrections as needed in the respective chapter.

Given the importance of resilience, I have provided additional resilience vignettes – aligned with the transformation hotspots – in Chapter 16 online. You can also find a dedicated page for the Appendix A: Guide for tools for further action, and Appendix B: Career guide in sustainable development. We also have a separate site with some links for SDG data for you to explore.

Beyond my site, you might also enjoy reading the IISD SDG Knowledge Hub – Generation 2030 blog. Young professionals from around the world, including my own students, regularly publish their analyses at https://sdg.iisd.org/generation-2030/. The overall IISD SDG Knowledge Hub itself features thousands of news articles, guest posts and policy briefs sorted by SDGs at https://sdg.iisd.org/sdgs/.

If you have any suggestions on additional resources to add, don't hesitate to send me your suggestion using the Contact Form.

You can also sign up for my bi-annual newsletter. Around May/June and November/December each year, I aim to send a brief update on resources added to the website, as well as any recent development of relevance on *How to Achieve Sustainable Development*.

Index